THE DŽARYLGAČ SURVEY PROJECT

VOLUME 2

BLACK SEA STUDIES

14

THE DANISH NATIONAL RESEARCH FOUNDATION'S
CENTRE FOR BLACK SEA STUDIES
&
GRONINGEN INSTITUTE OF ARCHAEOLOGY,
UNIVERSITY OF GRONINGEN

THE DŽARYLGAČ SURVEY PROJECT

VOLUME 2: Plates

Edited by
Pia Guldager Bilde, Peter Attema & Kristina Winther-Jacobsen

Aarhus University Press

THE DŽARYLGAČ SURVEY PROJECT, VOL. 2

Cover design Grafisk SIGNS
Photos by Pia Guldager Bilde and Kristina Winther-Jacobsen
Typeset and printed by Narayana Press, Gylling

ISBN 978 87 7934 521 8

Printed in Denmark 2012

Aarhus University Press
Langelandsgade 177
DK-8200 Aarhus N
www.unipress.dk

White Cross Mills
Hightown, Lancaster, LA1 4XS
United Kingdom
www.gazellebooks.co.uk

ISD
70 Enterprise Drive, Suite 2
Bristol, CT 06010
USA
www.isdistribution.com

The publication of this volume has been made possible by a generous grant from The Danish National Research Foundation, Aarhus University's Research Foundation, and the Augustinus Foundation.

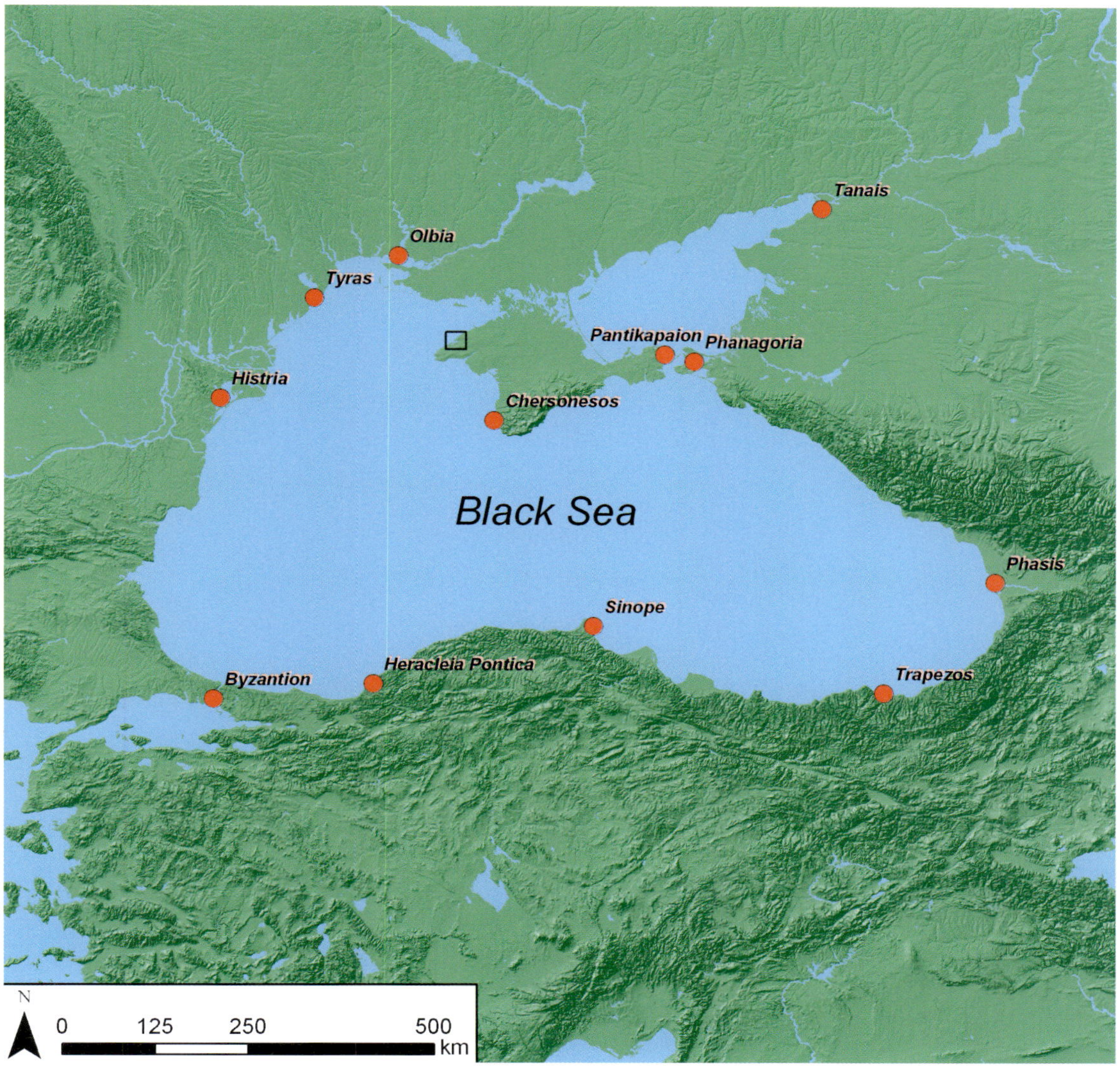

Fig. 0.1 • General map of Black Sea area.

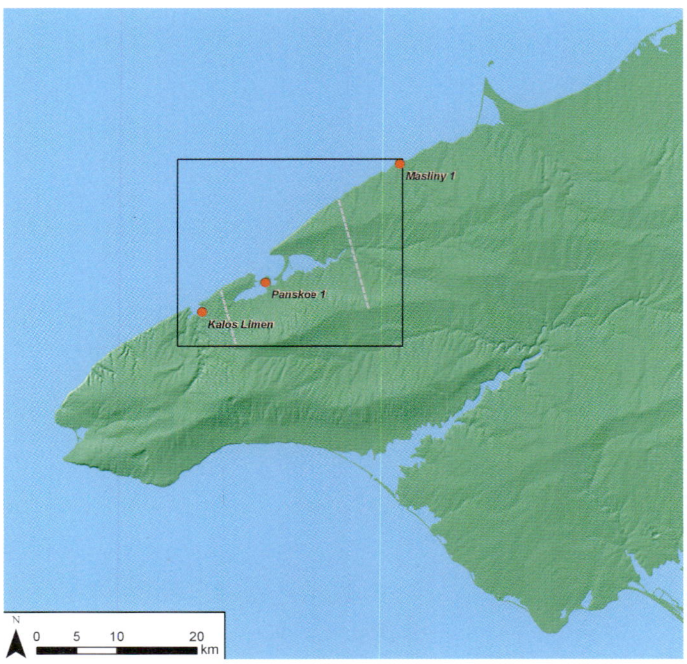

Fig. 0.2 • General map of the Tarchankut Peninsula.

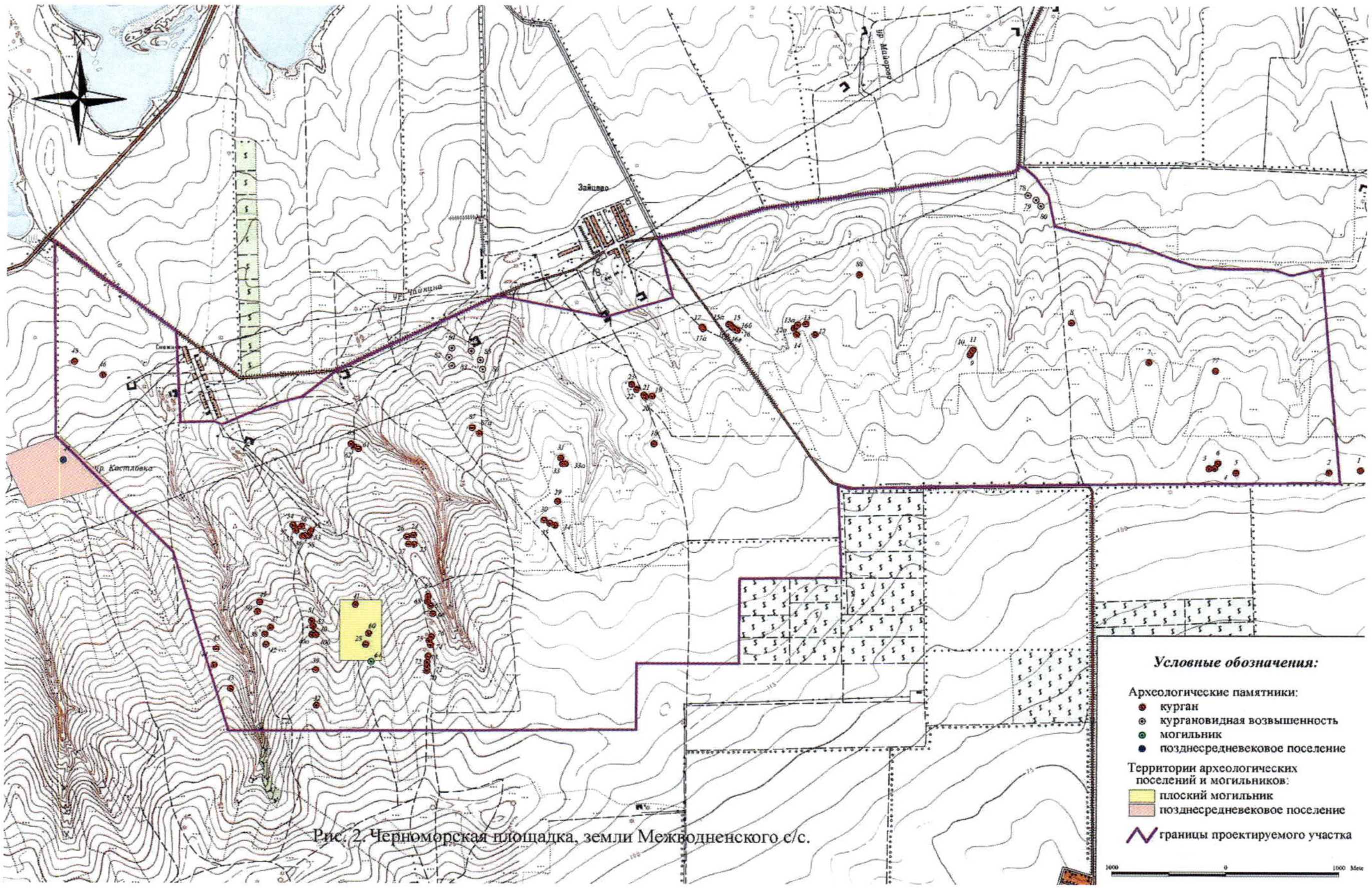

Fig. 0.3 • Investigations by S. Koltuchov before the DSP; the purple line gives the area of the planned wind power plant.

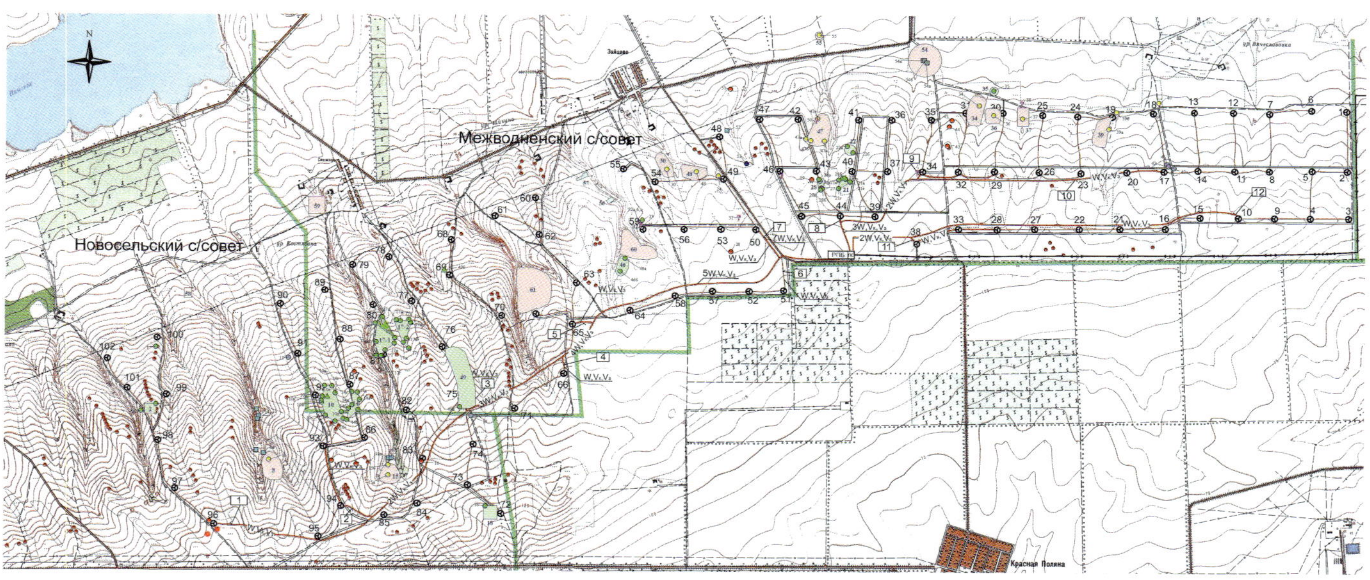

Fig. 0.4 • Map showing locations of proposed windmills for the wind power plant.

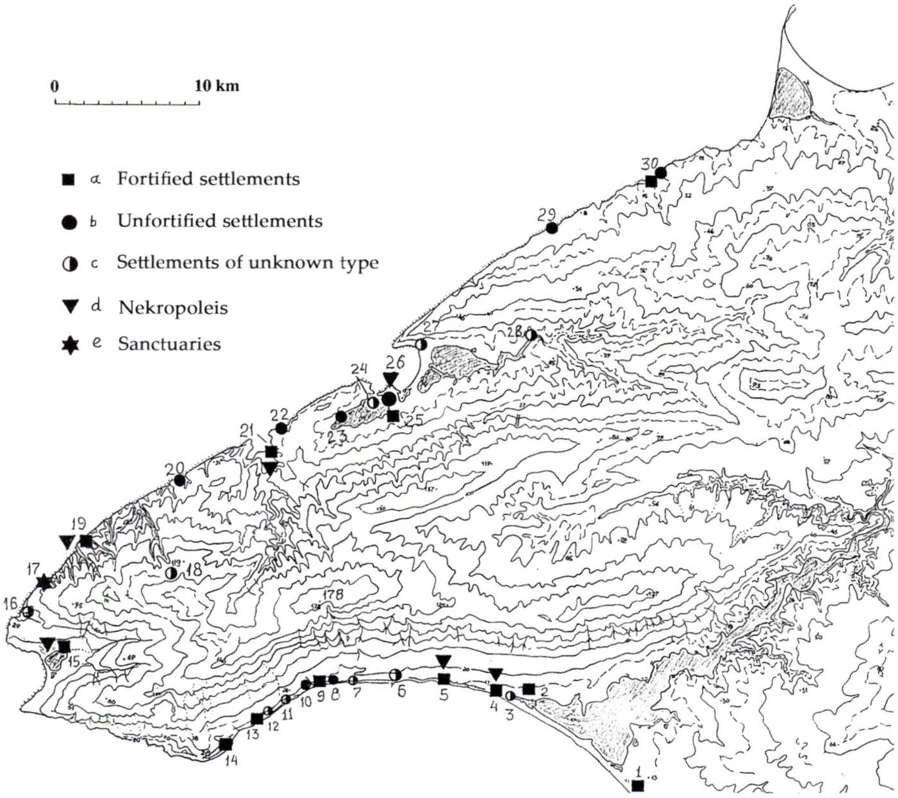

Fig. 1.1 • Summary map of ancient sites (after Ščeglov 2002, pl. 4.1).

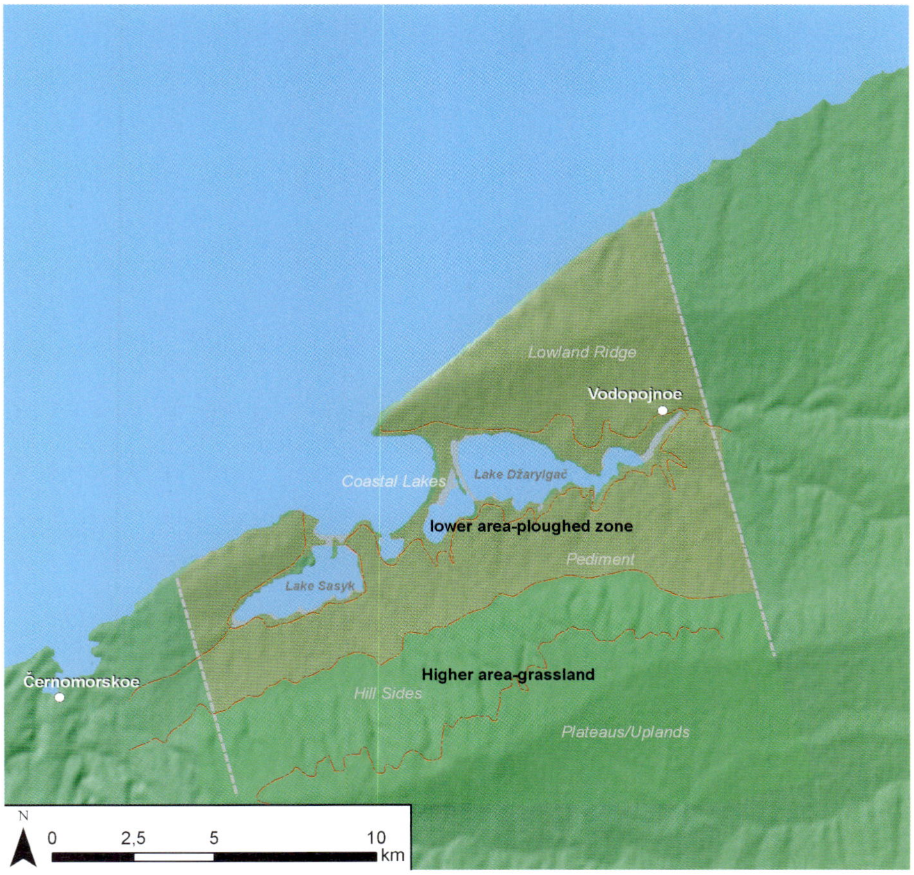

Fig. 2.1 • Area of intensive study and visibility zones.

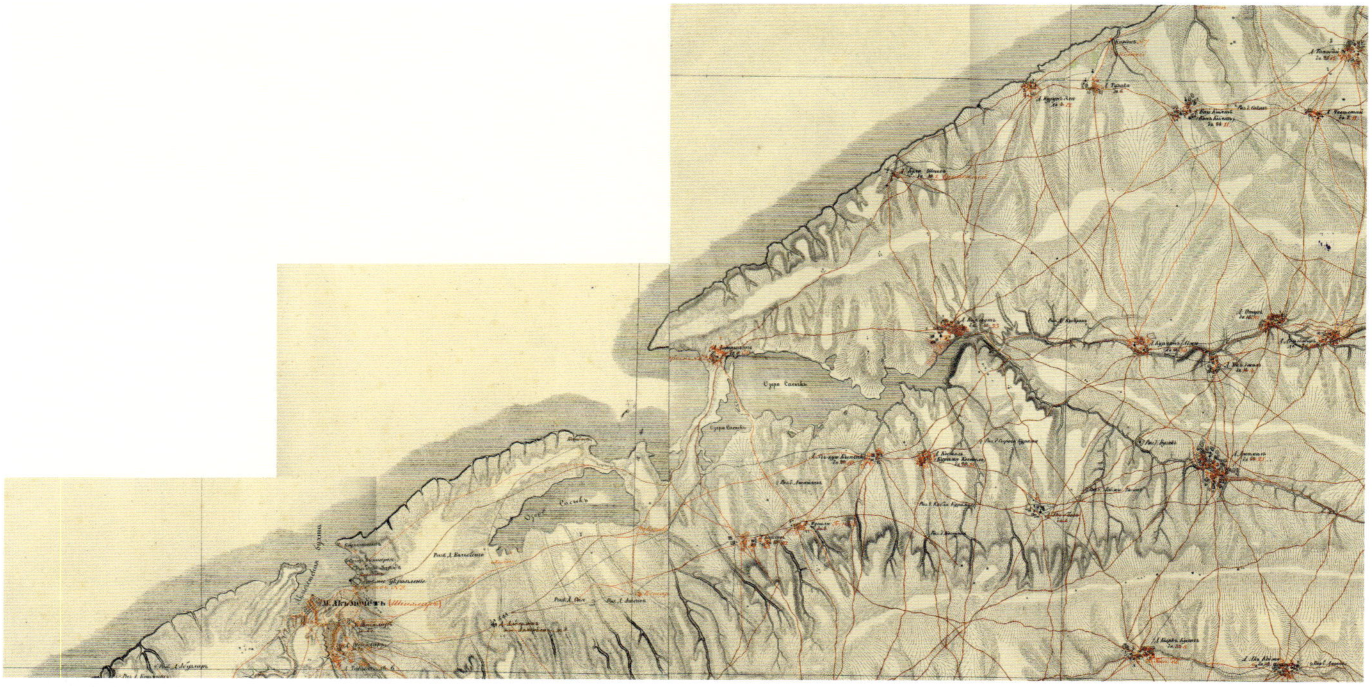

Fig. 2.2 • Betev's map of 1837 with emendations of 1865.

Fig. 2.3 • Map from the late 19th century in half-verst scale (1:21,000).

Fig. 2.4 • Map from the late 19th century in half-verst scale, detail of survey area.

Fig. 2.5 • Map of the mid-20th century, scale 1:25,000; detail of Panskoe area.

Fig. 2.6 • Map of 1989, scale 100,000; Tarchankut Peninsula.

Fig. 2.7 • Survey team in the field.

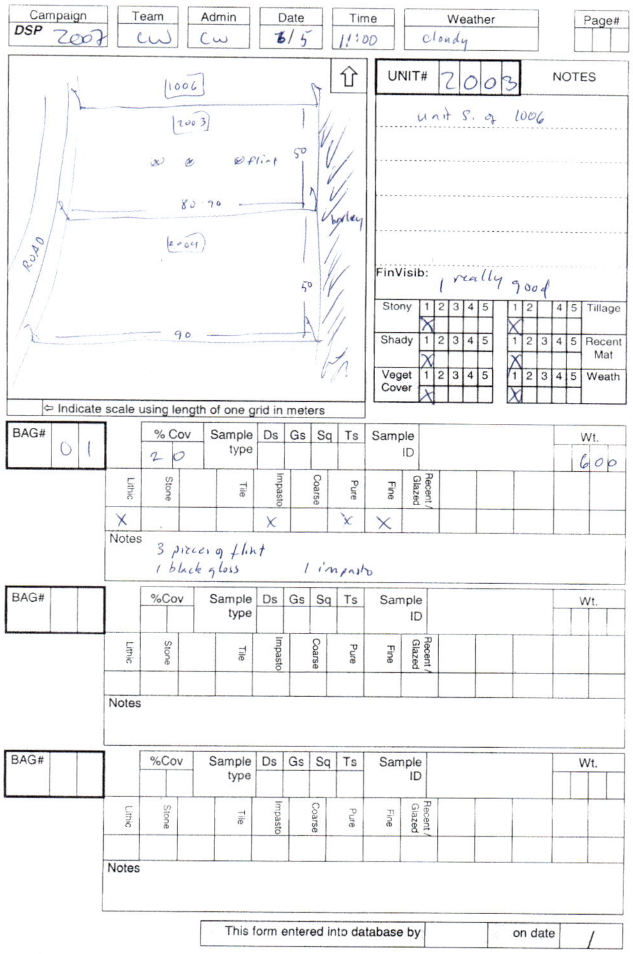
Campaign DSP 2007 | Team CW | Admin CW | Date 6/5 | Time 11:00 | Weather cloudy | Page#

UNIT# 2003 NOTES
unit S. of 1006

1006
1003
flint
50
80–90
hedge
1004
Road
50
90

FinVisib: really good

Stony	1	2	3	4	5		1	2	3	4	5	Tillage
Shady	1	2	3	4	5		1	2	3	4	5	Recent Mat
Veget Cover	1	2	3	4	5		1	2	3	4	5	Weath

⇦ Indicate scale using length of one grid in meters

BAG# 01 | % Cov 20 | Sample type | Ds | Gs | Sq | Ts | Sample ID | Wt. 60
Lithic X | Stone | Tile X | Impasto | Coarse X | Pure X | Fine | Recent / Glazed
Notes: 3 pieces of flint, 1 black gloss, 1 impasto

BAG# | %Cov | Sample type | Ds | Gs | Sq | Ts | Sample ID | Wt.
Lithic | Stone | Tile | Impasto | Coarse | Pure | Fine | Recent / Glazed
Notes

BAG# | %Cov | Sample type | Ds | Gs | Sq | Ts | Sample ID | Wt.
Lithic | Stone | Tile | Impasto | Coarse | Pure | Fine | Recent / Glazed
Notes

This form entered into database by | on date /

Fig. 2.8 • Example of recording forms for the block survey (2007 Unit 2003).

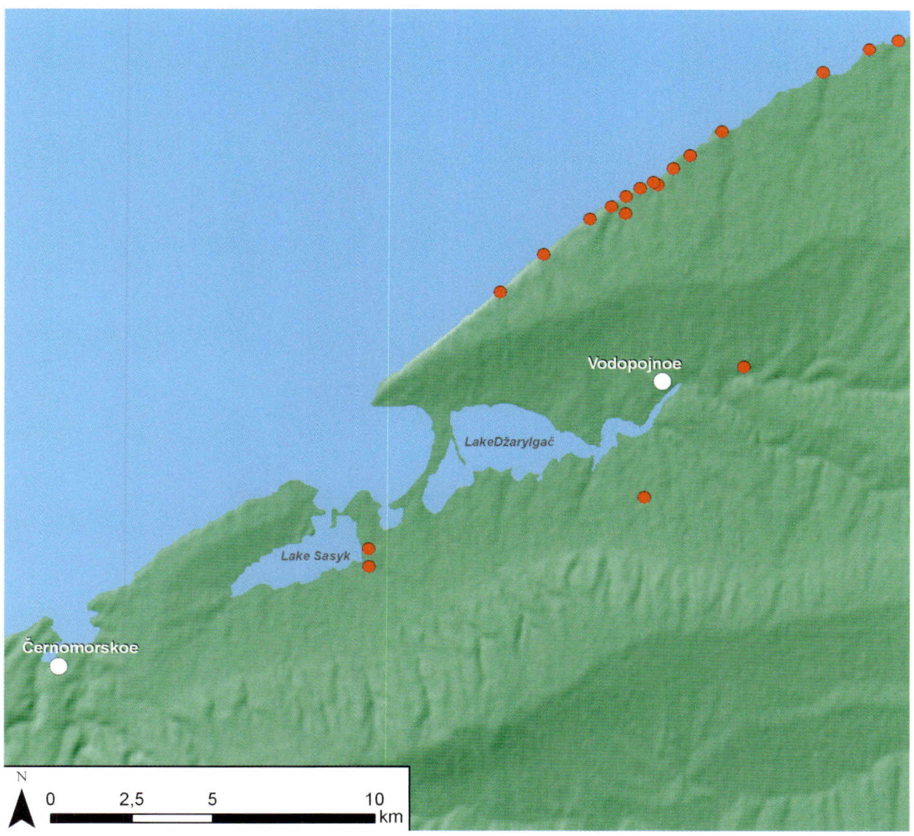

Fig. 2.9 • Location of sites visited and found during the DSP extensive survey (2007).

Fig. 2.10 • T. Smekalova surveying with hand-held probe.

Fig. 2.11 • Probes mounted on a cart, as used by Eastern Atlas.

Fig. 2.12 • Eastern Atlas using kite for photography.

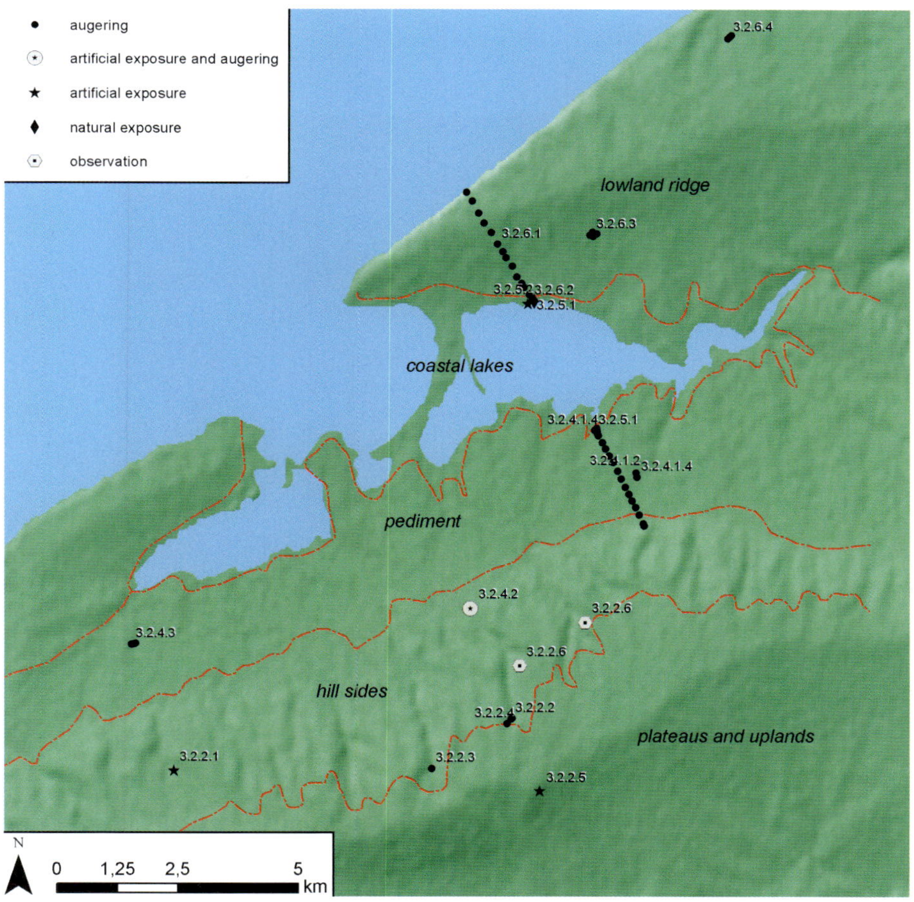

Fig. 2.13 • Map showing locations of augerings and geological observations.

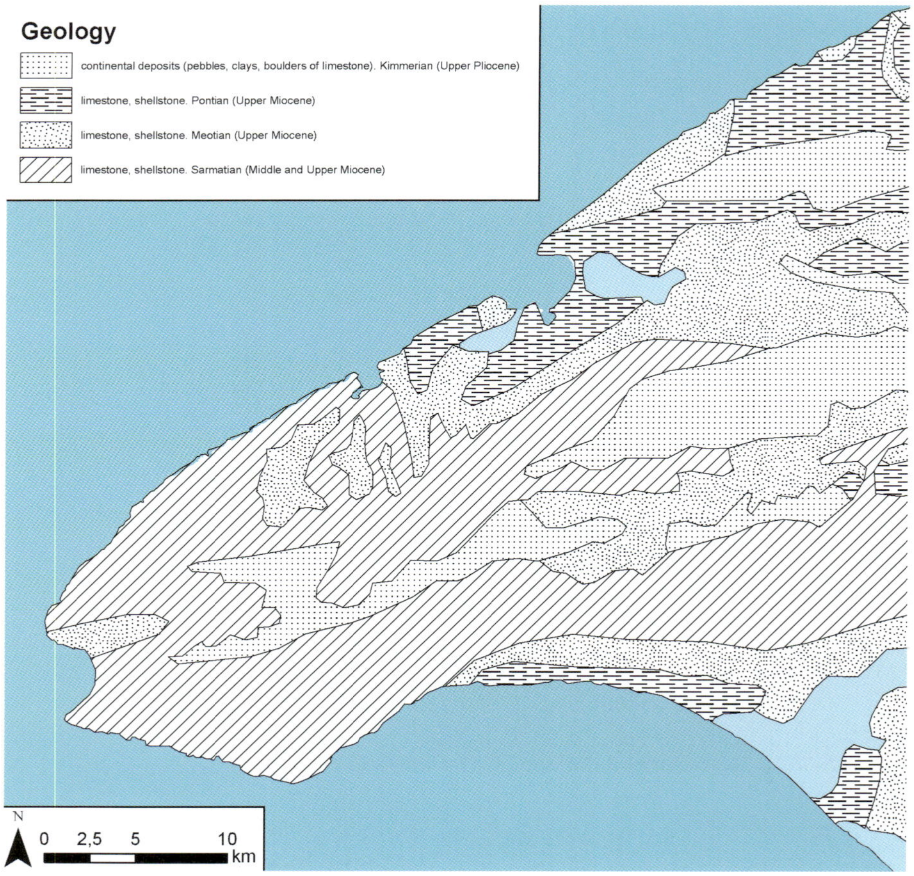

Fig. 3.1 • Map with main geological units drawn 1967, scale 1:250,000 (after the Geology of USSR, 1970)

Fig. 3.2 • Sinkhole in horizontally bedded limestone surface.

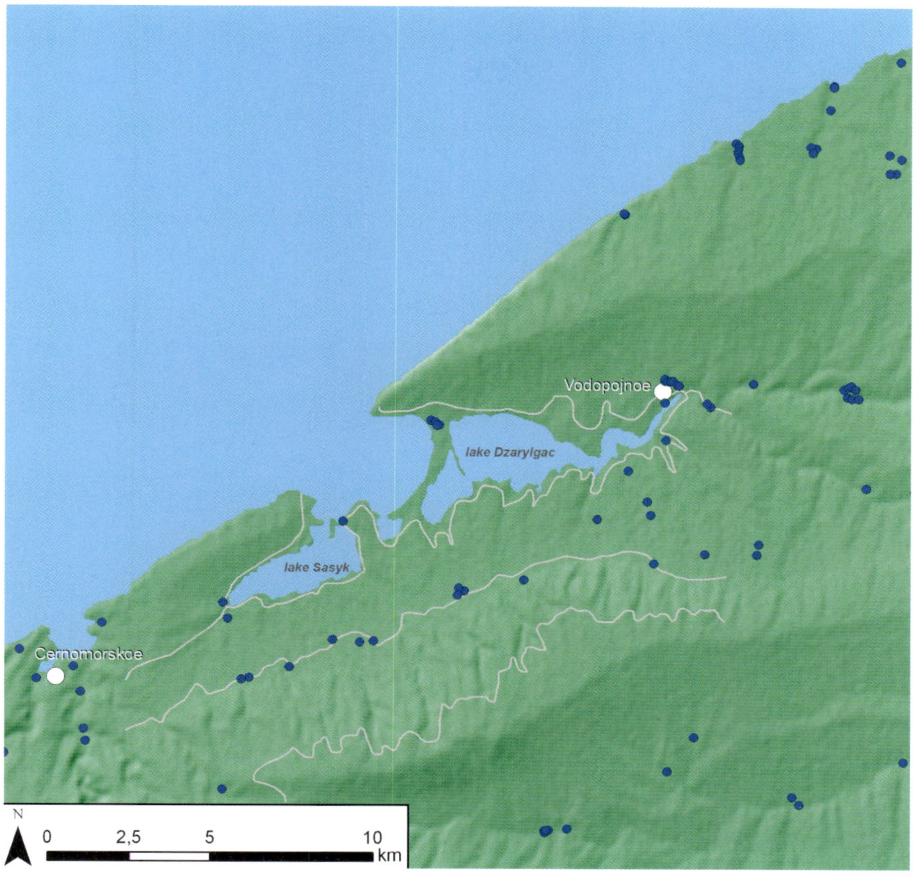

Fig. 3.3 • Map of the study area showing location of wells.

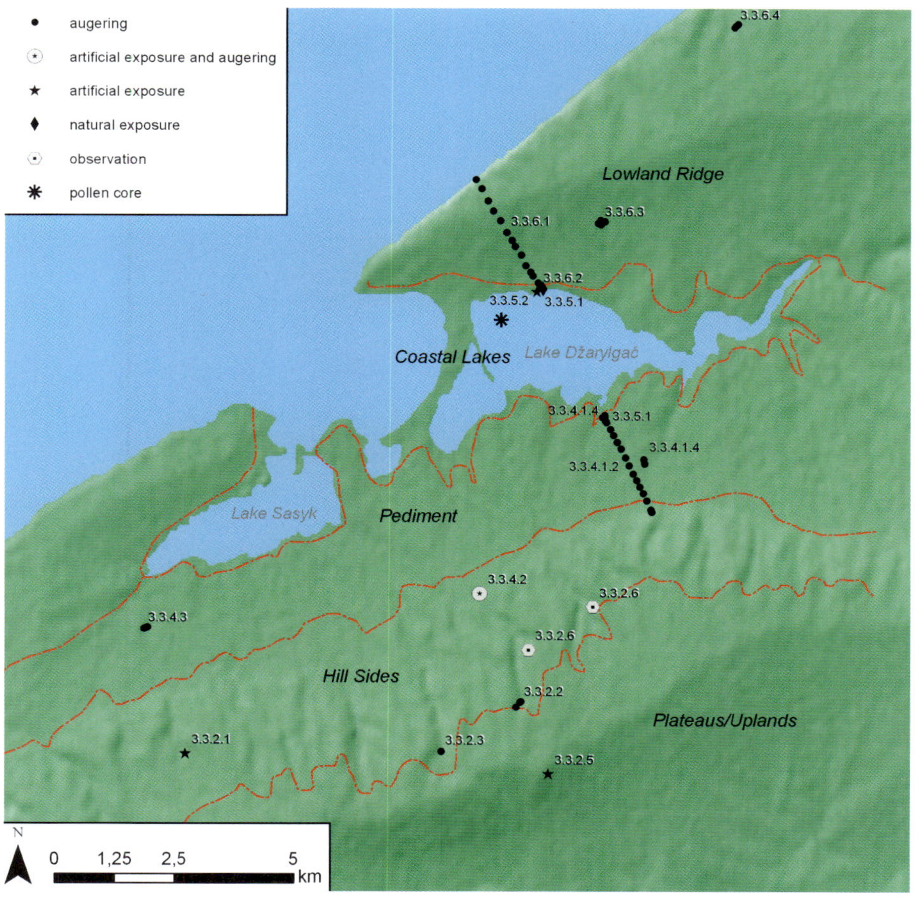

Fig. 3.4 • Map of the study area showing the five land classes and location of observations referred to in the text. The numbers correspond to paragraphs in the text.

Fig. 3.5 • Limestone plateau with kurgans.

Fig. 3.6 • Solifluction forms on the higher loess-covered parts of the landscape.

Fig. 3.7 • Ravines on the peninsular hillsides below the plateau which is controlled by a resistant limestone bed (visible as a thin white line in the background).

Fig. 3.8 • Agriculture on the pediment showing patches of different colours.

Fig. 3.9 • Loess quarry at the lower end of the hillsides.

Fig. 3.10 • Loess soils exposed in the northern cliff side of Lake Ďzarylgač.

Fig. 3.11 • Loess-free soils on the top of the Lowland Ridge.

Fig. 3.12 • Loess quarry near Mežvodnoe on the Lowland Ridge.

Fig. 3.13 • Soviet farm in ruins.

Fig. 4.1 • Baj Kijat 1 towards north.

Fig. 4.2 • The Baj Kijat sites viewed from Baj Kijat 3 eastwards towards Baj Kijat 2 on the opposite side of small bay; Baj Kijat 1 is just outside the picture to the right.

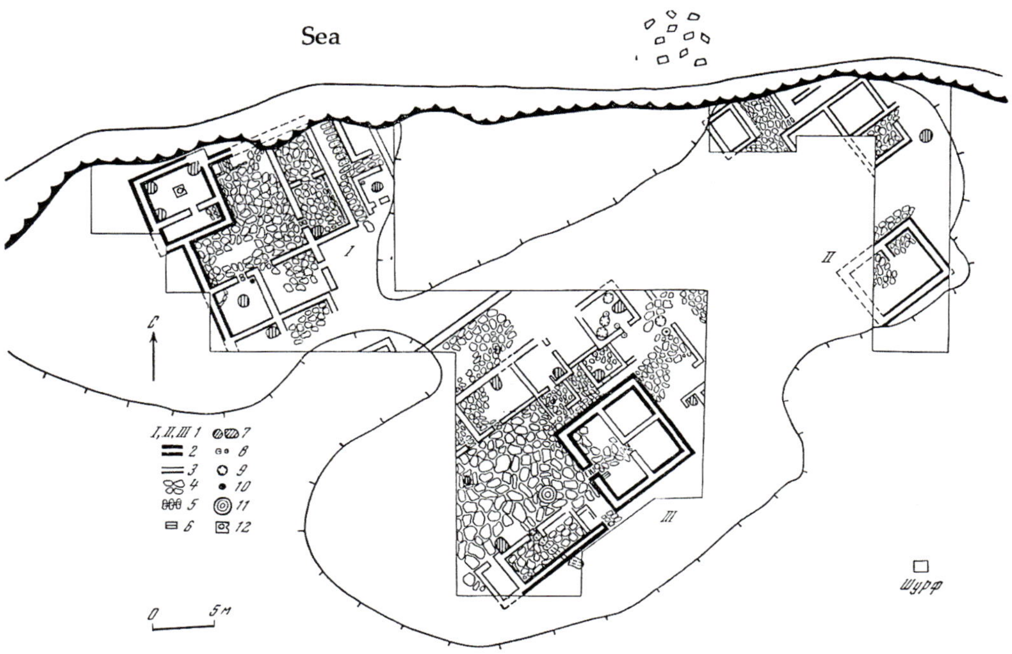

Fig. 4.3 • Plan of Masliny 1, modified after Latyčeva 1978, fig. 2.

Fig. 4.4 • Masliny 1. Tower.

Fig. 4.5 • Masliny 1. Farmhouse in the process of being eroded away.

Fig. 4.6 • Groty 1 seen from W across small ravine.

Fig. 4.7 • View from Burun Eli 2 towards east with fishing station.

Fig. 4.8 • Burun Eli 2 seen from east; a large ash hill marks the site.

Fig. 4.9 • Burun Eli 2. Pavement in soil profile.

Fig. 4.10 • Burun Eli 2. A road (over the shadow of the photographer) leading from the site aiming between the two kurgans in the horizon.

Fig. 4.11 • Skalistoe sites seen towards west; to the left Skalistoe 5, in the centre by fishing station Skalistoe 2, to the right Skalistoe 8.

Fig. 4.12 • Skalistoe 5 and 2 seen from Skalistoe 8.

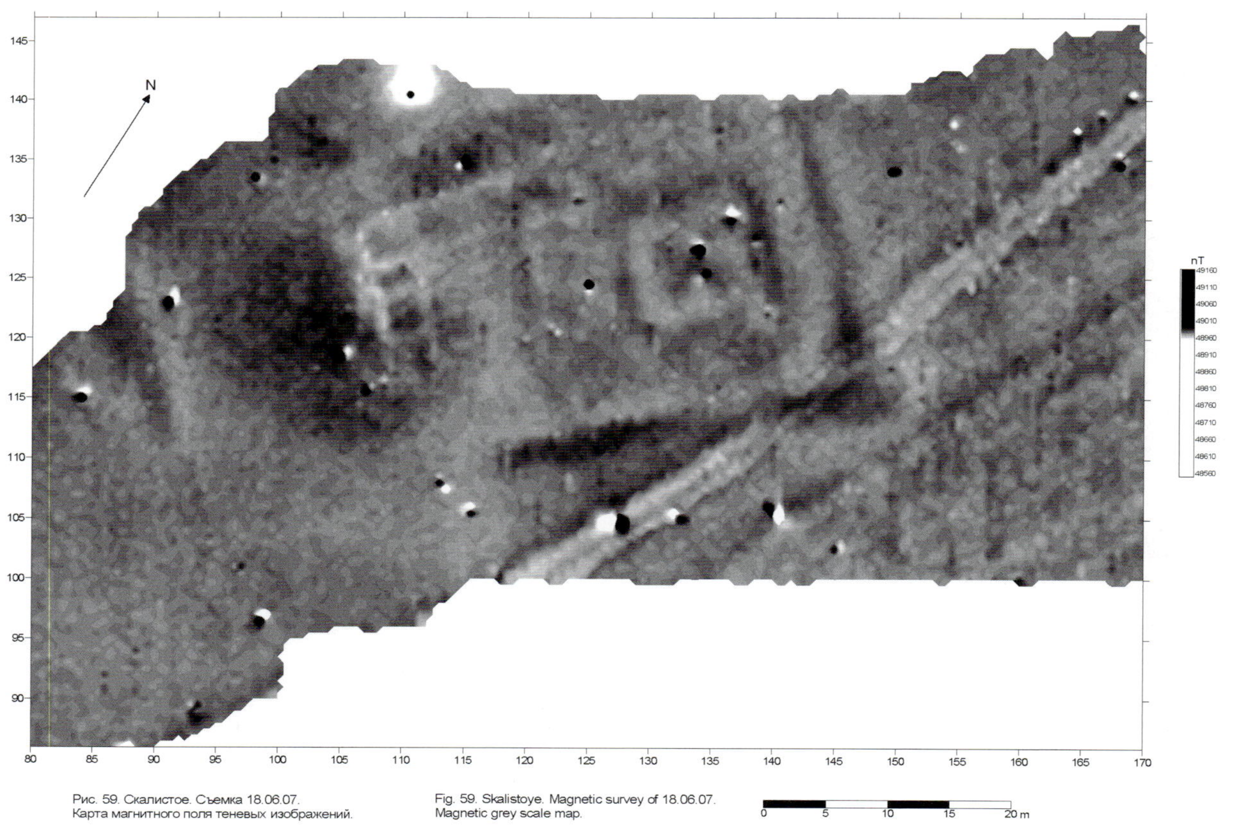

Fig. 4.13 • Skalistoe 5, geomagnetic measurements.

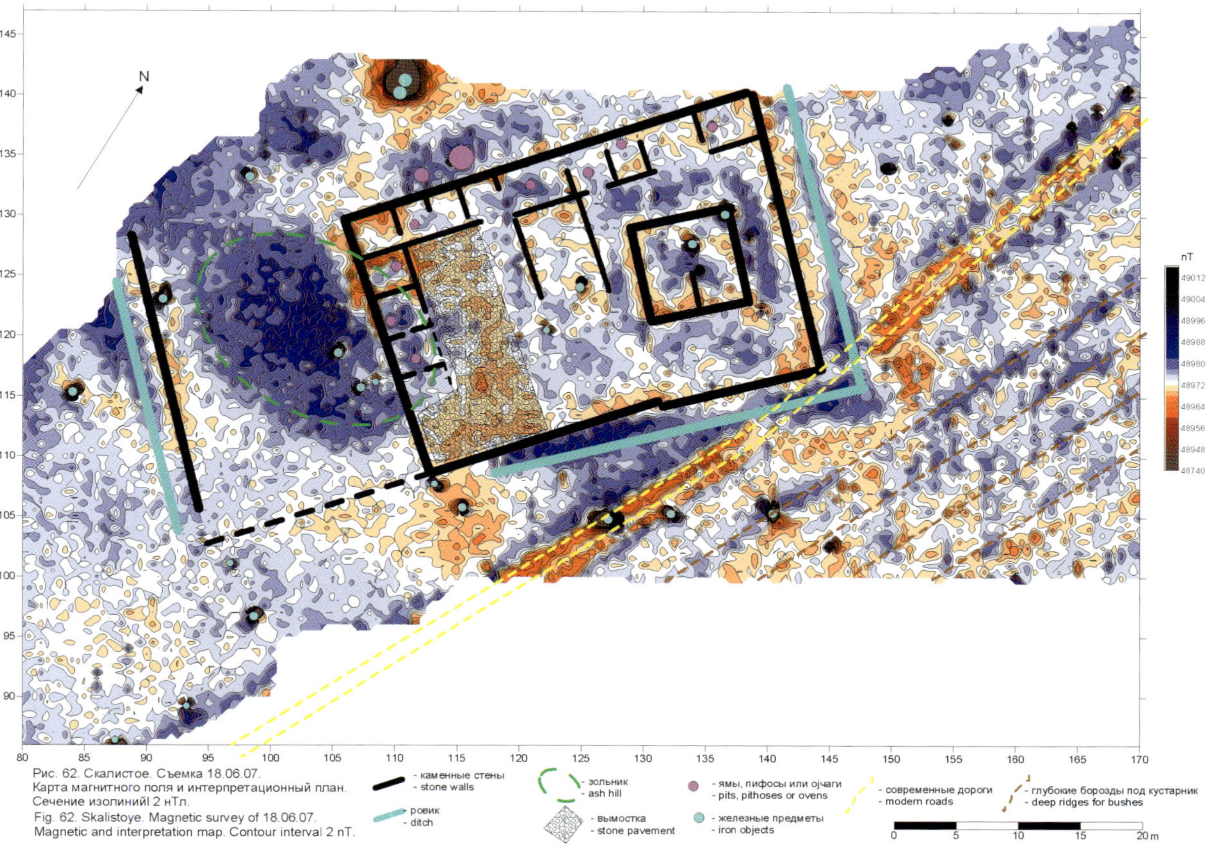

Fig. 4.14 • Skalistoe 5, geomagnetic measurements, interpretation.

Fig. 4.15 • Skalistoe 5, trial trench from southeast.

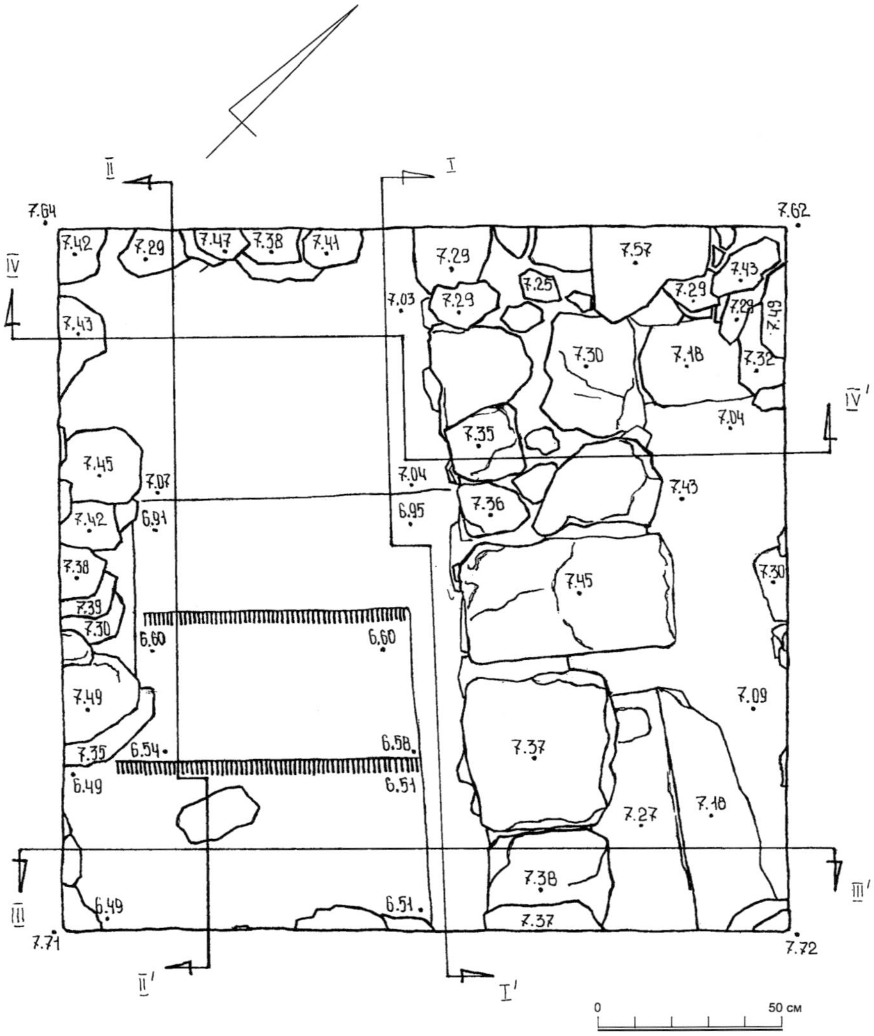

Fig. 4.16 • Skalistoe 5, trial trench, plan.

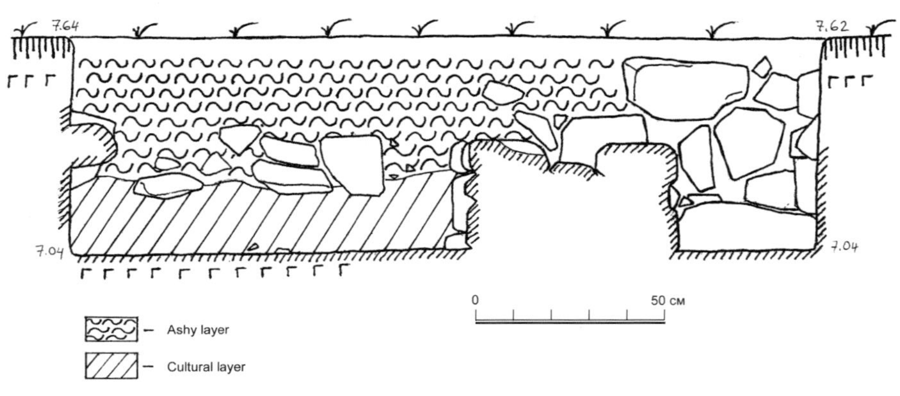

Fig. 4.17 • Skalistoe 5, trial trench, northwest profile.

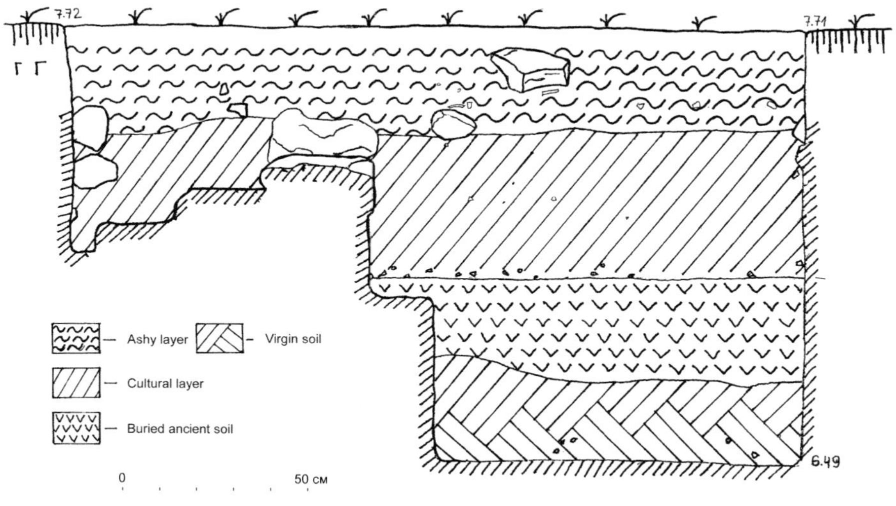

Fig. 4.18 • Skalistoe 5, trial trench, southeast profile.

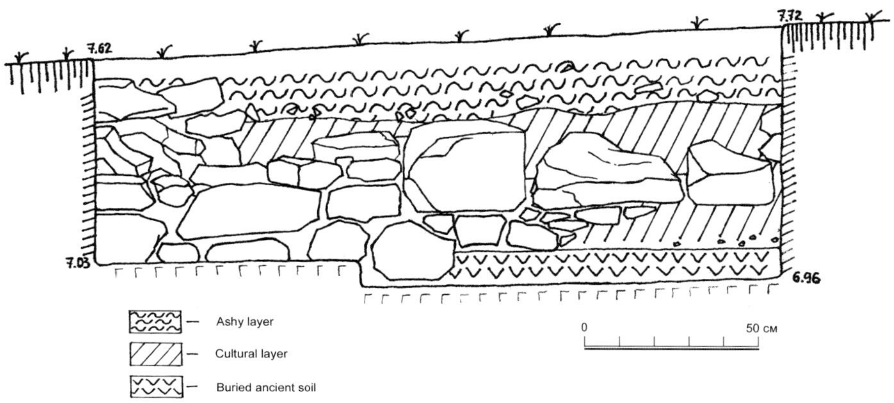

Fig. 4.19 • Skalistoe 5, trial trench, southwest façade.

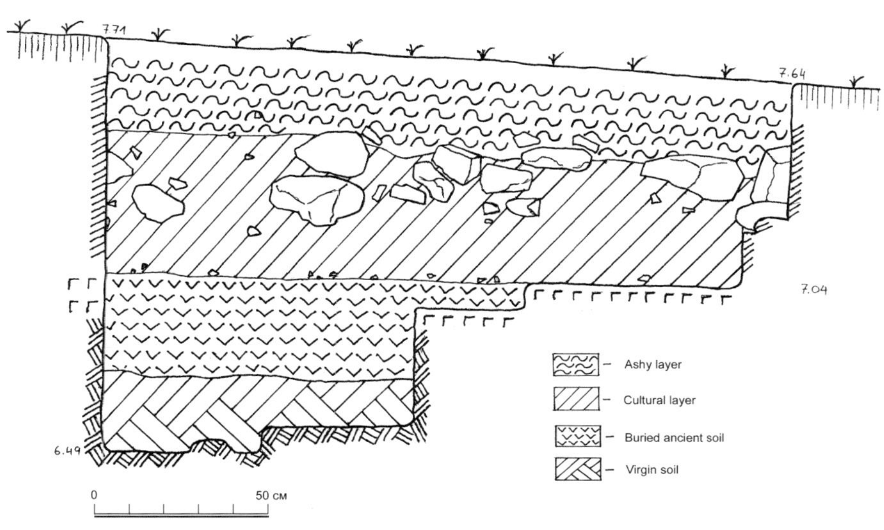

Fig. 4.20 • Skalistoe 5, trial trench, southwest profile.

Fig. 4.21 • Modern village (Kalinovka) enjoying protection from the wind.

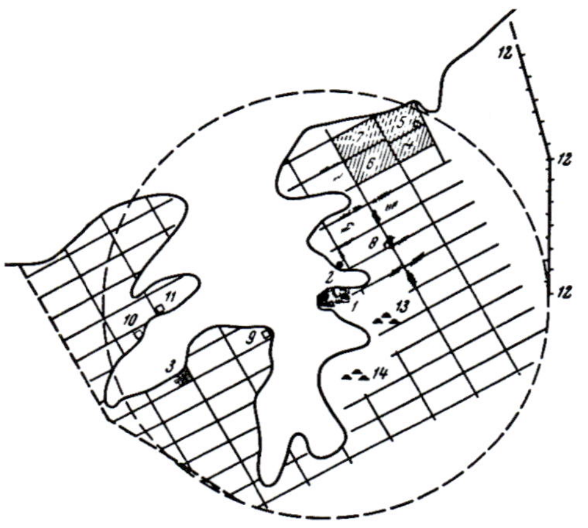

Fig. 4.22 • Uzkaja Bay (after Ščeglov 1967, fig. 2) the numbers 8-11 mark the location of farmhouses.

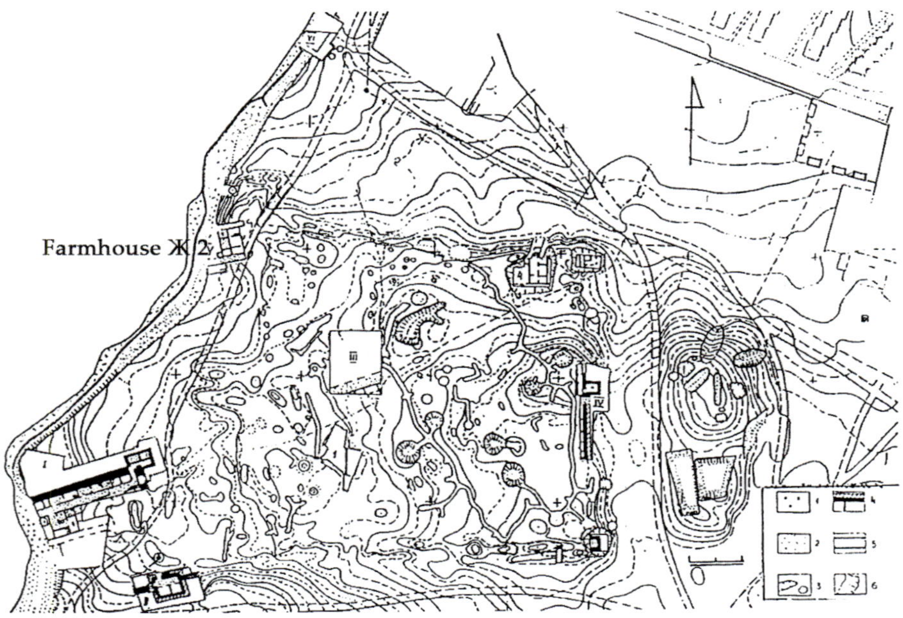

Fig. 4.23 • The Černomorskoe Settlement, plan; Ж marks farmhouse (modified after Užencev 2006, fig. 4).

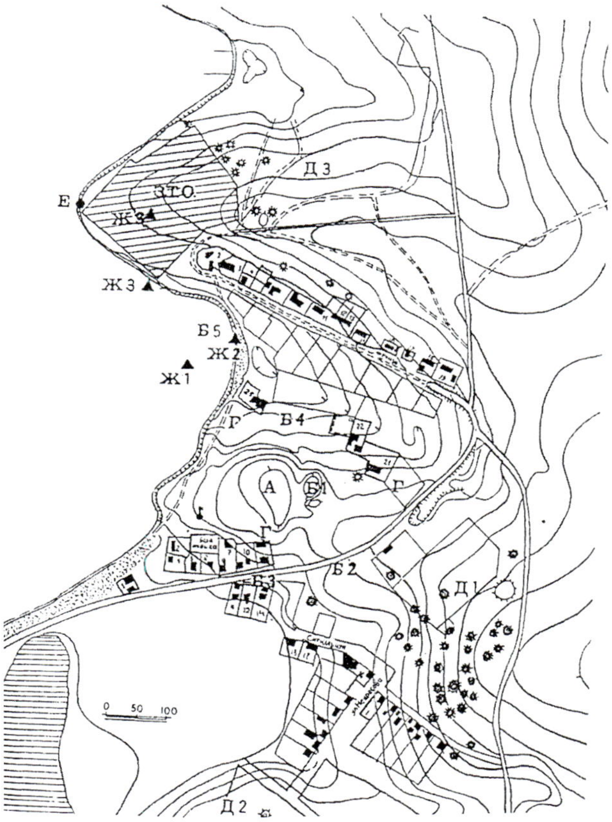

Fig. 4.24 • The Černomorskoe Settlement, plan of immediate surroundings; Ж marks farmhouses (after Užencev 2006, fig. 4).

Fig. 4.25 • The Černomorskoe Settlement, tower with 'anti-ramming' belt.

Fig. 4.26 • The Černomorskoe Settlement, farmhouse Ж 2 lying on the beach.

Fig. 4.27 • Well head found during construction work at the Florya Pension in Ul. Revolucii.

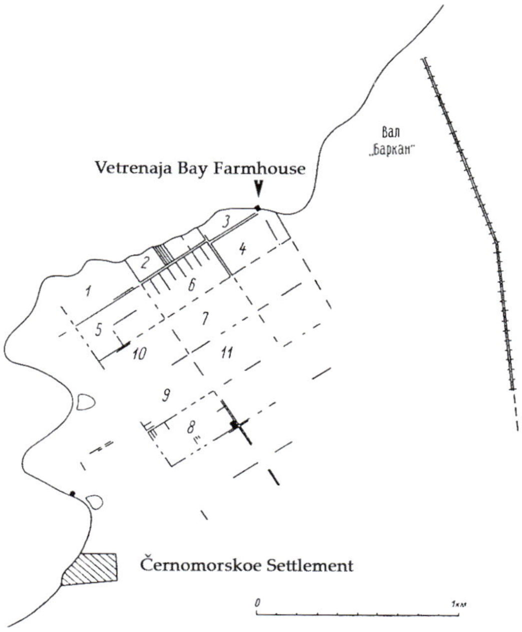

Fig. 4.28 • Cadastral system northeast of the Černomorskoe Settlement and the location of the Vetrenaja Bay farmhouse (modified after Ščeglov 1978, fig. 43).

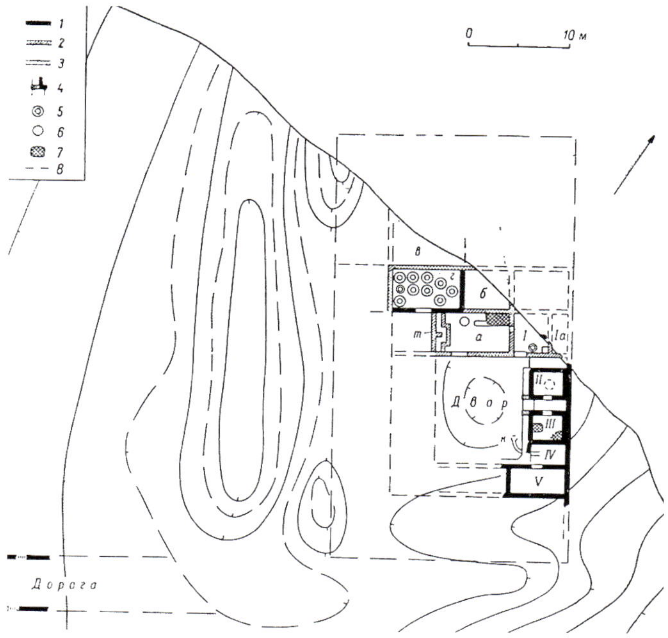

Fig. 4.29 • Vetrenaja Bay farmhouse (after Ščeglov 1978, fig. 48).

Fig. 4.30 • Vetrenaja Bay.

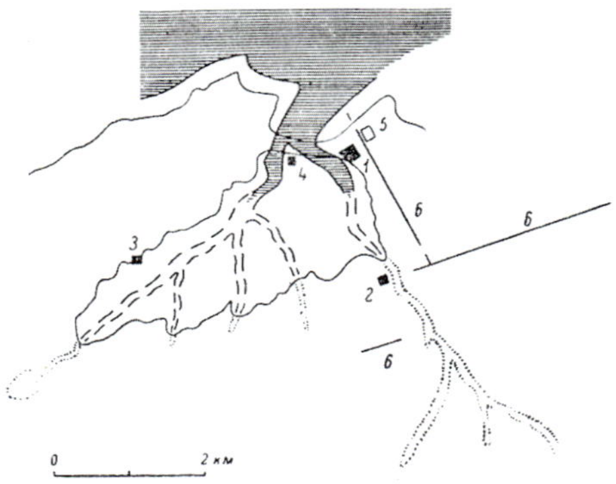

Fig. 4.31 • Panskoe sites (after Ščeglov 1987, fig. 2).

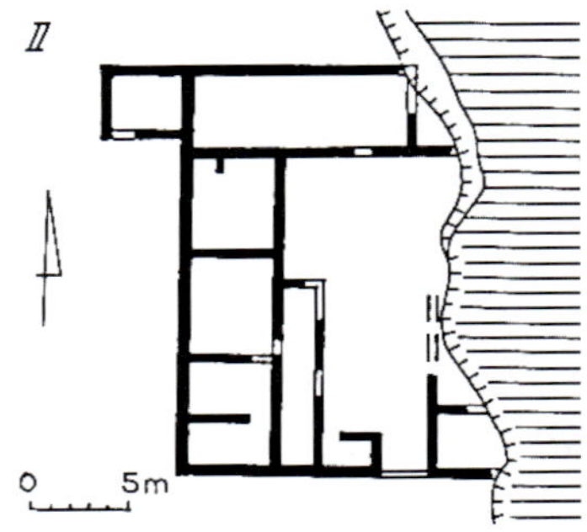

Fig. 4.32 • Panskoe III (after Ščeglov 1987, fig. 2).

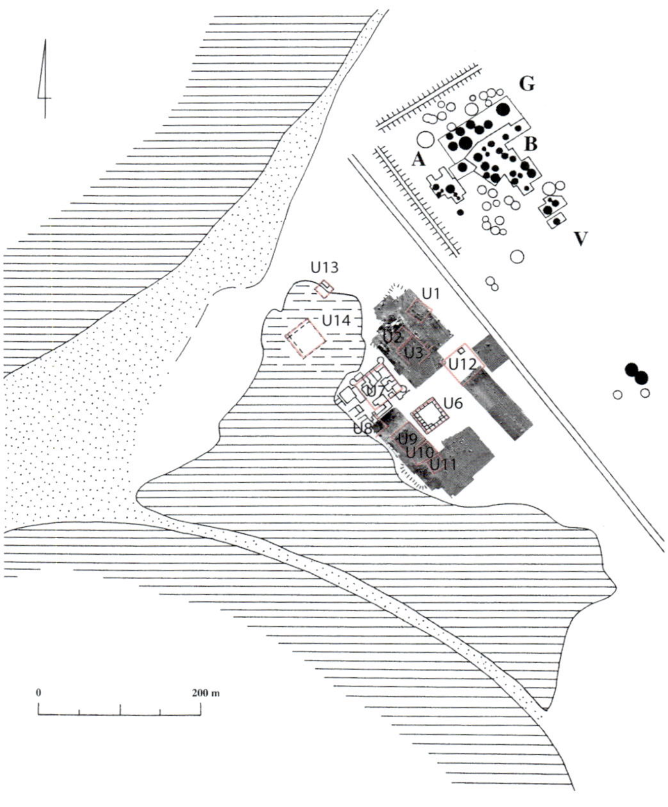

Fig. 4.33 • Panskoe I, plan (modified after Ščeglov 2002, pl. 6 with addition of geomagnetic measurements).

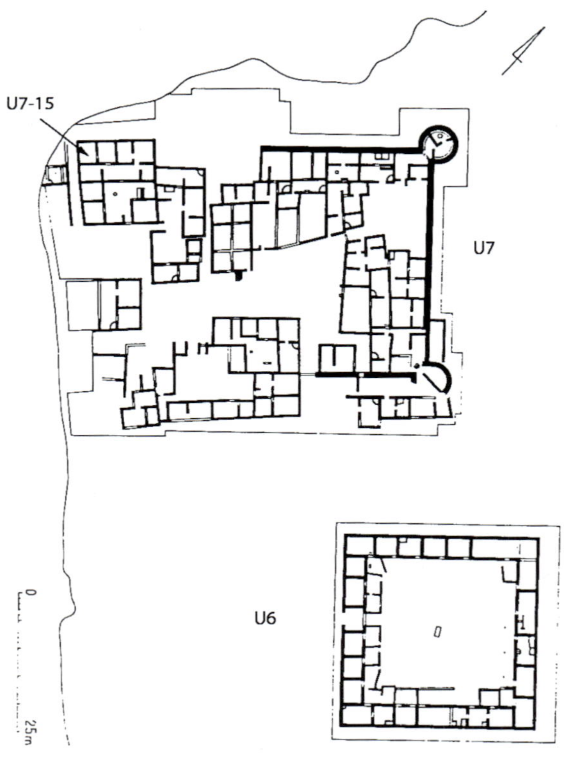

Fig. 4.34 • Panskoe I, plan of U7 and U6 (after Ščeglov 2002, pl. 7).

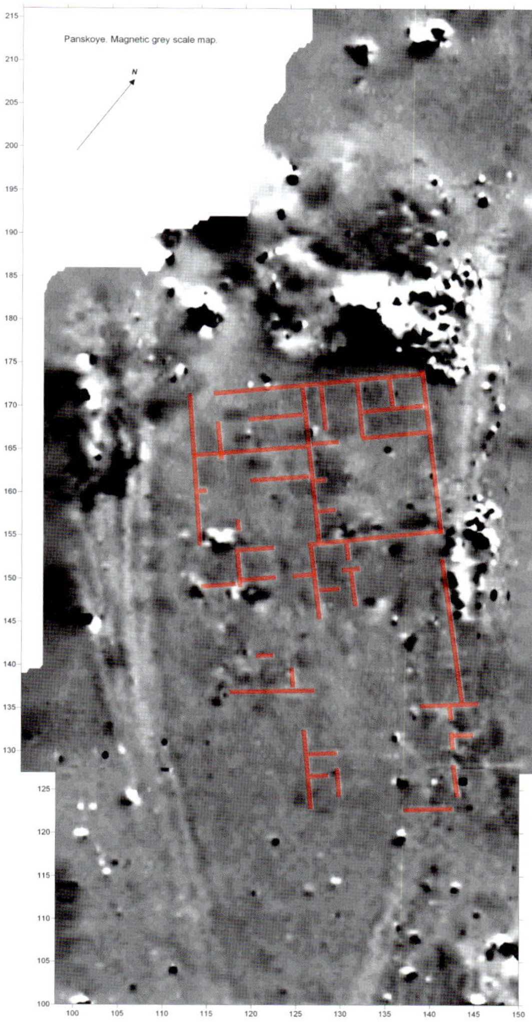

Fig. 4.35 • Panskoe I, geomagnetic measurements of U2 and U3, interpretation.

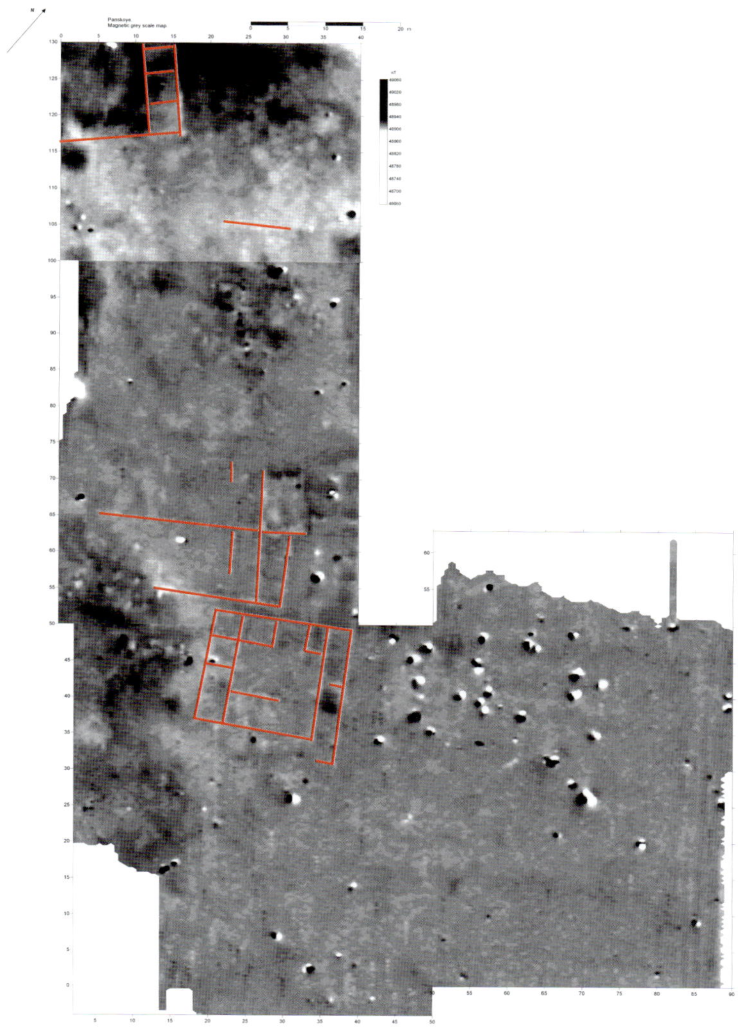

Fig. 4.36 • Panskoe I, geomagnetic measurements of U8, U9, U10, and U11, interpretation.

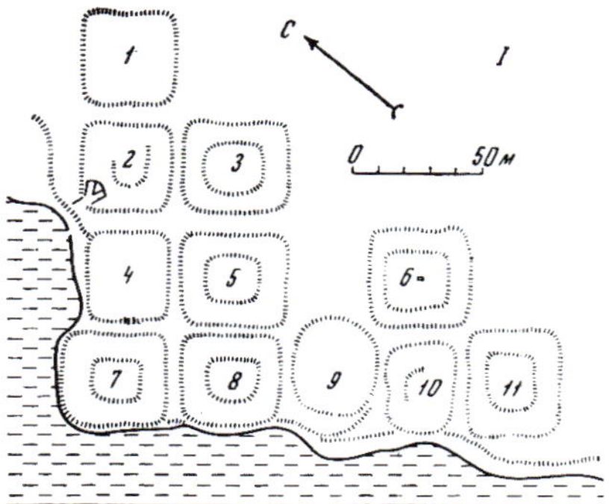

Fig. 4.37 • Panskoe I, sketch plan (after Wasowic 1972, fig. 12.I).

Fig. 4.38 • The Tatarian village of Vodopojnoe.

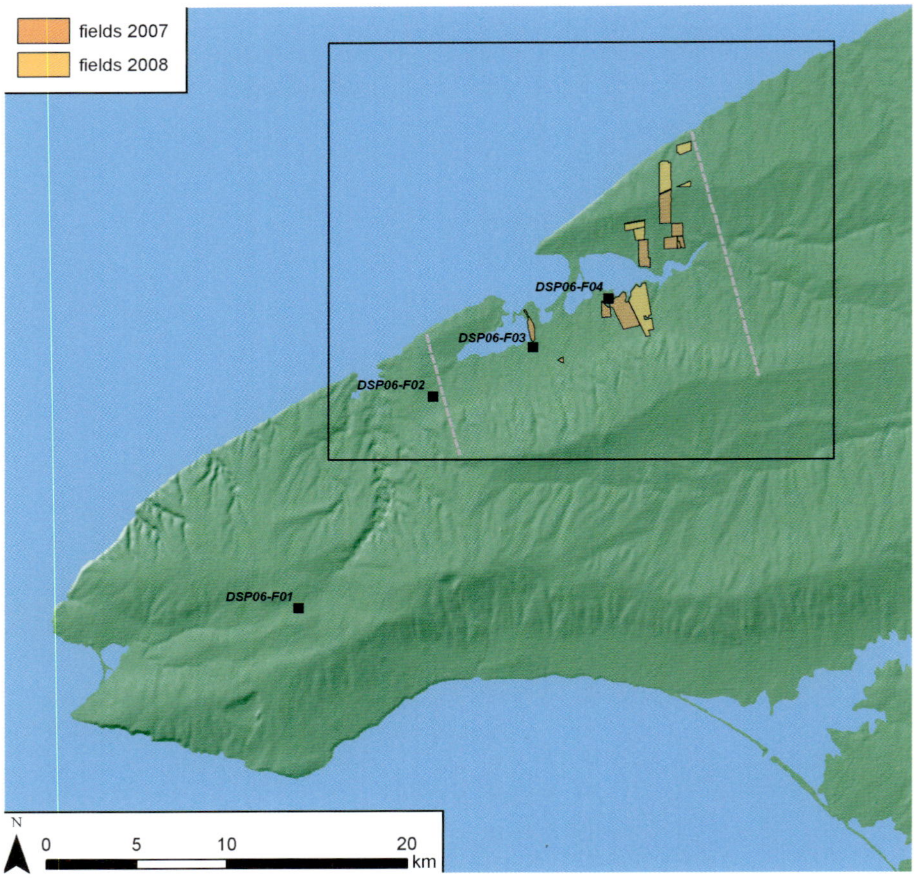

Fig. 4.39 • Map of fields surveyed in 2006.

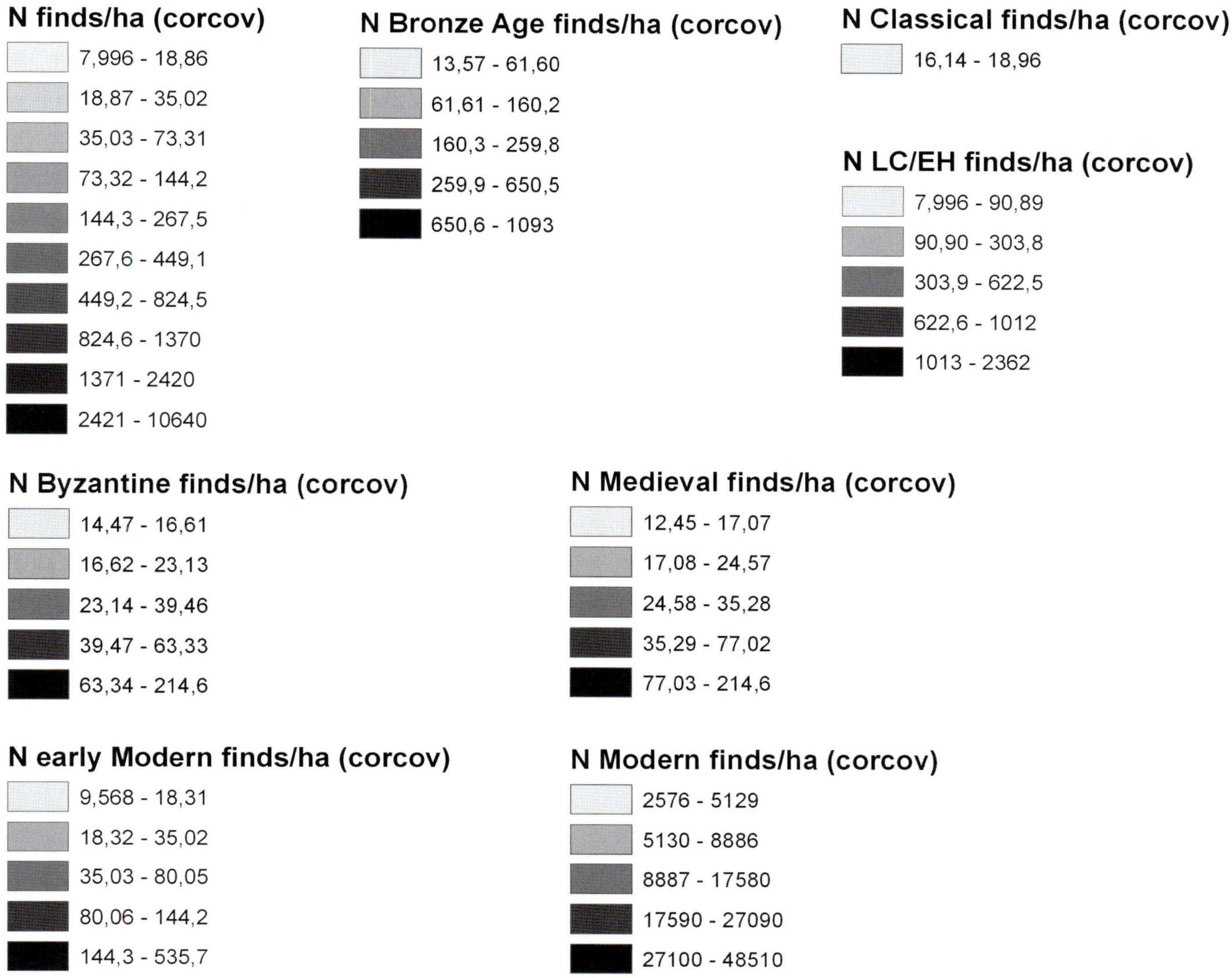

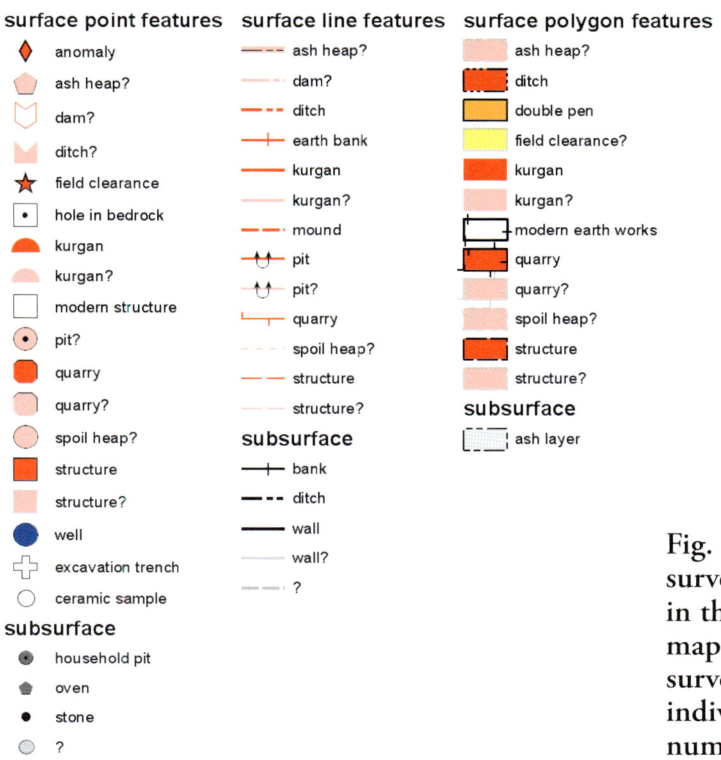

Fig. 4.40 • Above period-specific scales for the block survey distribution maps on the Lowland Ridge and in the Pediment zone. Below legends for distribution maps on the Hillsides and Plateaus/ Uplands. In block survey maps raw artefact counts are displayed in the individual site maps as well as, in white circles, the number of sherds from total samples.

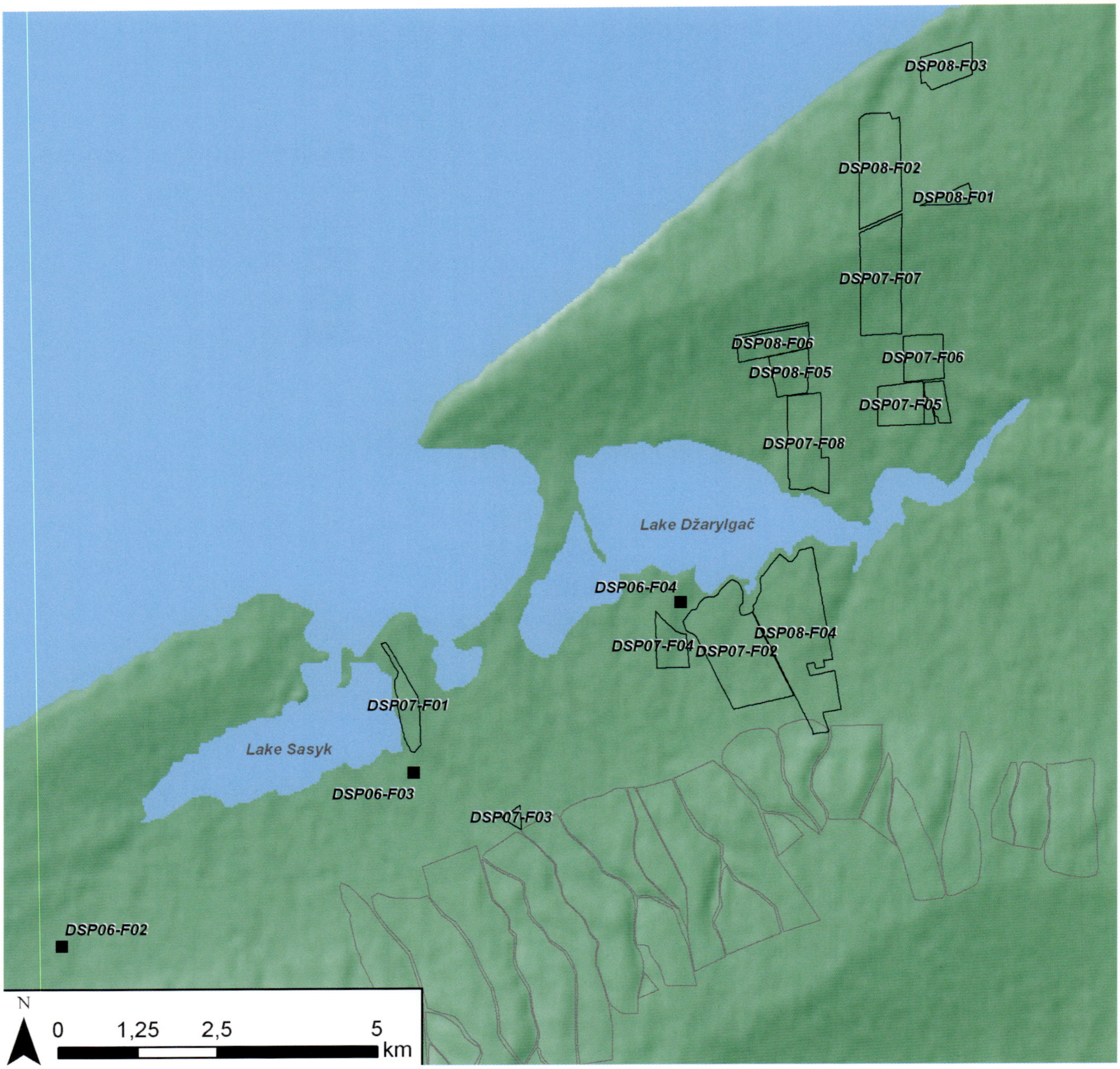

Fig. 4.41 • Map of fields surveyed in 2007 and 2008.

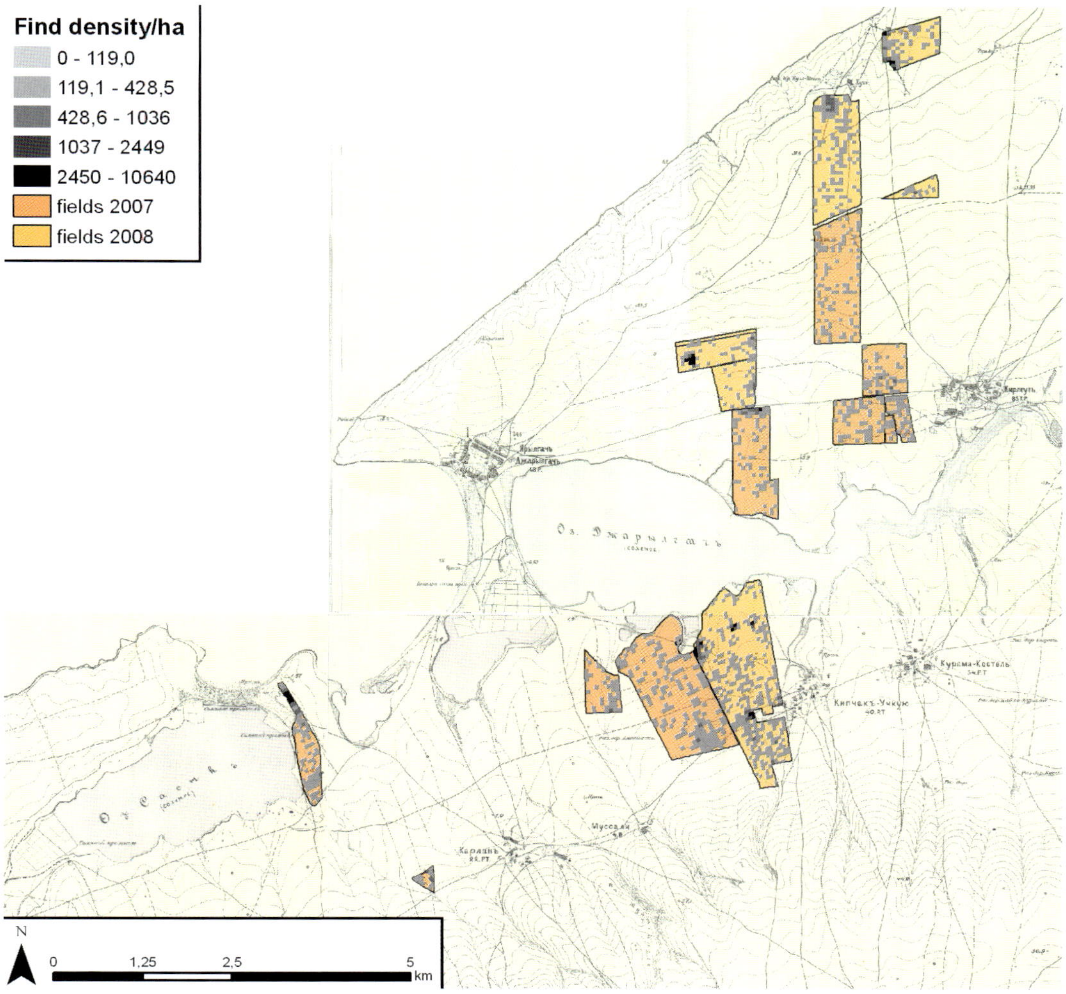

Fig. 4.42 • The distribution of finds (modern material excluded) over the survey area (corrected for coverage and unit size).

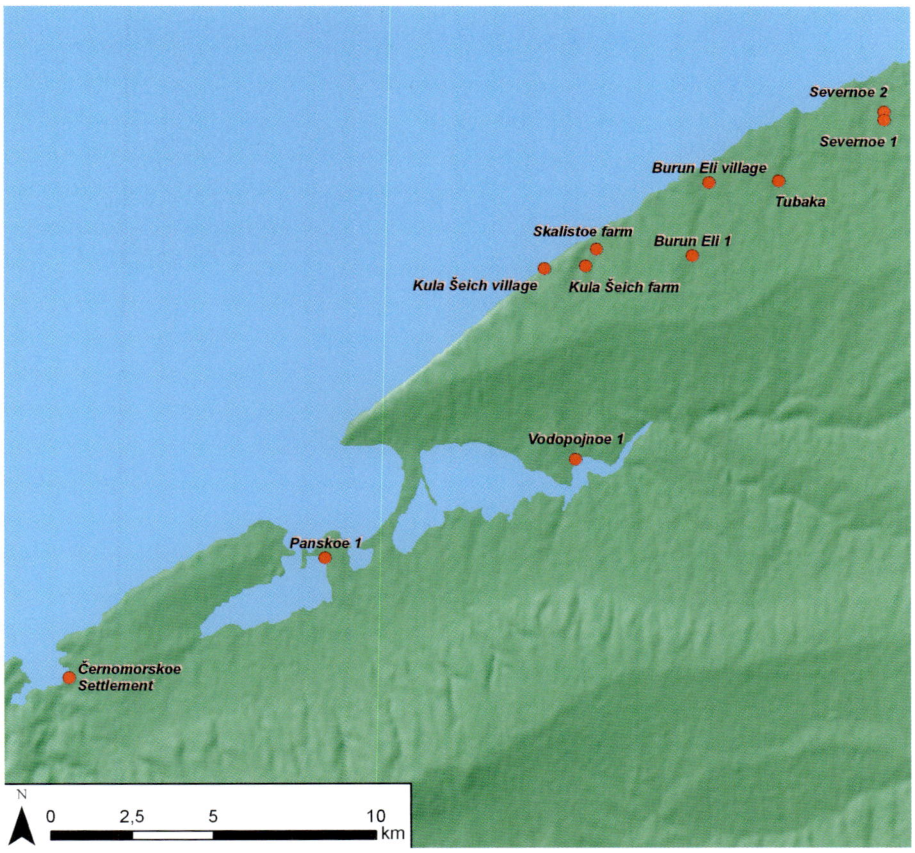

Fig. 4.43 • Map of sites previously known in the area as discussed in the text.

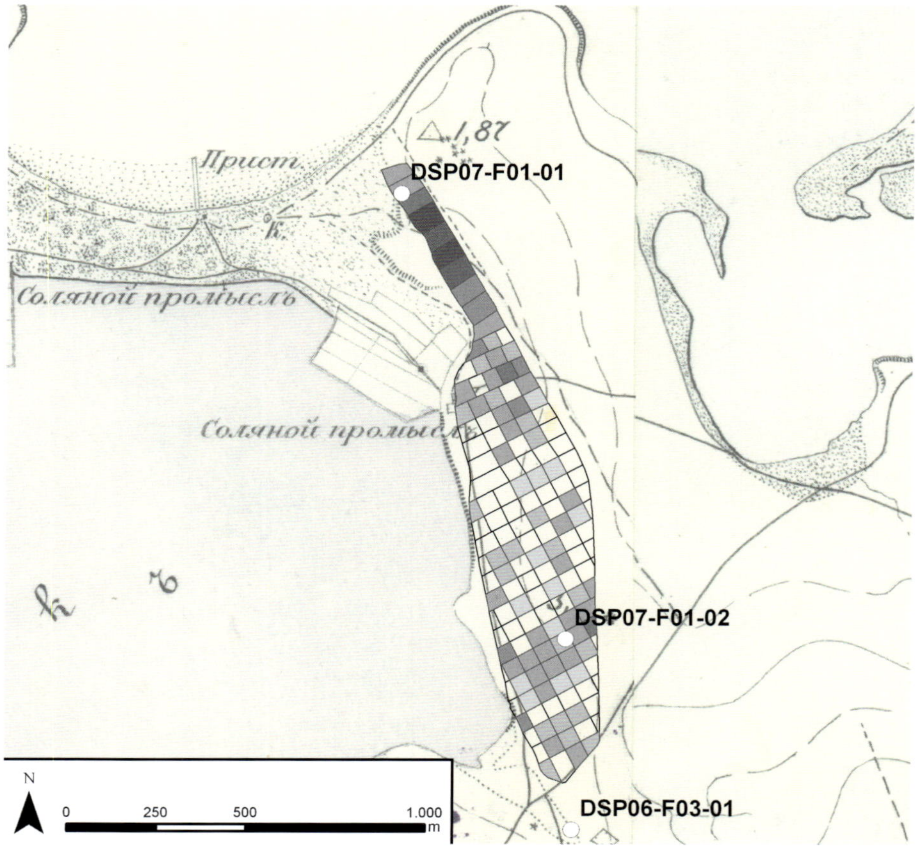

Fig. 4.44 • Distribution map of all finds from DSP07-F01. For scales see fig. 4.40.

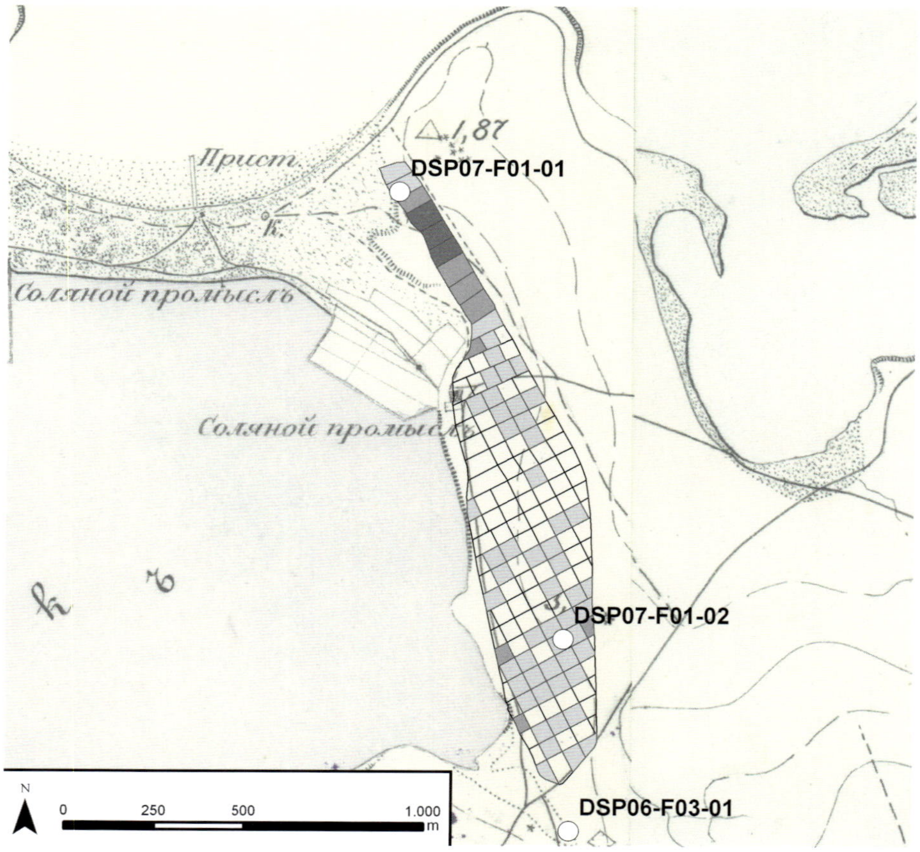

Fig. 4.45 • Distribution map of LCEH finds from DSP07-F01.

Fig. 4.46 • Distribution map of DSP07-F01-01 (LCEH).

Fig. 4.47 • Distribution map of DSP07-F01-02 (LCEH).

Fig. 4.48 • Across DSP07-F01-02 from the southwest with the kurgan in the background right.

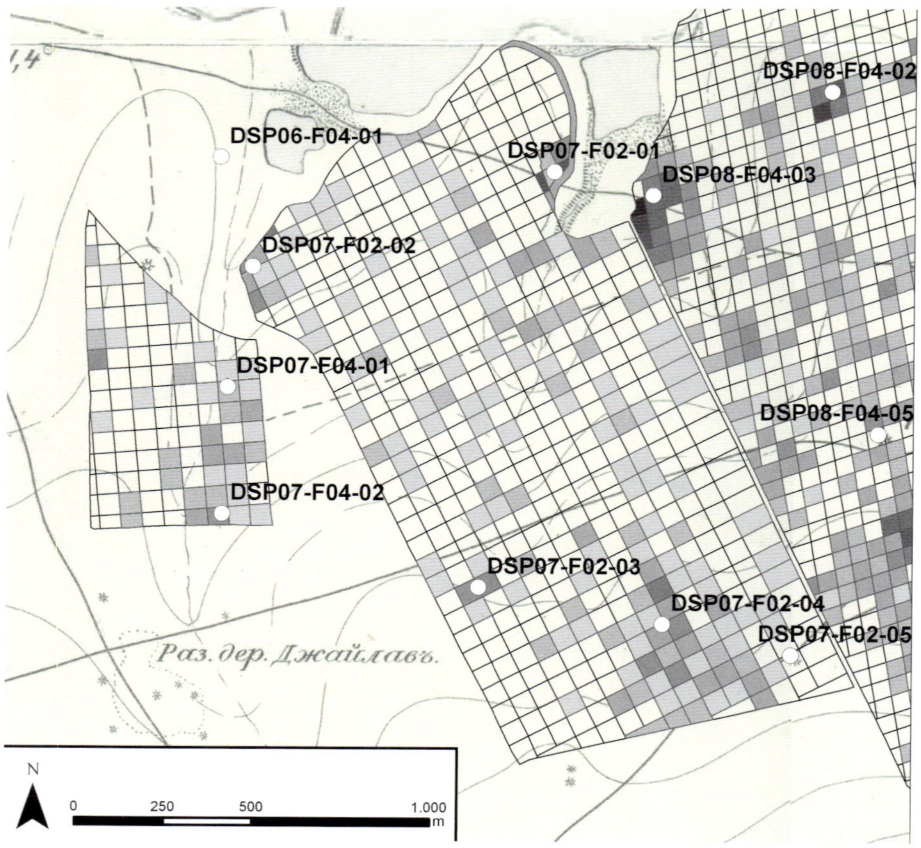

Fig. 4.49 • Distribution map of all finds from DSP07-F02.

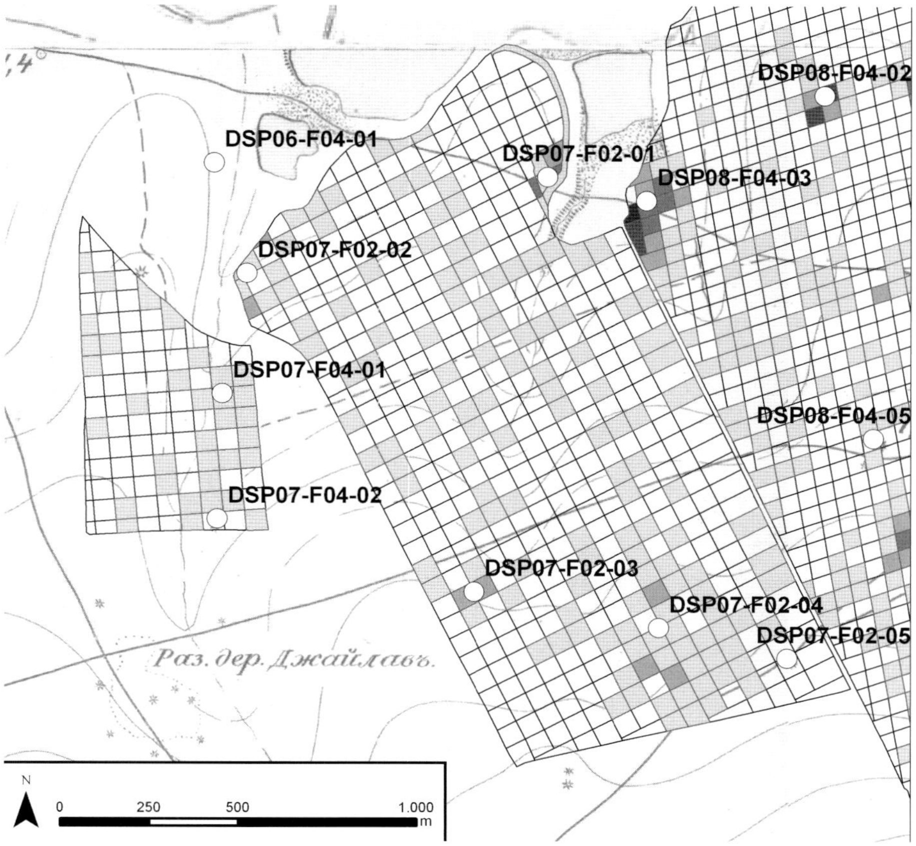

Fig. 4.50 • Distribution map of LCEH finds from DSP07-F02.

Fig. 4.51 • Distribution map of DSP07-F02-01 (LCEH).

Fig. 4.52 • Across DSP07-F02-01 from the northwest.

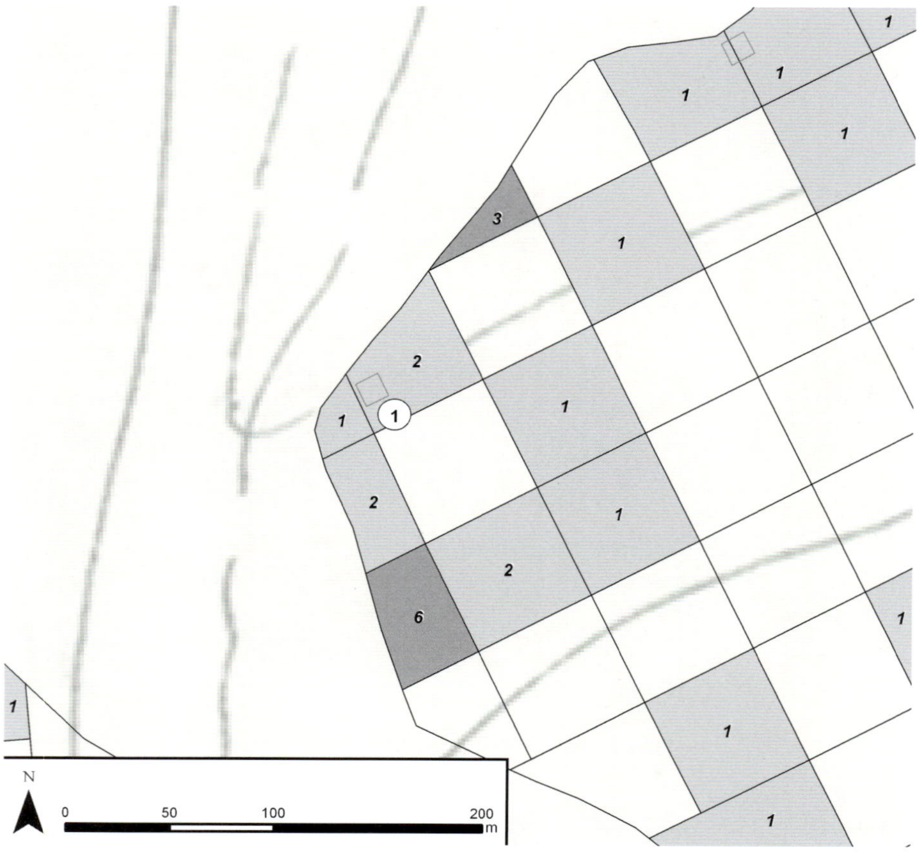

Fig. 4.53 • Distribution map of DSP07-F02-02 (LCEH).

Fig. 4.54 • DSP07-F02-02 from the west.

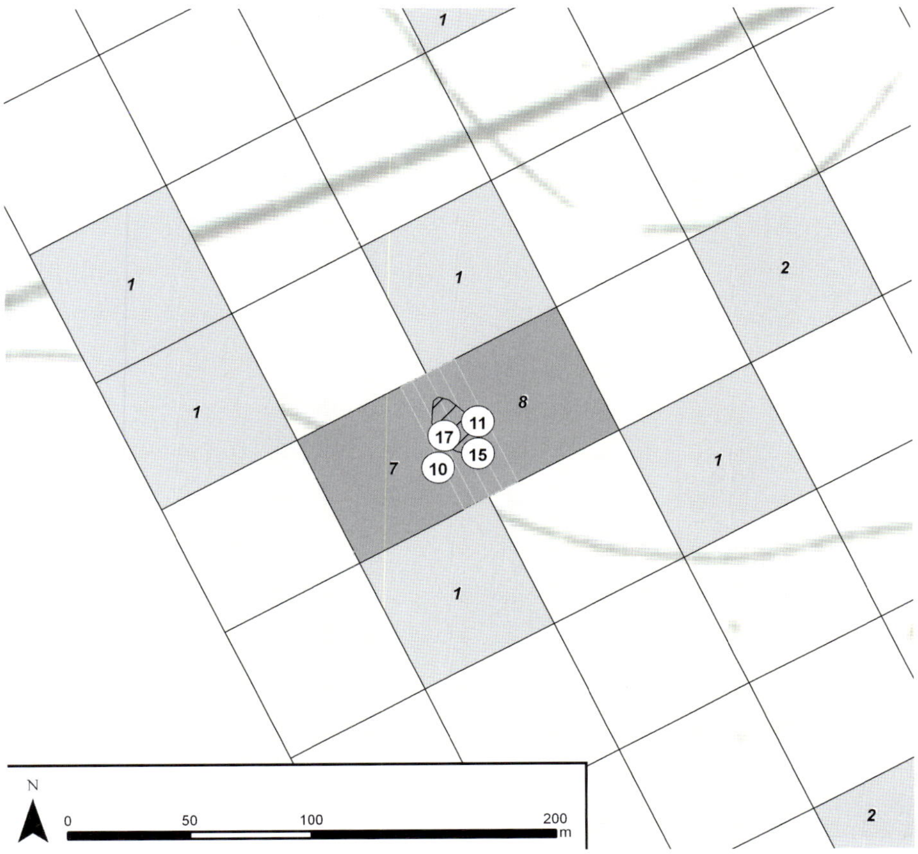

Fig. 4.55 • Distribution map of DSP07-F02-03 (LCEH).

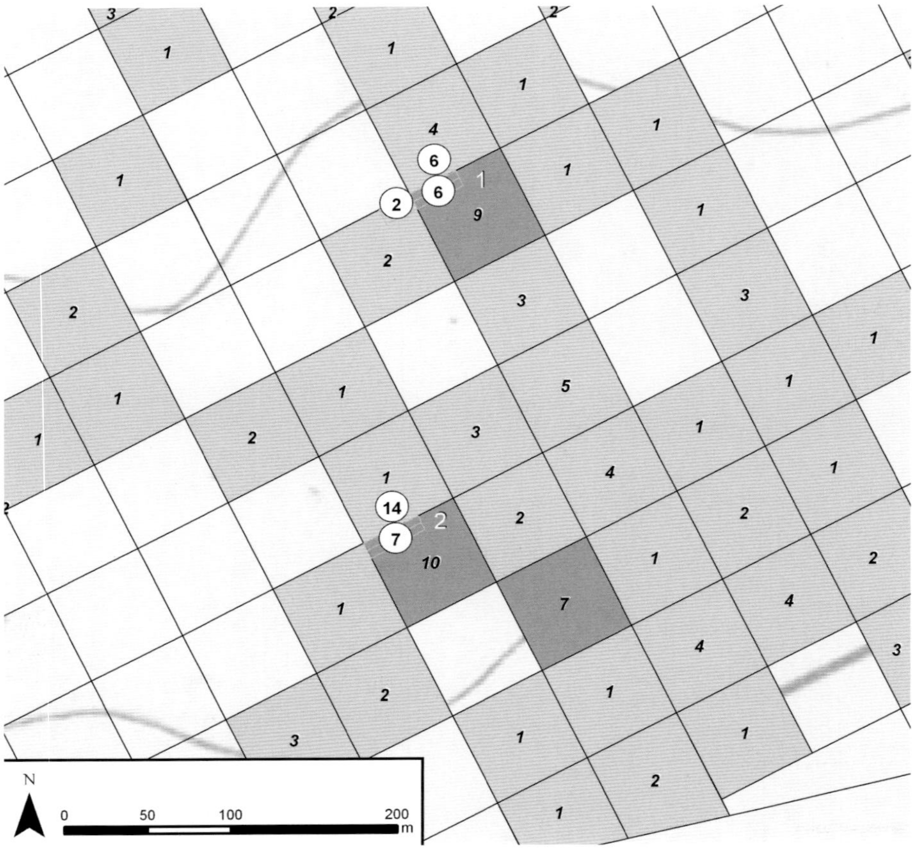

Fig. 4.56 • Distribution map of DSP07-F02-04 (LCEH).

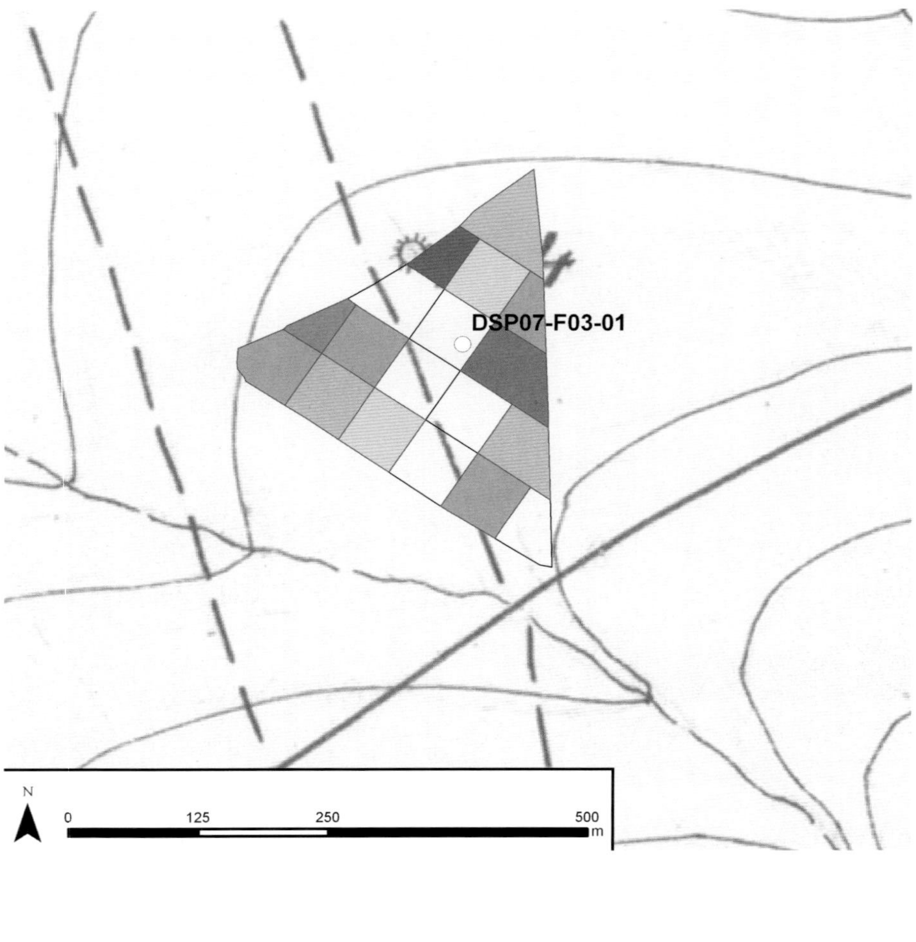

Fig. 4.57 • Distribution map of all finds from DSP07-F03.

Fig. 4.58 • Distribution map of LCEH finds from DSP07-F03.

Fig. 4.59 • Across DSP07-F03 from the northwest from the Snežnoe kurgan reused as a cemetery.

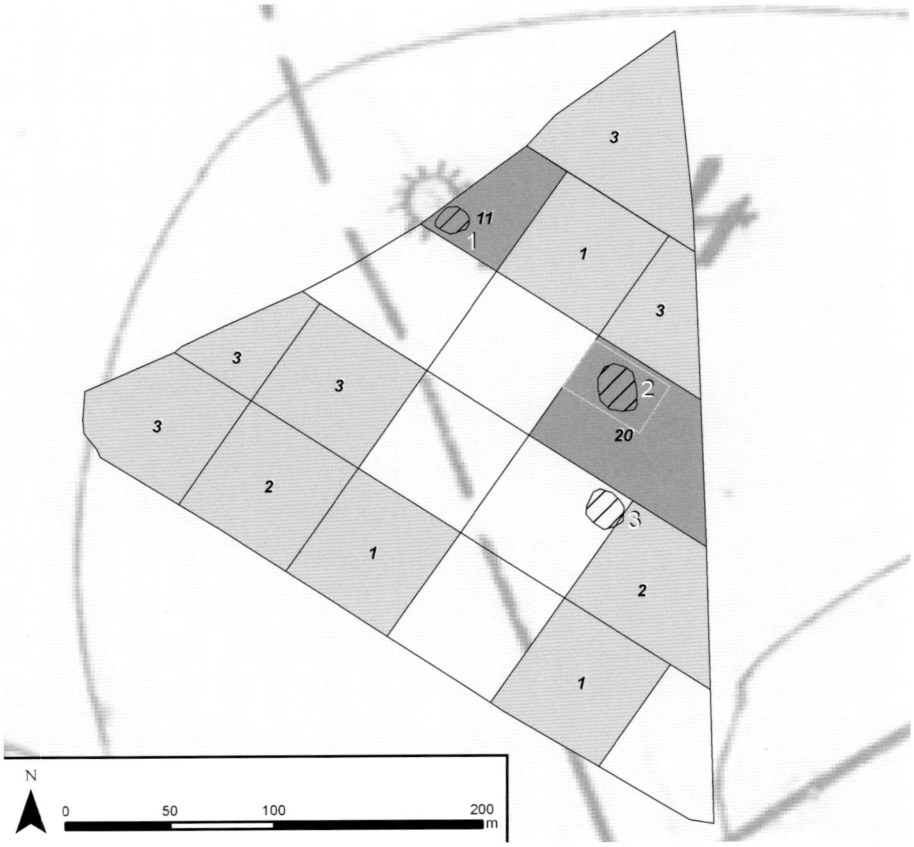

Fig. 4.60 • Distribution map of DSP07-F03-01 (LCEH).

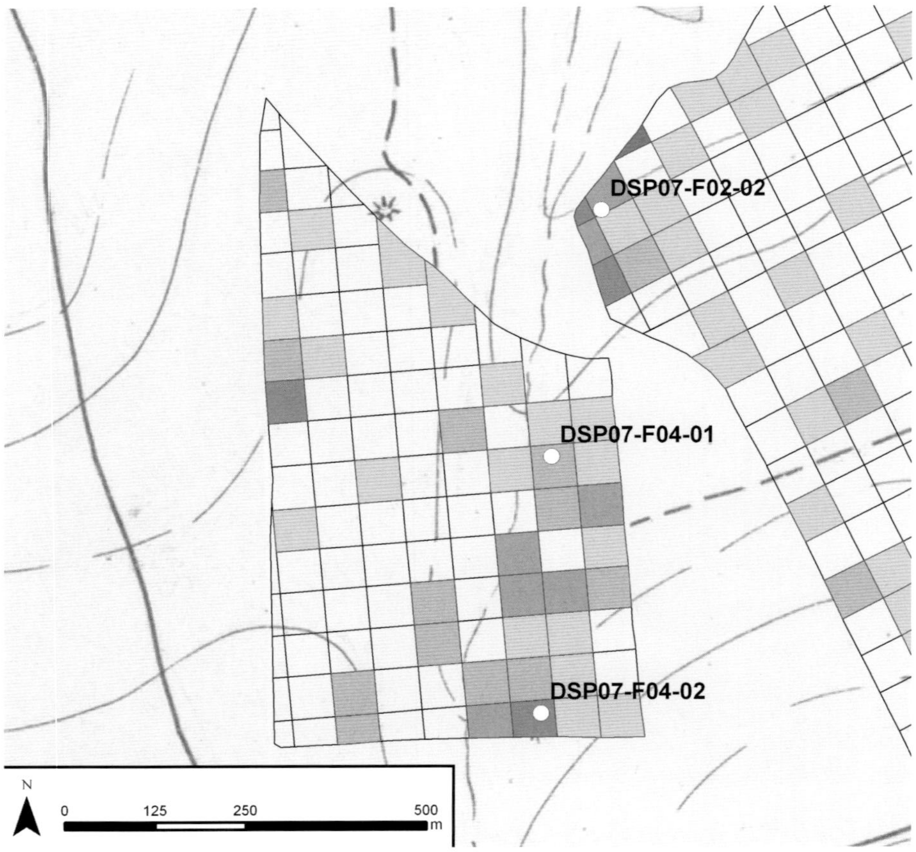

Fig. 4.61 • Distribution map of all finds from DSP07-F04.

Fig. 4.62 • Distribution map of BA finds from DSP07-F04.

Fig. 4.63 • Distribution map of LCEH finds from DSP07-F04.

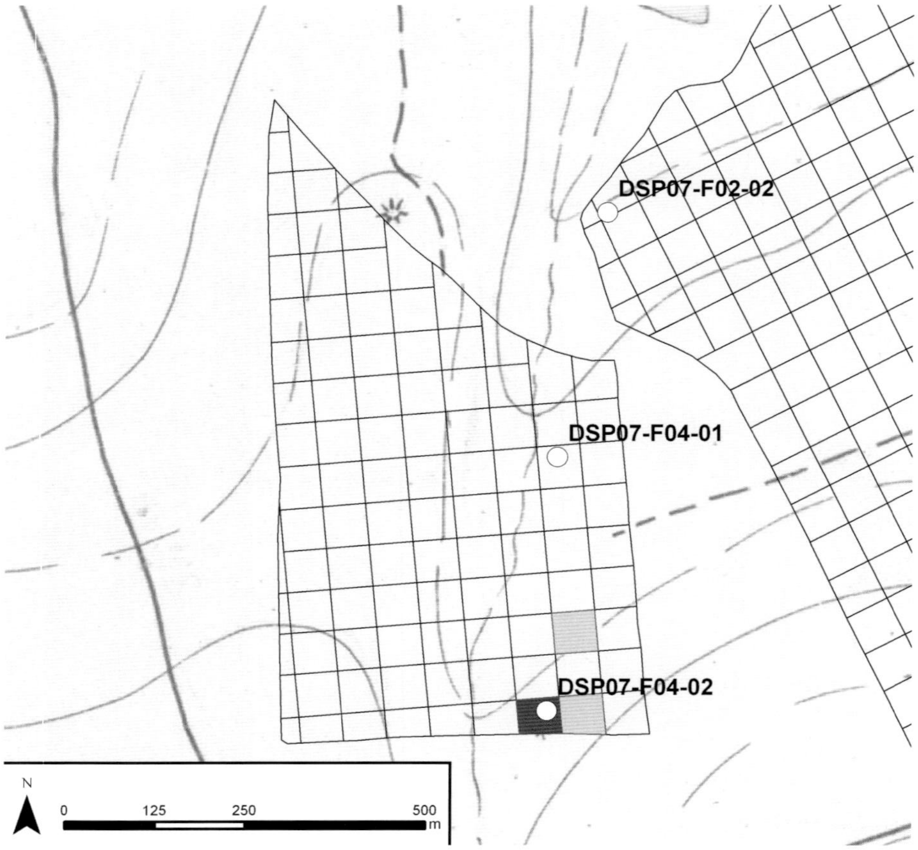

Fig. 4.64 • Distribution map of EM finds from DSP07-F04.

Fig. 4.65 • Distribution map of DSP07-F04-01 (BA). Triangle marks find location of hand axe.

Fig. 4.66 • Distribution map of DSP07-F04-02 (EM).

Fig. 4.67 • DSP07-F04-02 from the northeast.

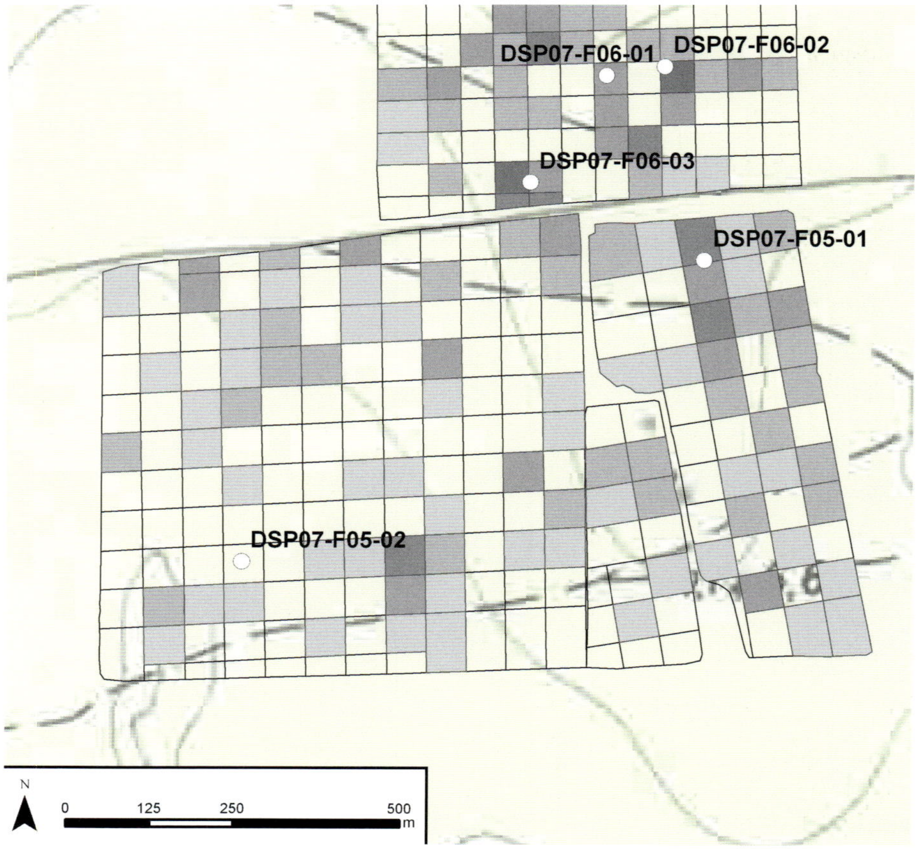

Fig. 4.68 • Distribution map of all finds from DSP07-F05.

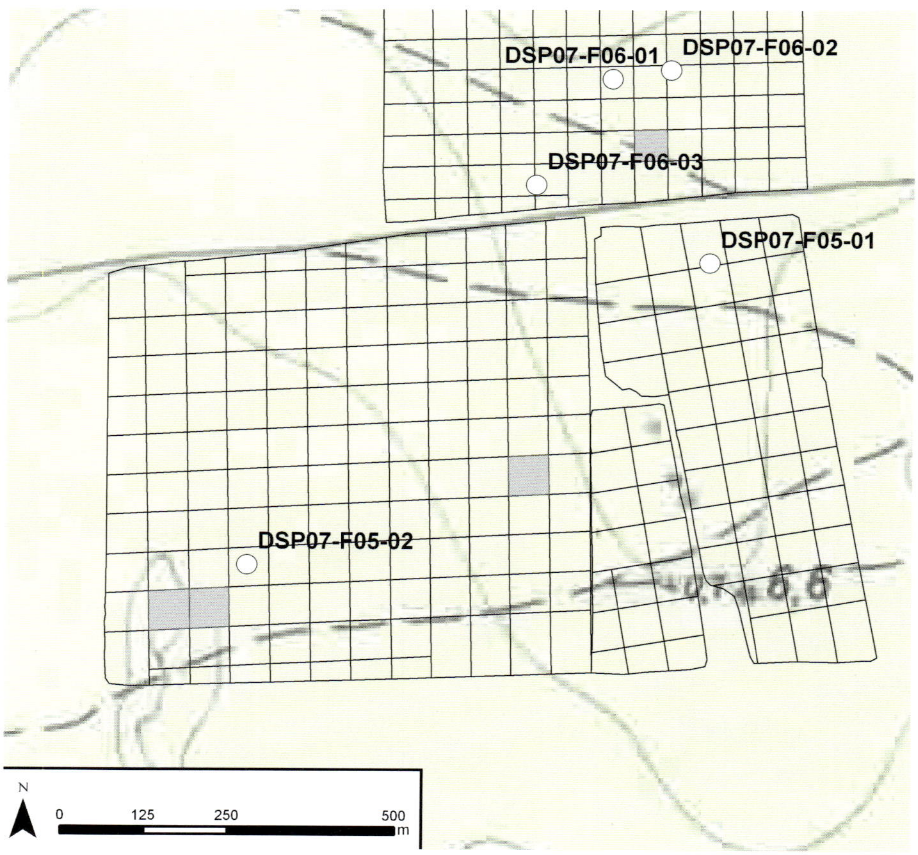

Fig. 4.69 • Distribution map of BA finds from DSP07-F05.

Fig. 4.70 • Distribution map of LCEH finds from DSP07-F05.

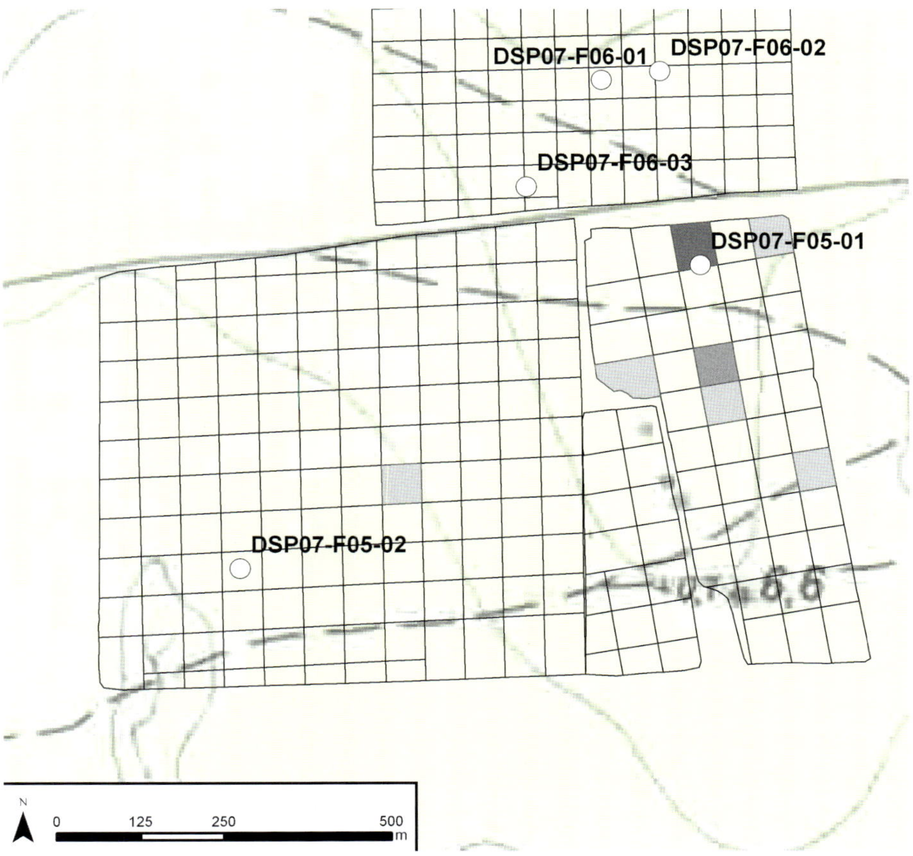

Fig. 4.71 • Distribution map of EM finds from DSP07-F05.

Fig. 4.72 • Distribution map of DSP07-F05-01 (EM).

Fig. 4.73 • Across DSP07-F05-01 to Vodopojnoe from the northwest in 2008.

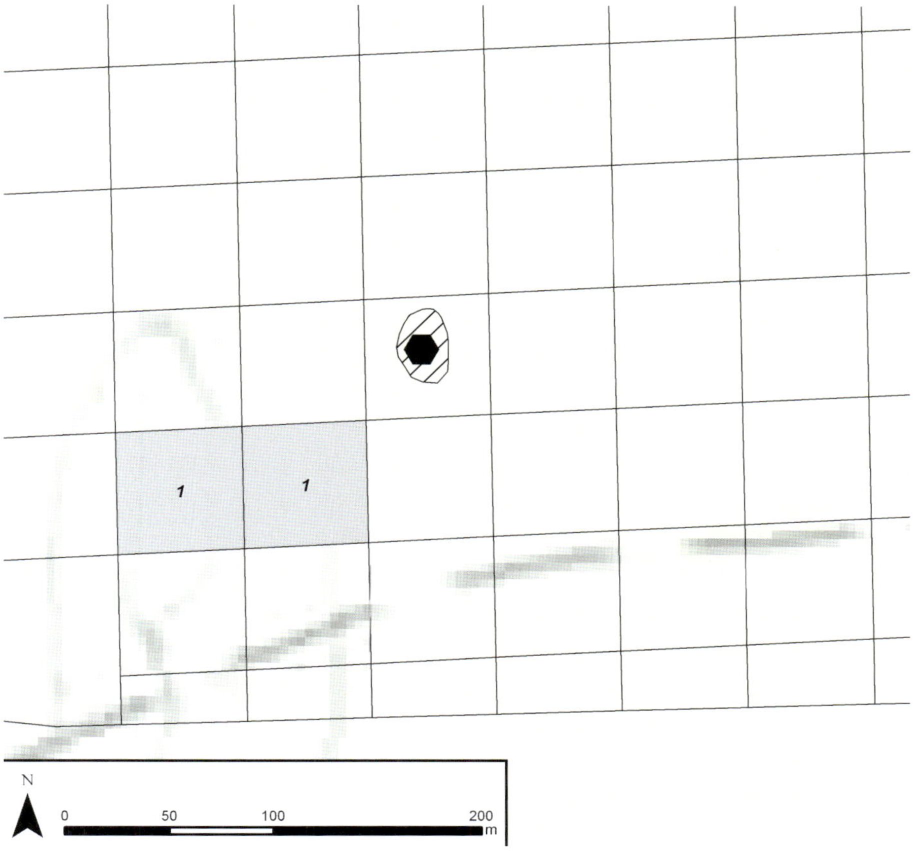

Fig. 4.74 • Distribution map of DSP07-F05-02 (BA). Symbol marks find location of stele.

Fig. 4.75 • Anthropomorphic limestone stele from DSP07-F05-02.

Fig. 4.76 • Distribution map of all finds from DSP07-F06.

Fig. 4.77 • Distribution map of LCEH finds from DSP07-F06.

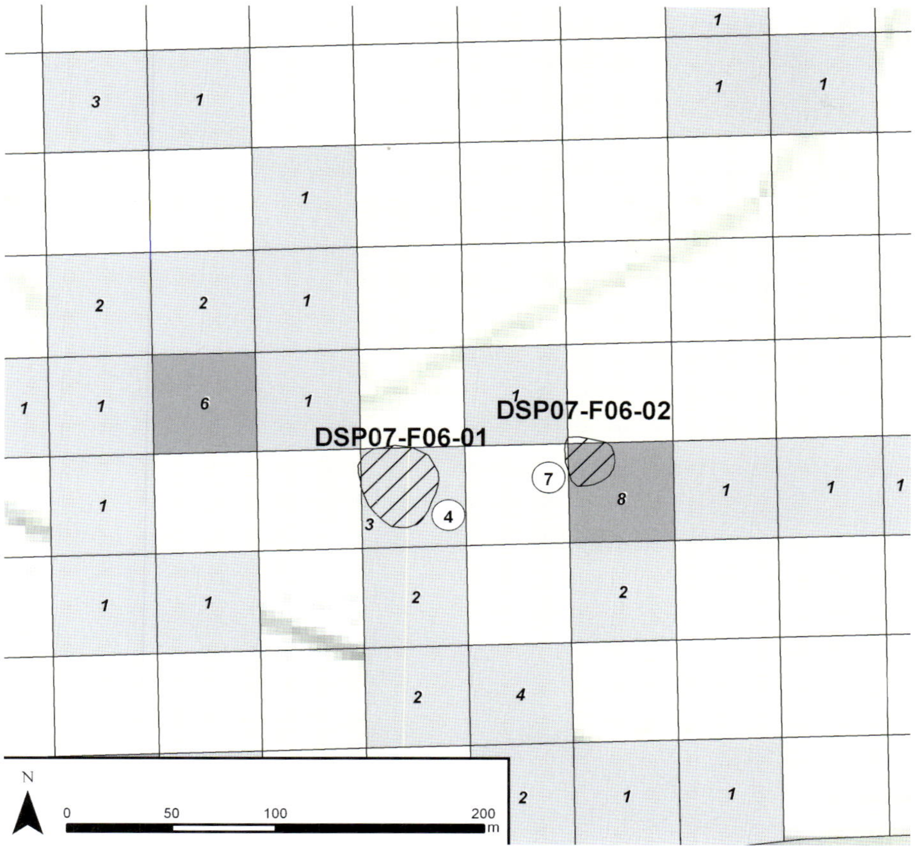

Fig. 4.78 • Distribution map of DSP07-F06-01-02 (LCEH).

Fig. 4.79 • Across DSP07-F06-01-02 from the southwest.

Fig. 4.80 • Distribution map of DSP07-F06-03 (LCEH).

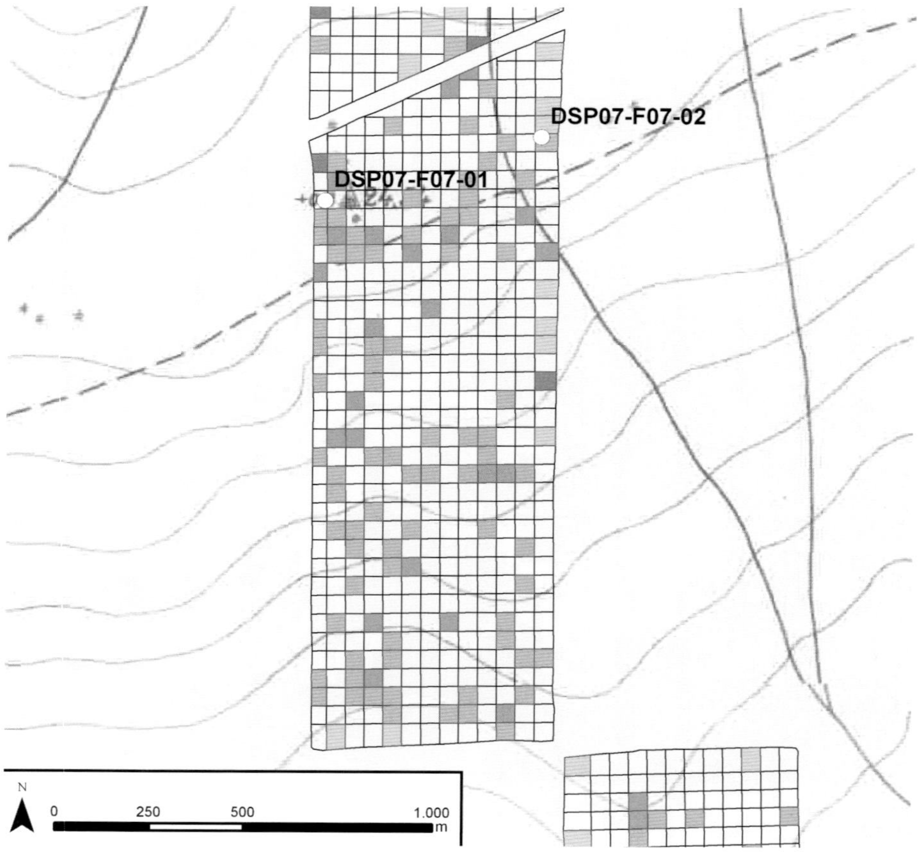

Fig. 4.81 • Distribution map of all finds from DSP07-F07.

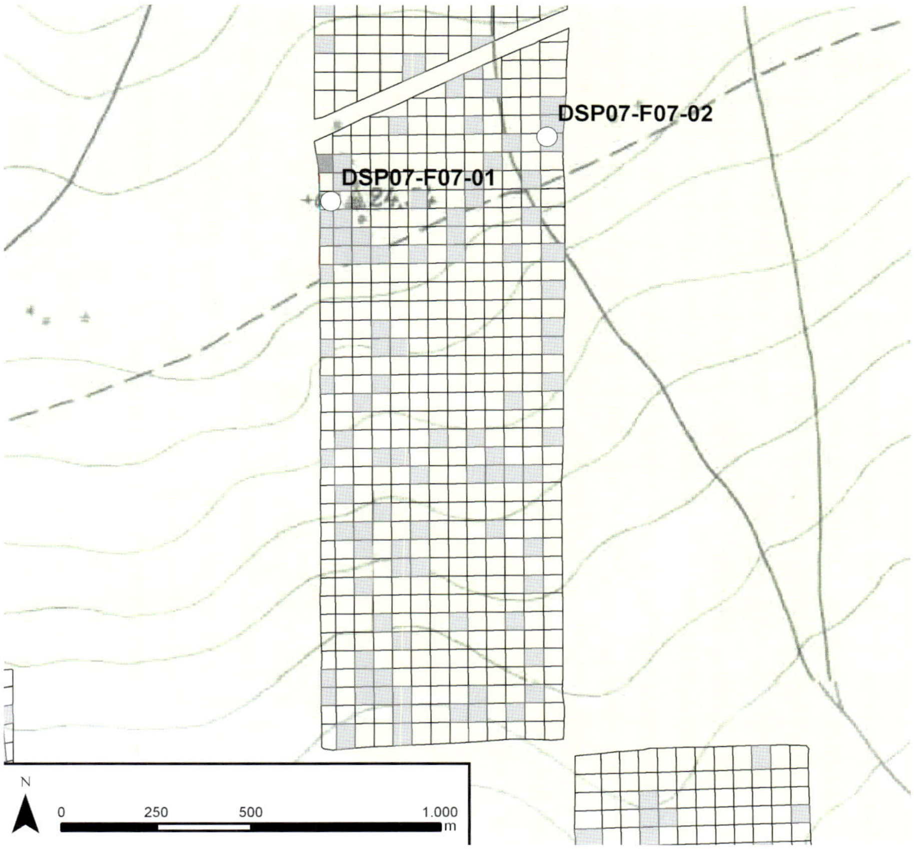

Fig. 4.82 • Distribution map of LCEH finds from DSP07-F07.

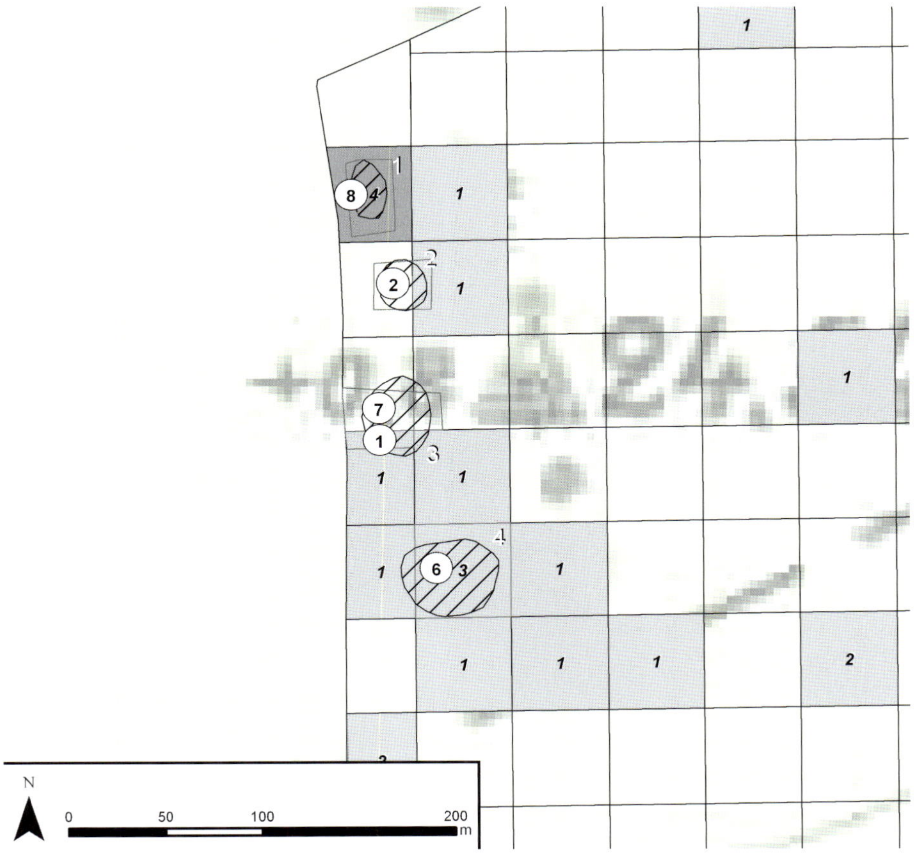

Fig. 4.83 • Distribution map of DSP07-F07-01 (LCEH).

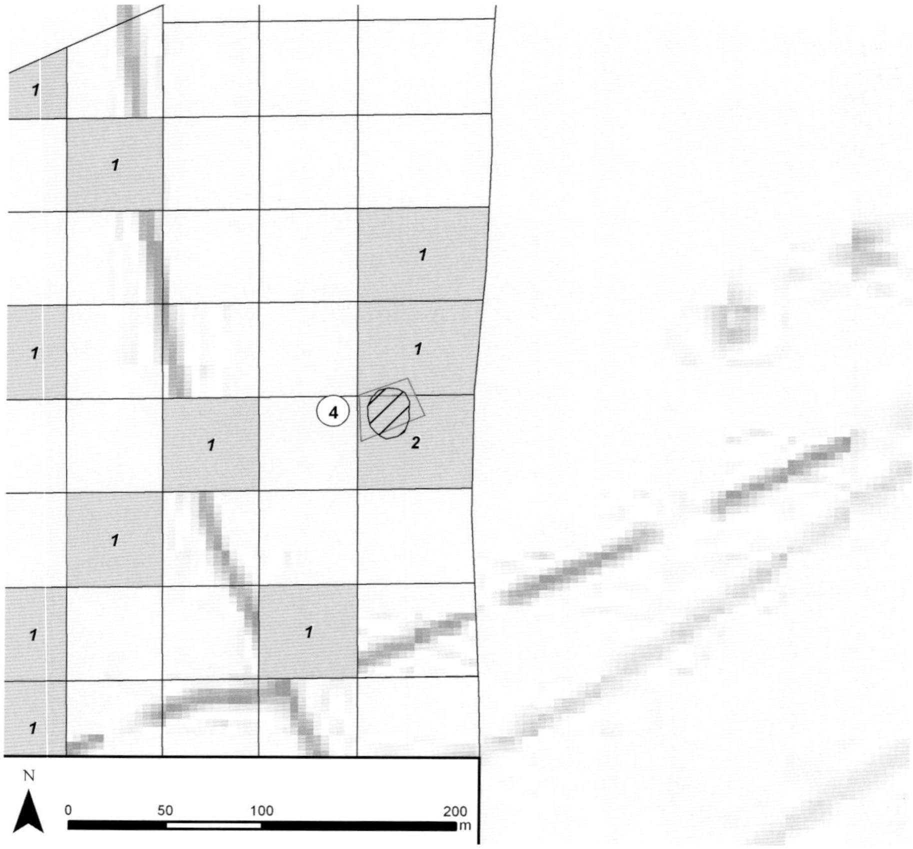

Fig. 4.84 • Distribution map of DSP07-F07-02 (LCEH).

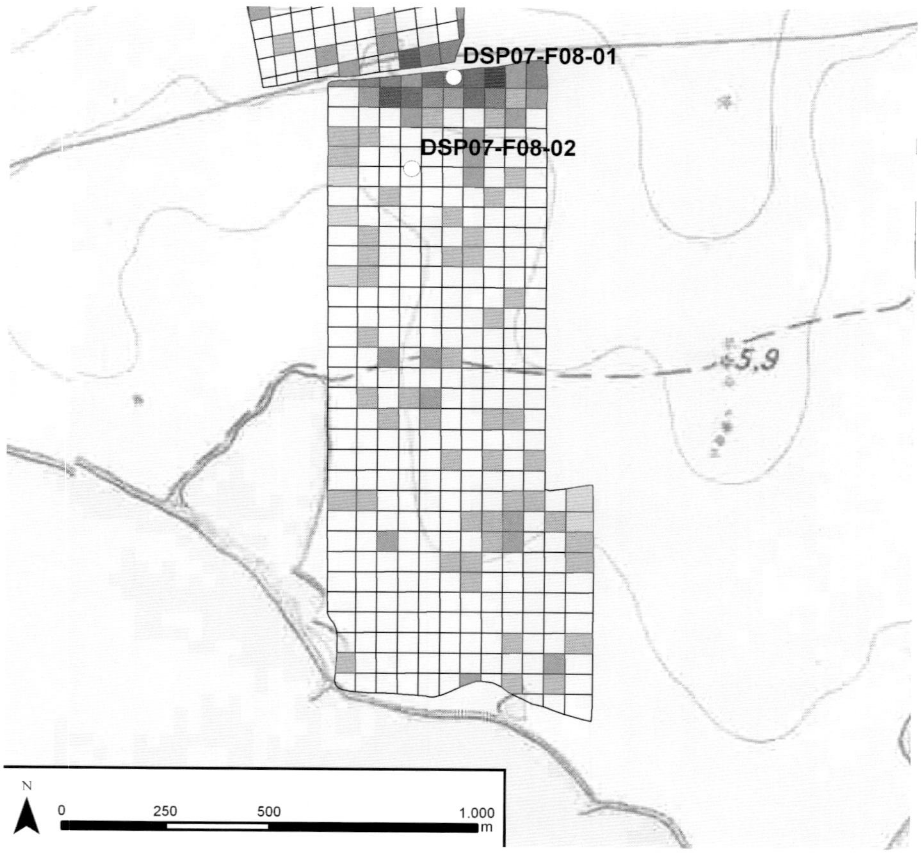

Fig. 4.85 • Distribution map of all finds from DSP07-F08.

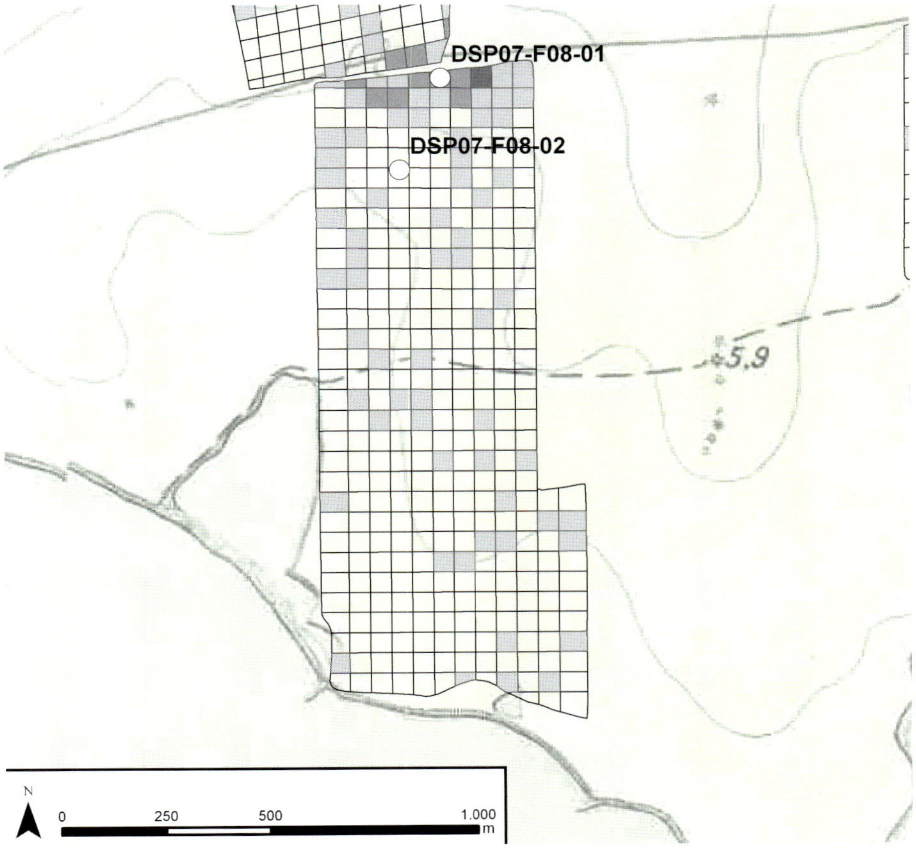

Fig. 4.86 • Distribution map of LCEH finds from DSP07-F08.

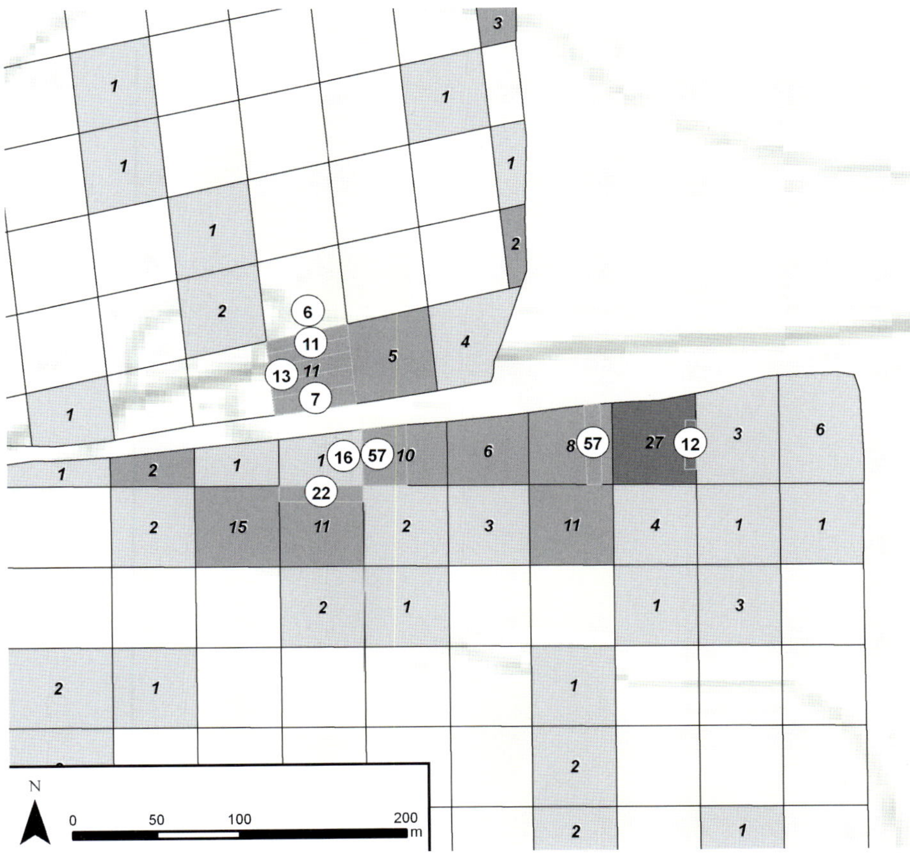

Fig. 4.87 • Distributions map of DSP07-F08-01 (LCEH).

Fig. 4.88 • Across DSP07-F08-01 from the east.

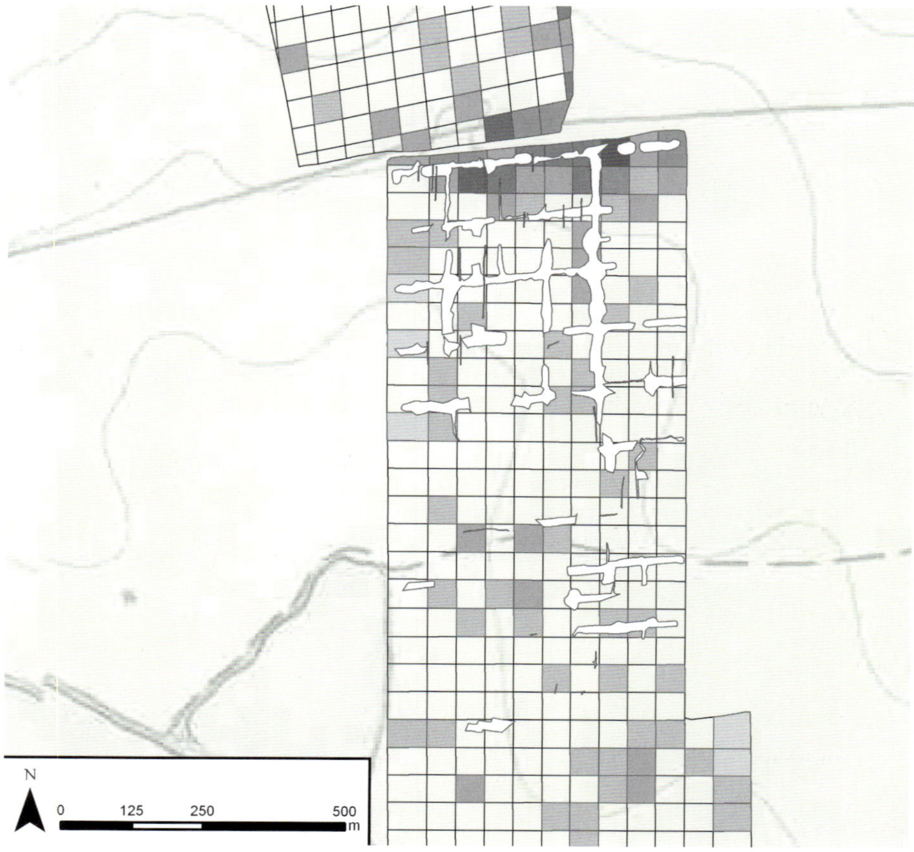

Fig. 4.89 • Distributions map of DSP07-F08-02 (all).

Fig. 4.90 • Distribution map of all finds from DSP08-F01.

Fig. 4.91 • Distribution map of LCEH finds from DSP08-F01.

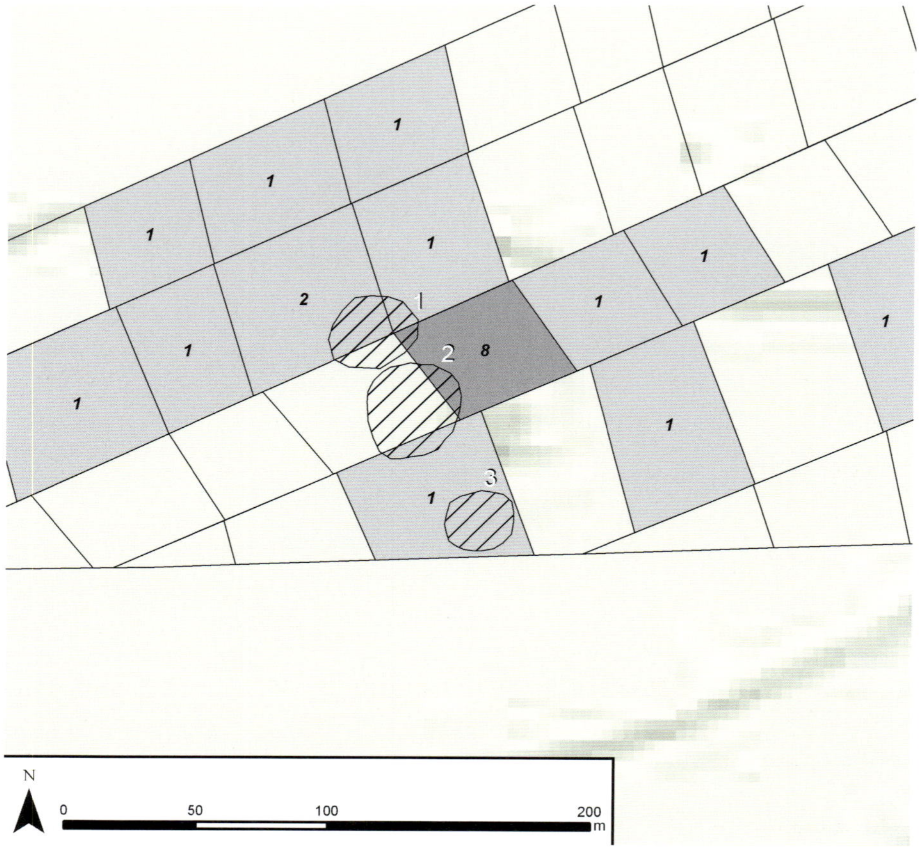

Fig. 4.92 • Distribution map of DSP08-F01-01.

Fig. 4.93 • Across DSP08-F01-01 from the north with Feature 3 in the background.

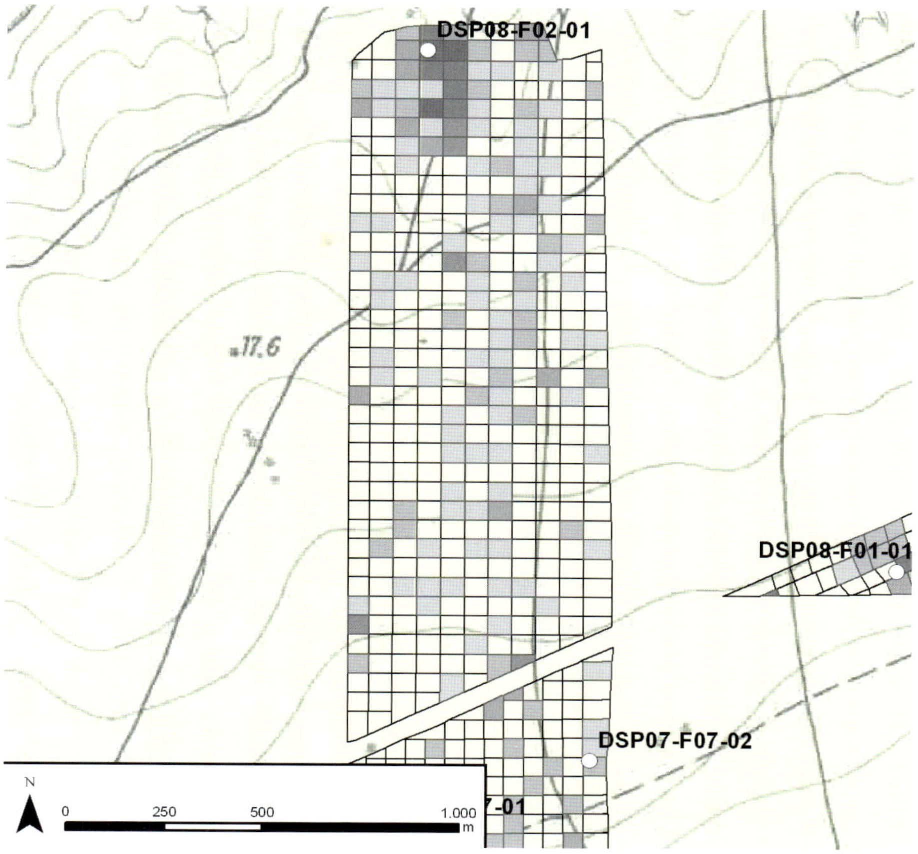

Fig. 4.94 • Distribution map of all finds from DSP08-F02.

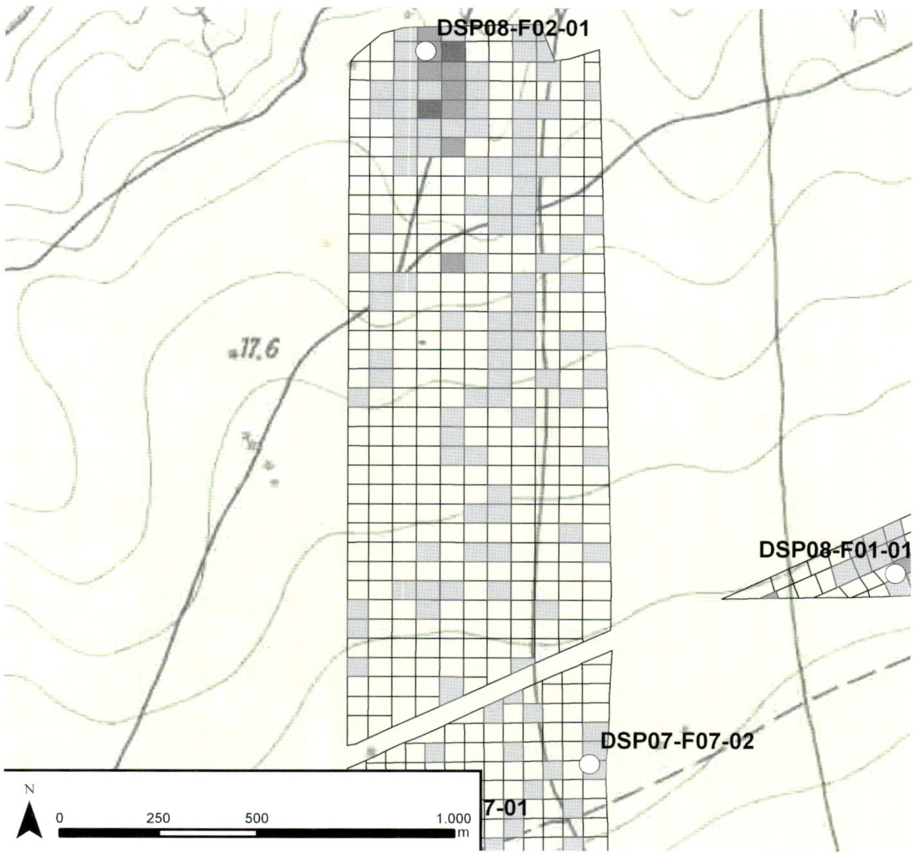

Fig. 4.95 • Distribution map of LCEH finds from DSP08-F02.

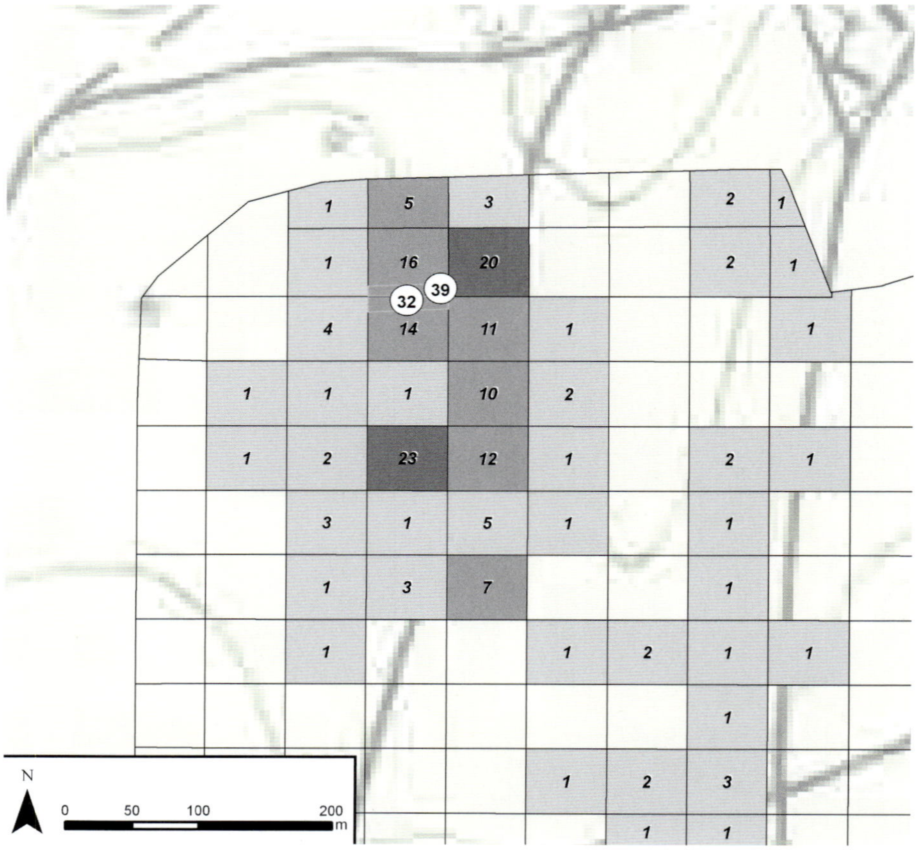

Fig. 4.96 • Distribution map of DSP08-F02-01.

Fig. 4.97 • DSP08-F02-01 from the southeast with the cemetery in the background.

Fig. 4.98 • Distribution map of all finds from DSP08-F03.

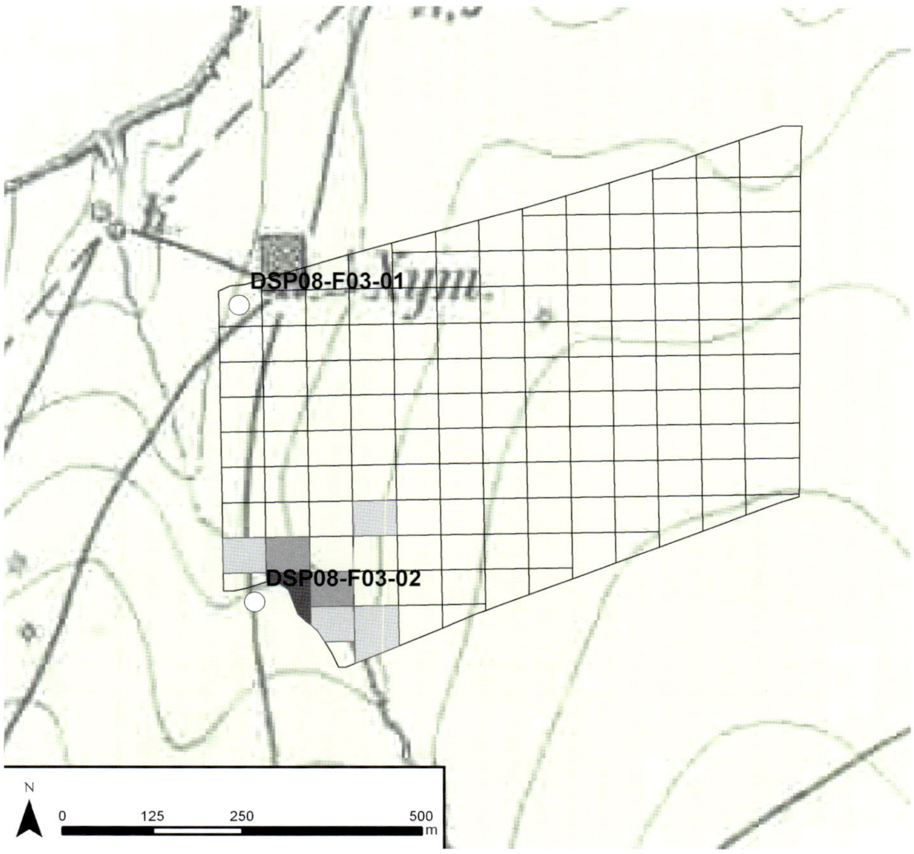

Fig. 4.99 • Distribution map of BA finds from DSP08-F03.

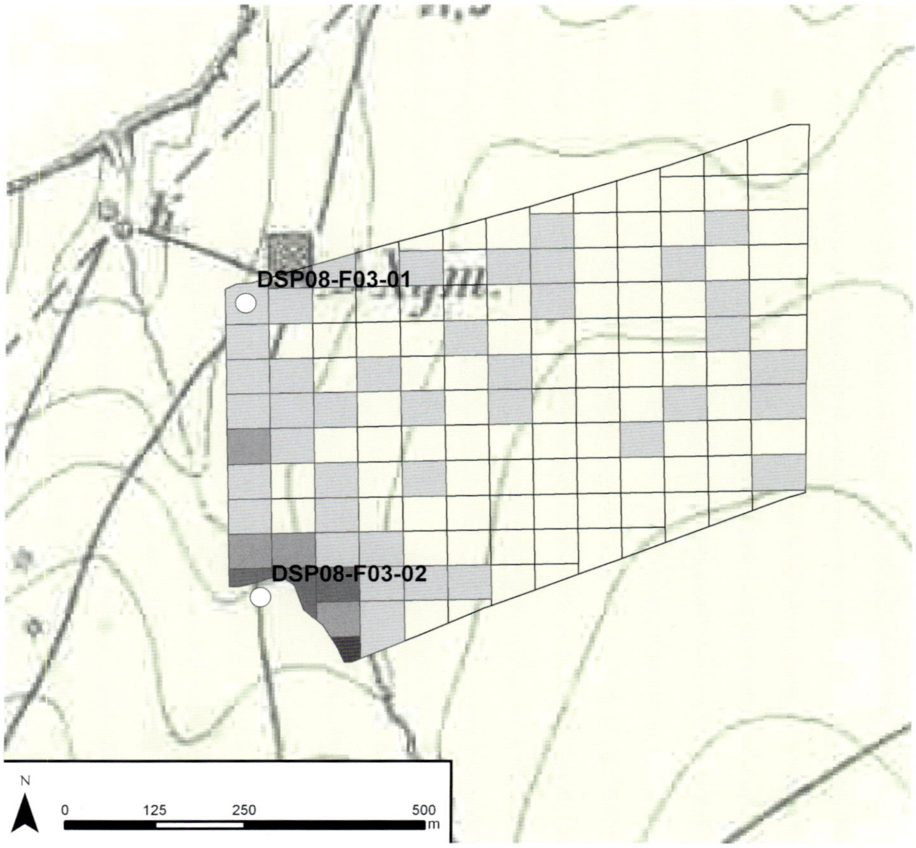

Fig. 4.100 • Distribution map of LCEH finds from DSP08-F03.

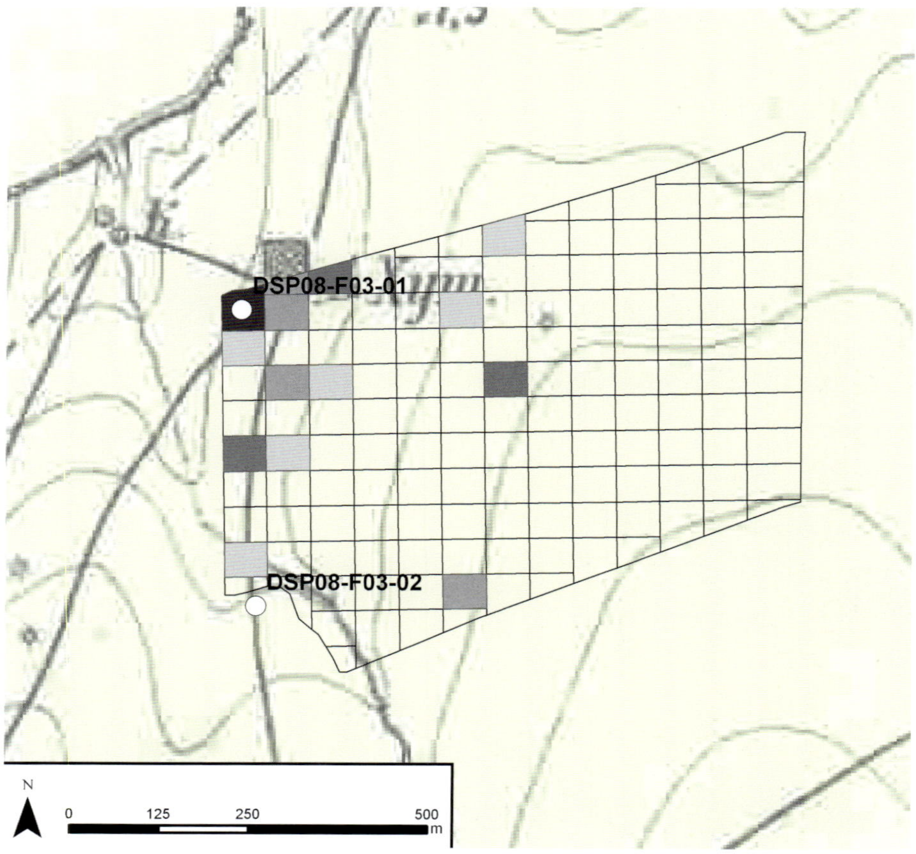

Fig. 4.101. • Distribution map of EM finds from DSP08-F03.

Fig. 4.102 • Distribution map of DSP08-F03-01 (LCEH).

Fig. 4.103 • Distribution map of DSP08-F03-01 (EM).

Fig. 4.104 • DSP08-F03-01 from the southwest.

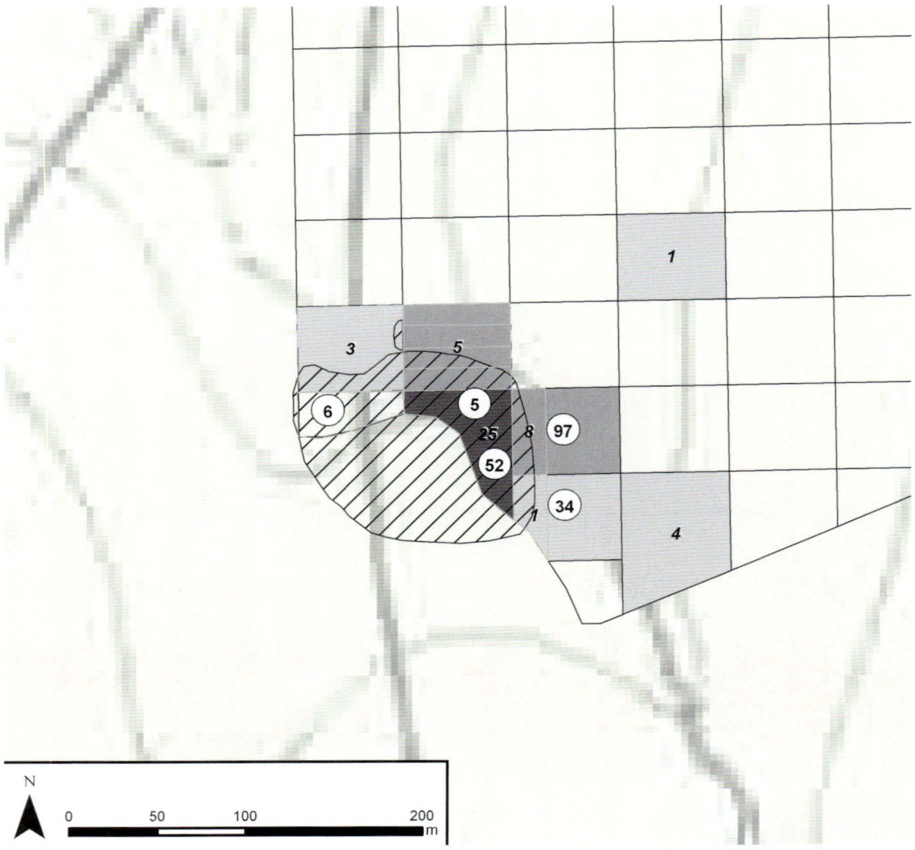

Fig. 4.105 • Distribution map of DSP08-F03-02/Skalistoe 2 (BA).

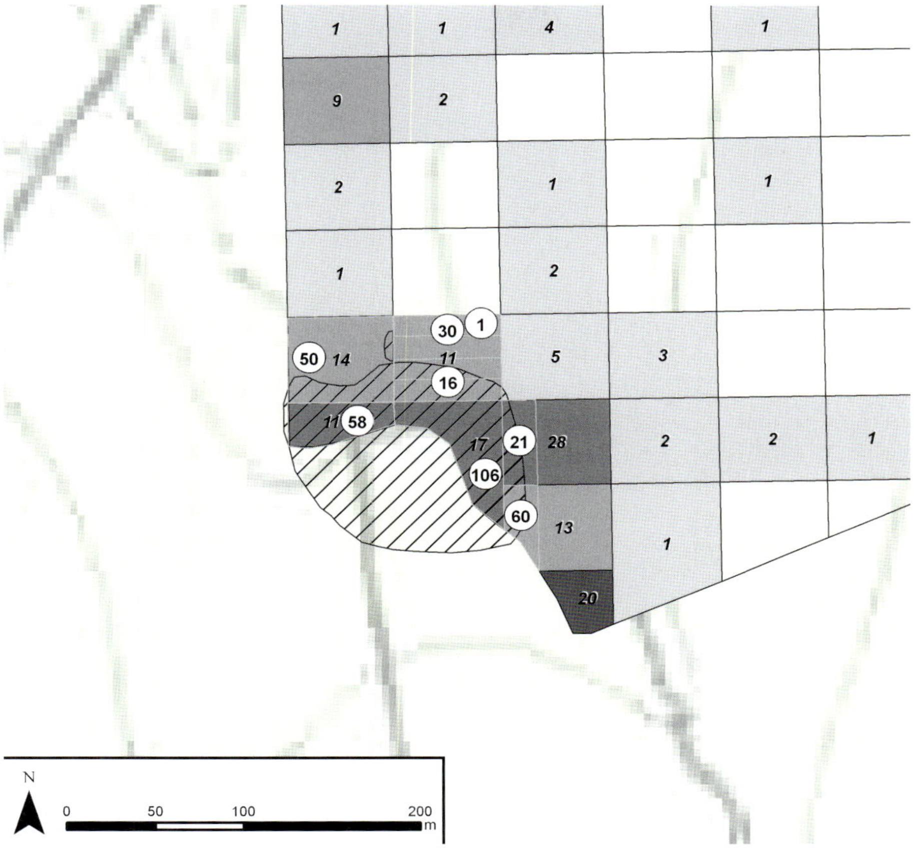

Fig. 4.106 • Distribution map of DSP08-F03-02/Skalistoe 2 (LCEH).

Fig. 4.107 • DSP08-F03-02/Skalistoe 2 from the northeast.

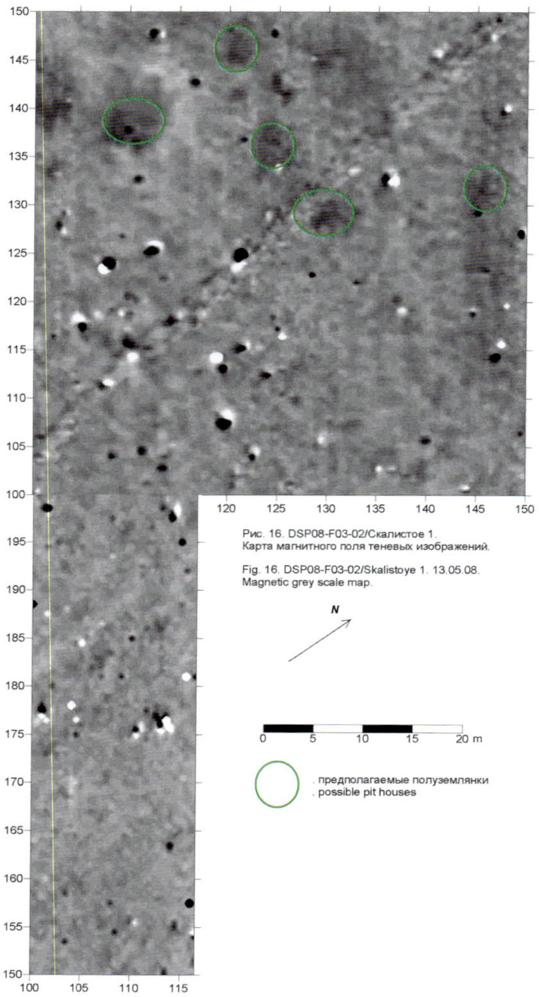

Fig. 4.108 • Geomagnetic map of DSP08-F03-02/ Skalistoe 2, interpretation.

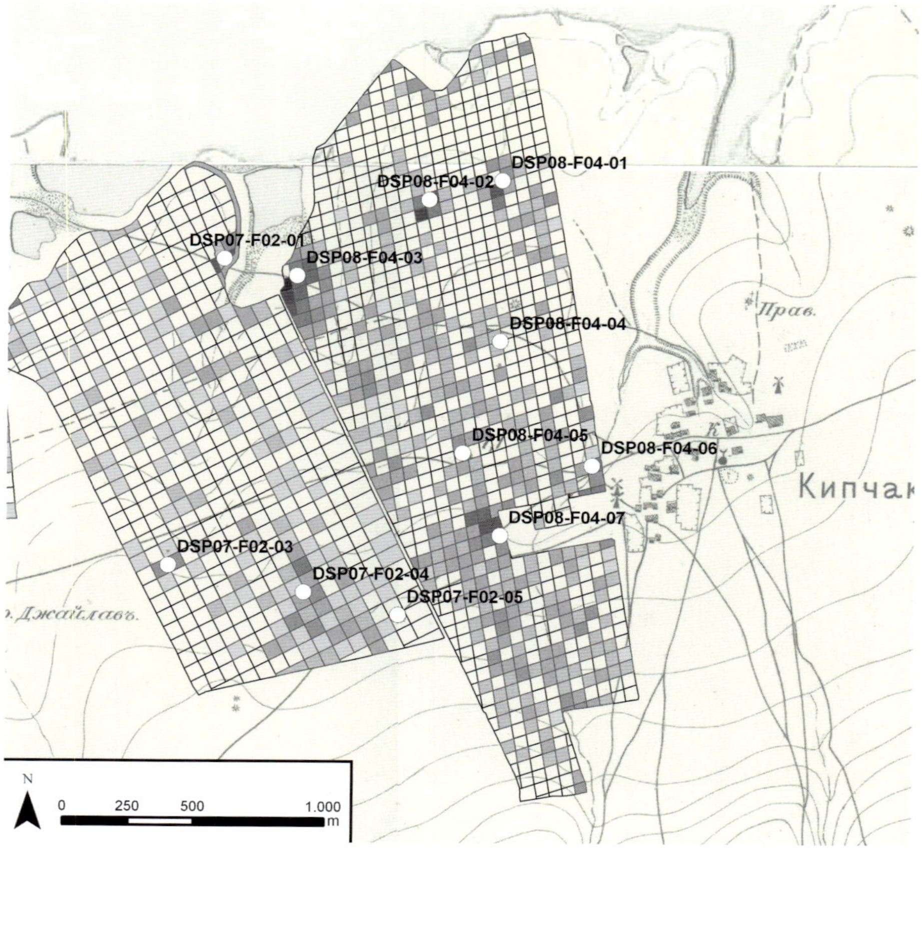

Fig. 4.109 • Distribution map of all finds from DSP08-F04.

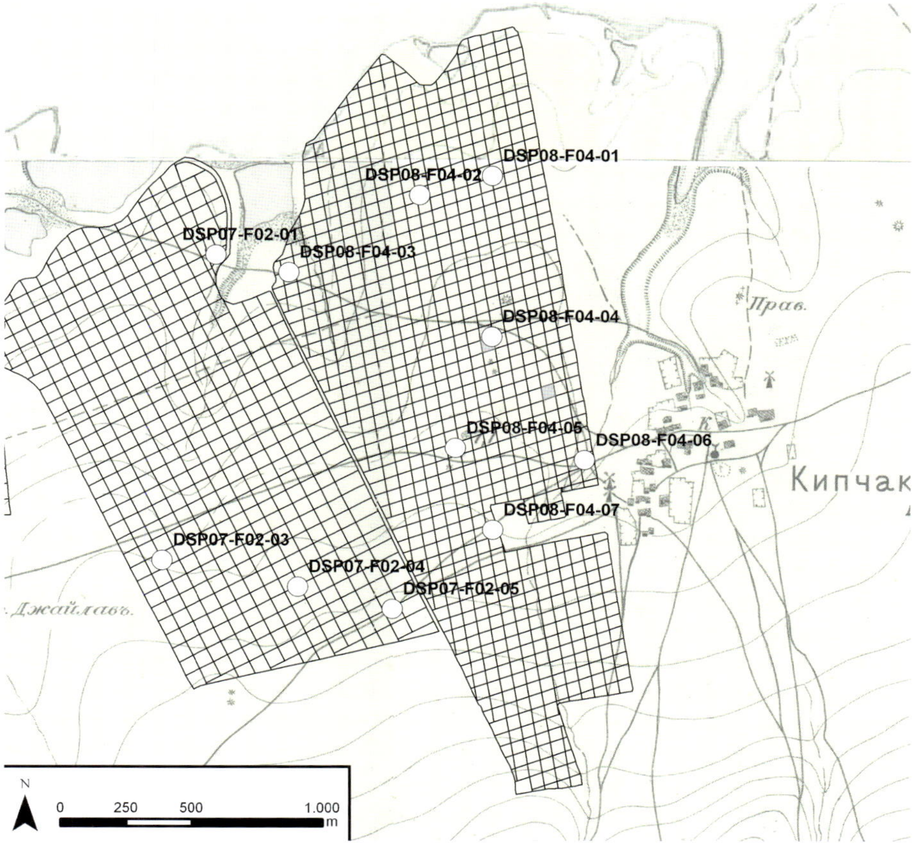

Fig. 4.110 • Distribution map of BA finds from DSP08-F04.

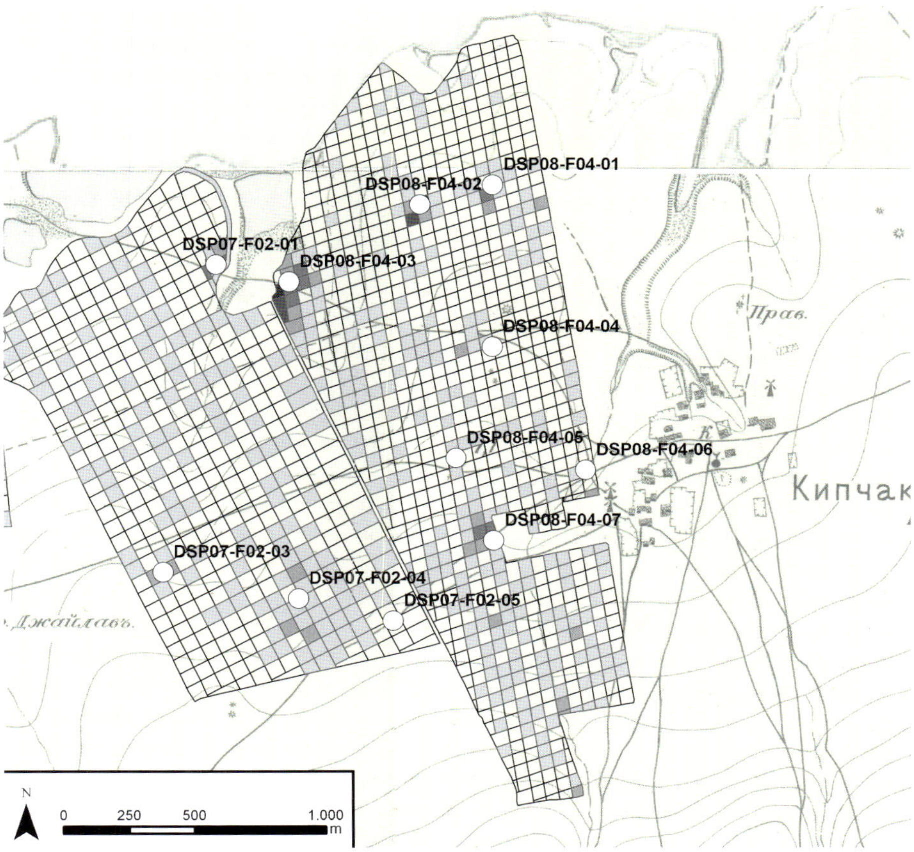

Fig. 4.111 • Distribution map of LCEH finds from DSP08-F04.

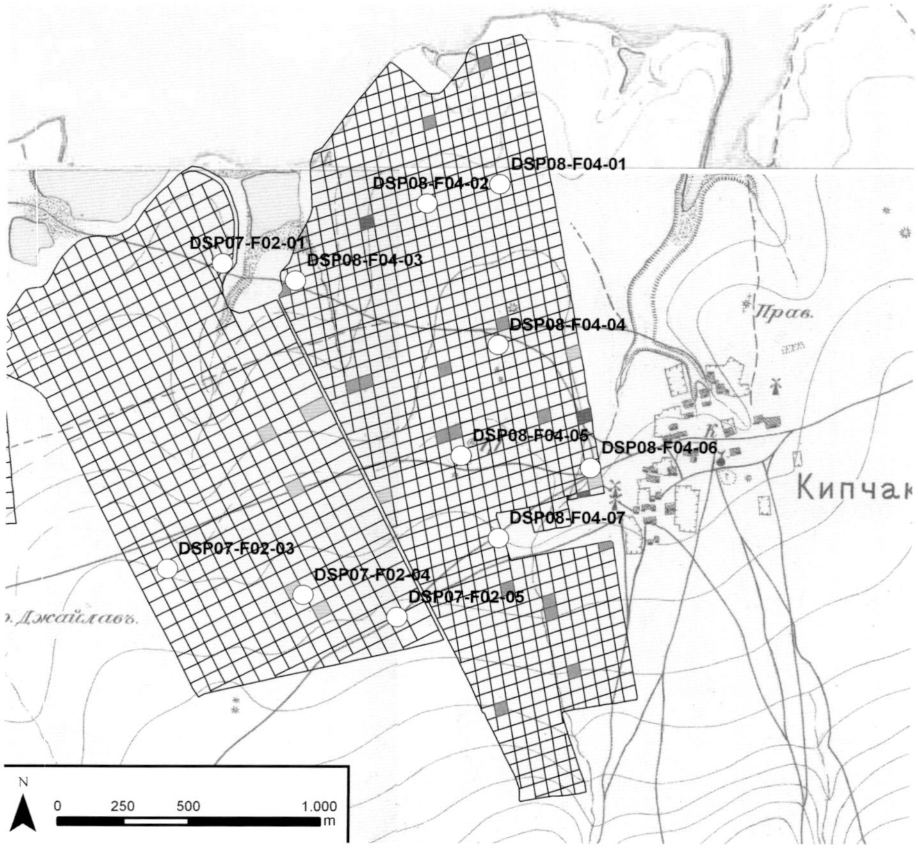

Fig. 4.112 • Distribution map of EM finds from DSP08-F04.

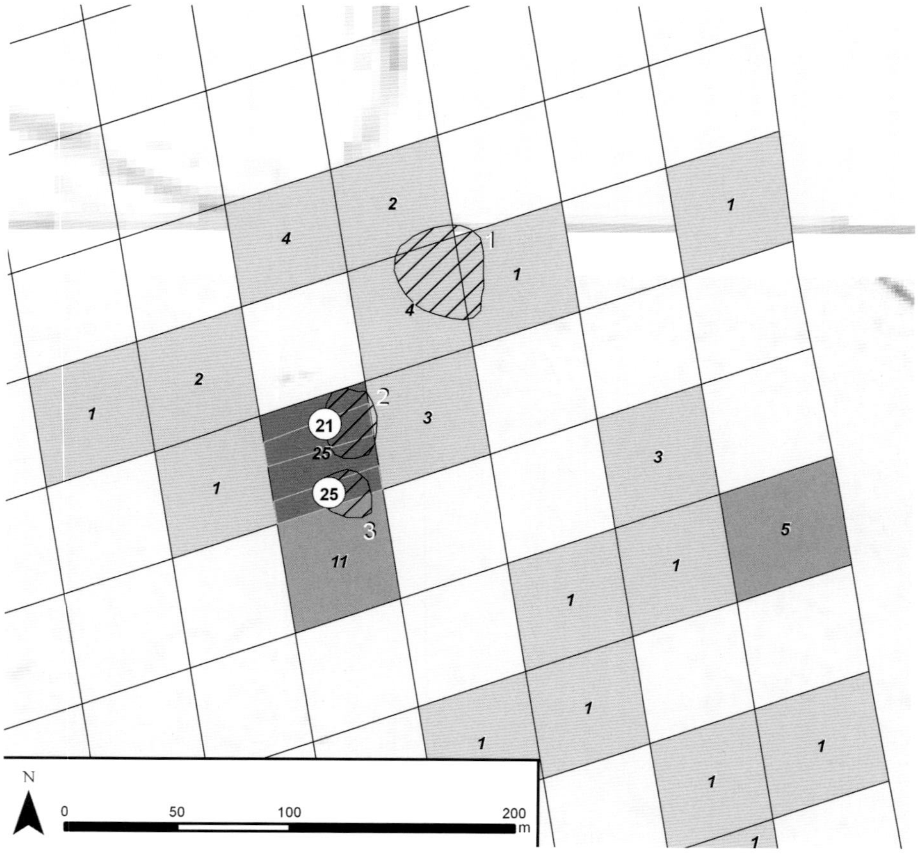

Fig. 4.113 • Distribution map of DSP08-F04-01 (LCEH).

Fig. 4.114 • DSP08-F04-01 from the southeast.

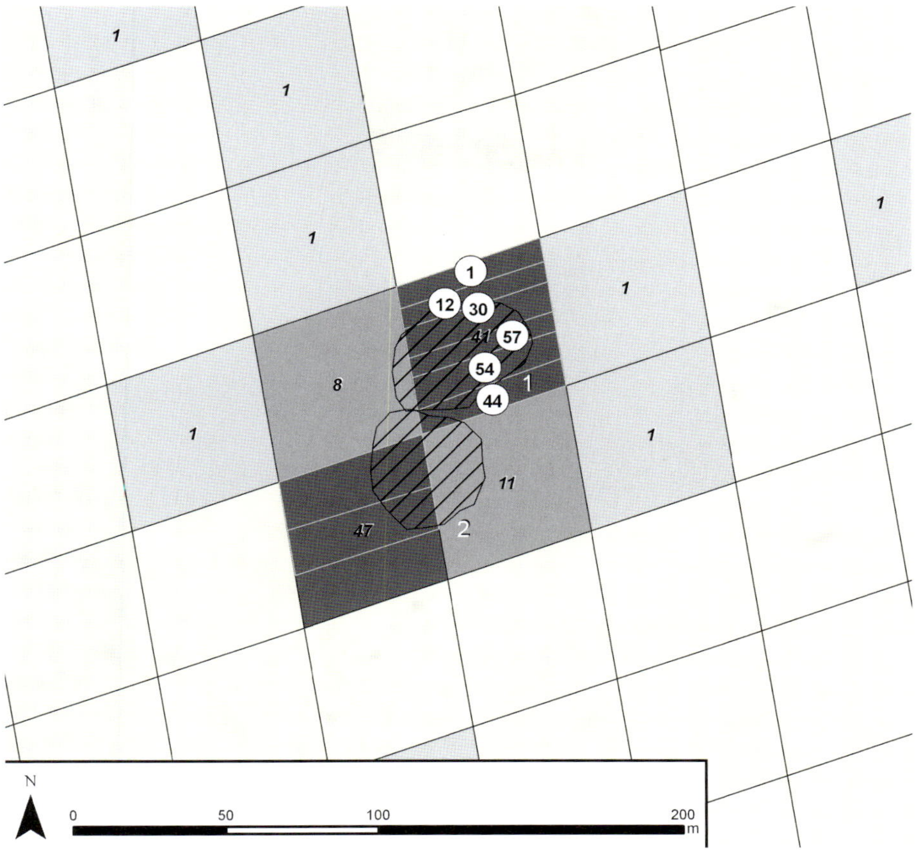

Fig. 4.115 • Distribution map of DSP08-F04-02 (LCEH).

Fig. 4.116 • DSP08-F04-02 from the southeast.

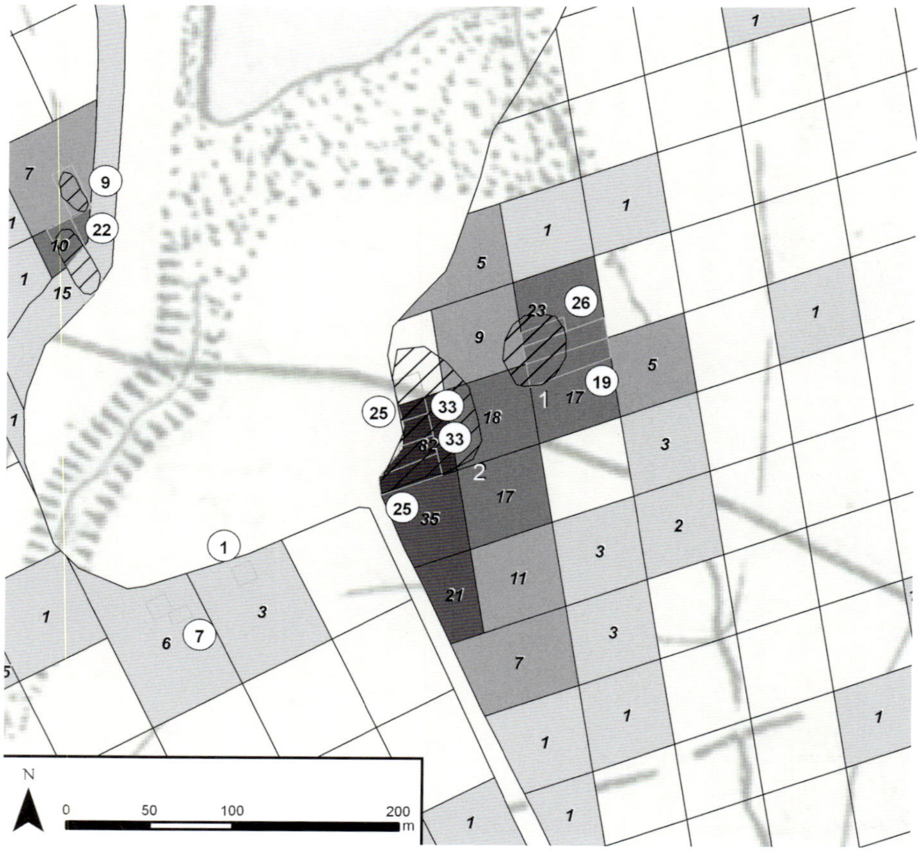

Fig. 4.117 • Distribution map of DSP08-F04-03 (LCEH).

Fig. 4.118 • DSP08-F04-03 from the southwest.

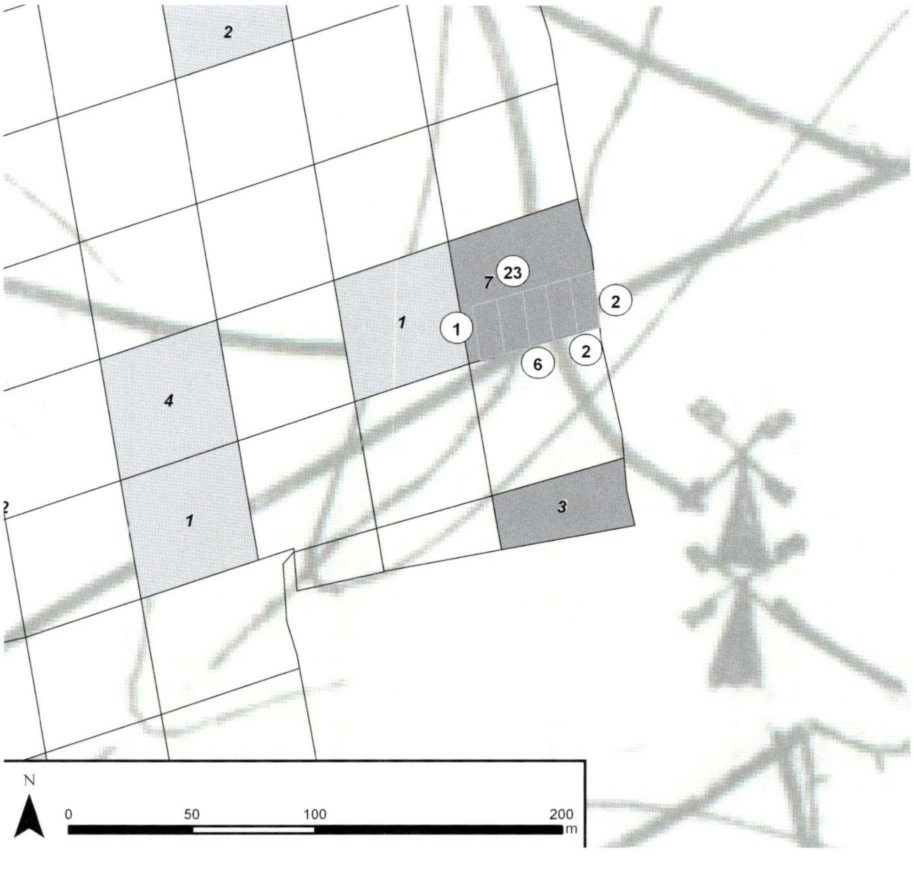

Fig. 4.119 • Distribution map of DSP08-F04-06 (LCEH).

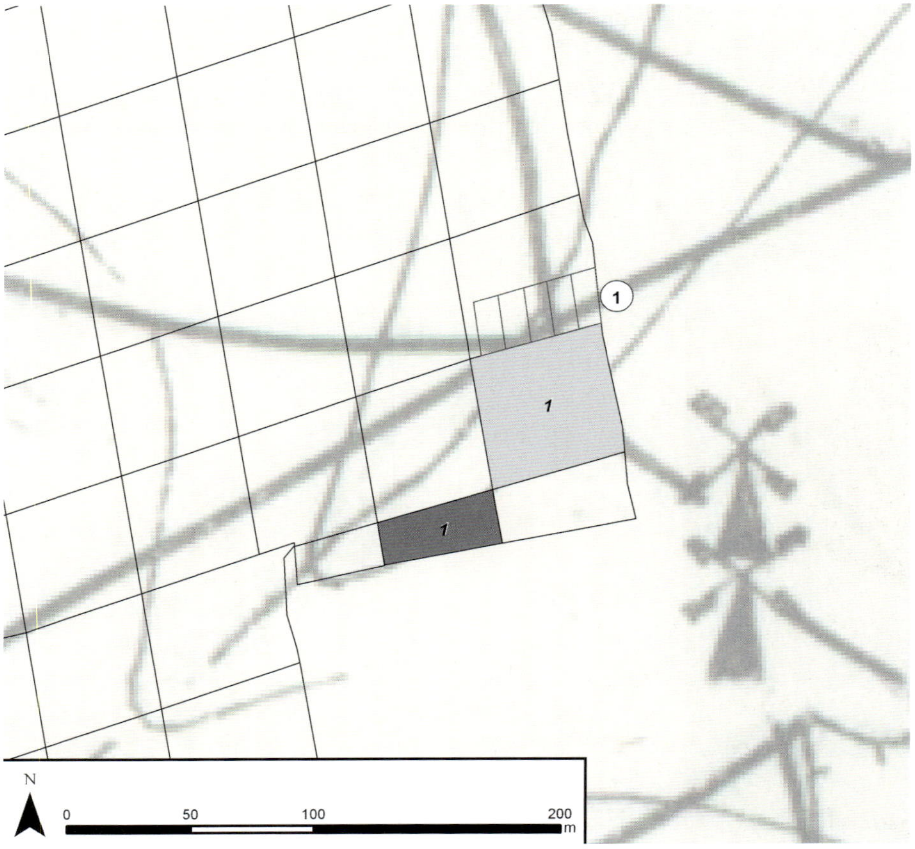

Fig. 4.120 • Distribution map of DSP08-F04-06 (EM).

Fig. 4.121 • DSP08-F04-06 from the southwest.

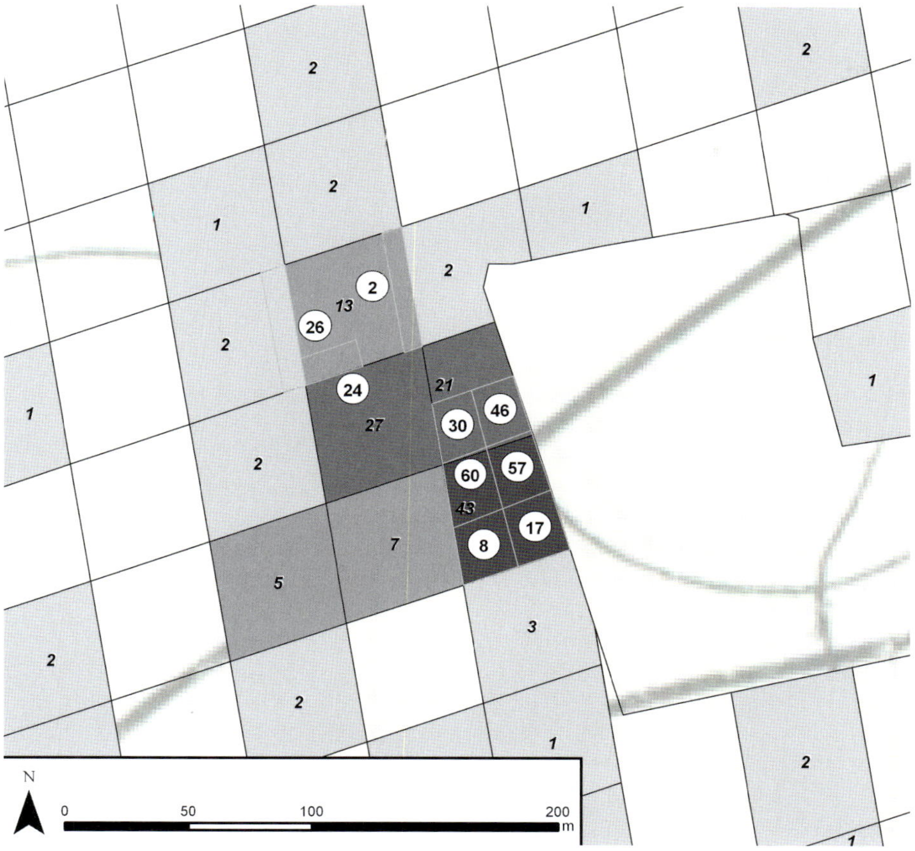

Fig. 4.122 • Distribution map of DSP08-F04-07 (LCEH).

Fig. 4.123 • DSP08-F04-07 from the north.

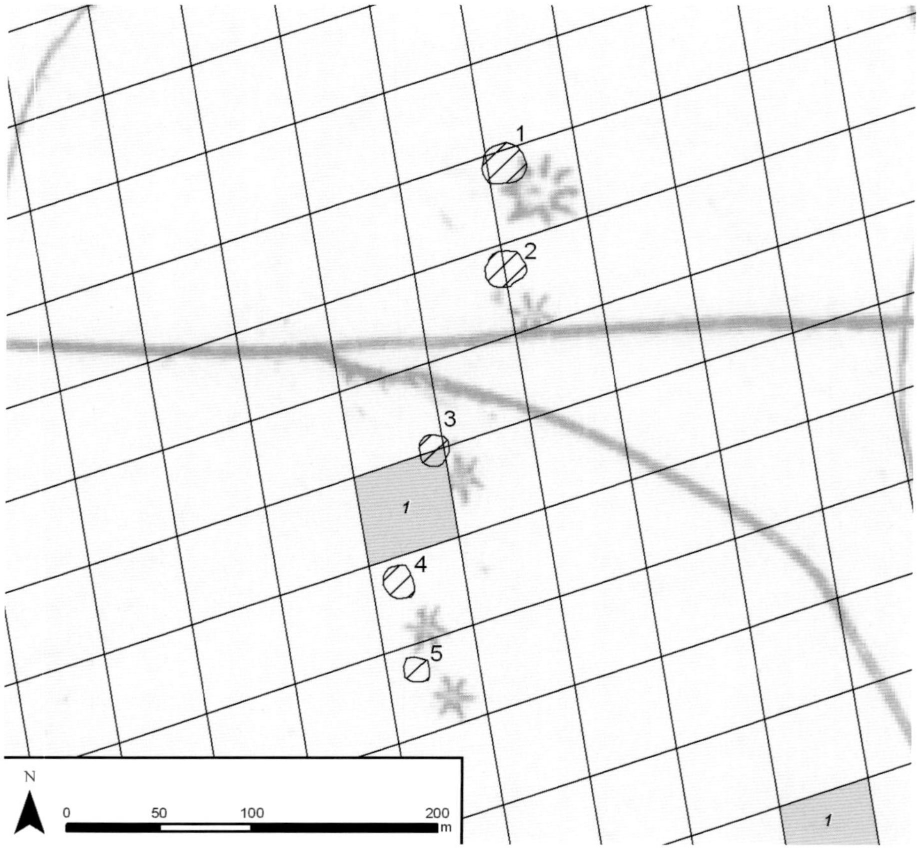

Fig. 4.124 • Distribution map of DSP08-F04-04 (BA).

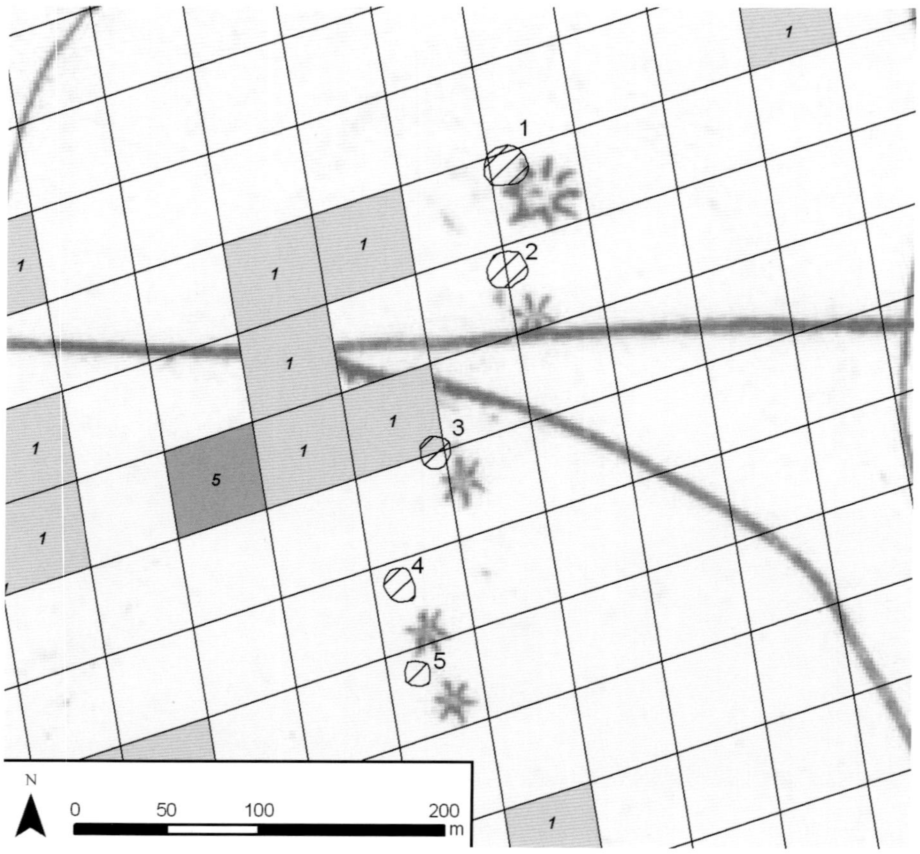

Fig. 4.125 • Distribution map of DSP08-F04-04 (LCEH).

Fig. 4.126 • DSP08-F04-04 from the southwest.

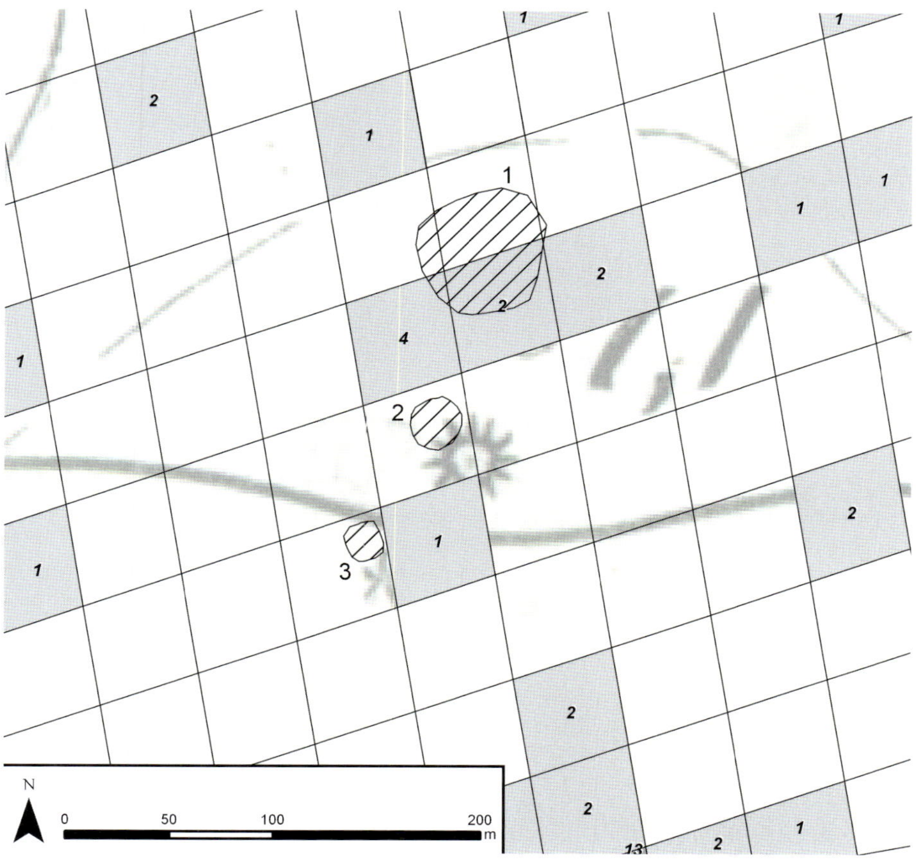

Fig. 4.127 • Distribution map of DSP08-F04-05 (LCEH).

Fig. 4.128 • DSP08-F04-05 from the south.

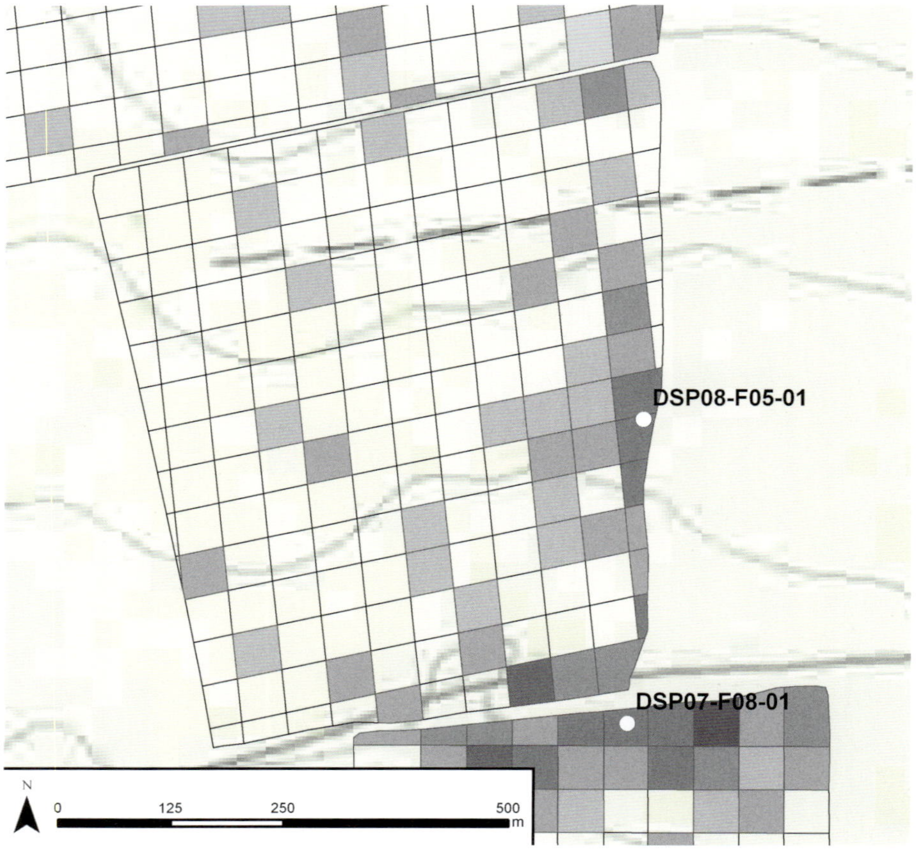

Fig. 4.129 • Distribution map of all finds from DSP08-F05.

Fig. 4.130 • Distribution map of LCEH finds from DSP08-F05.

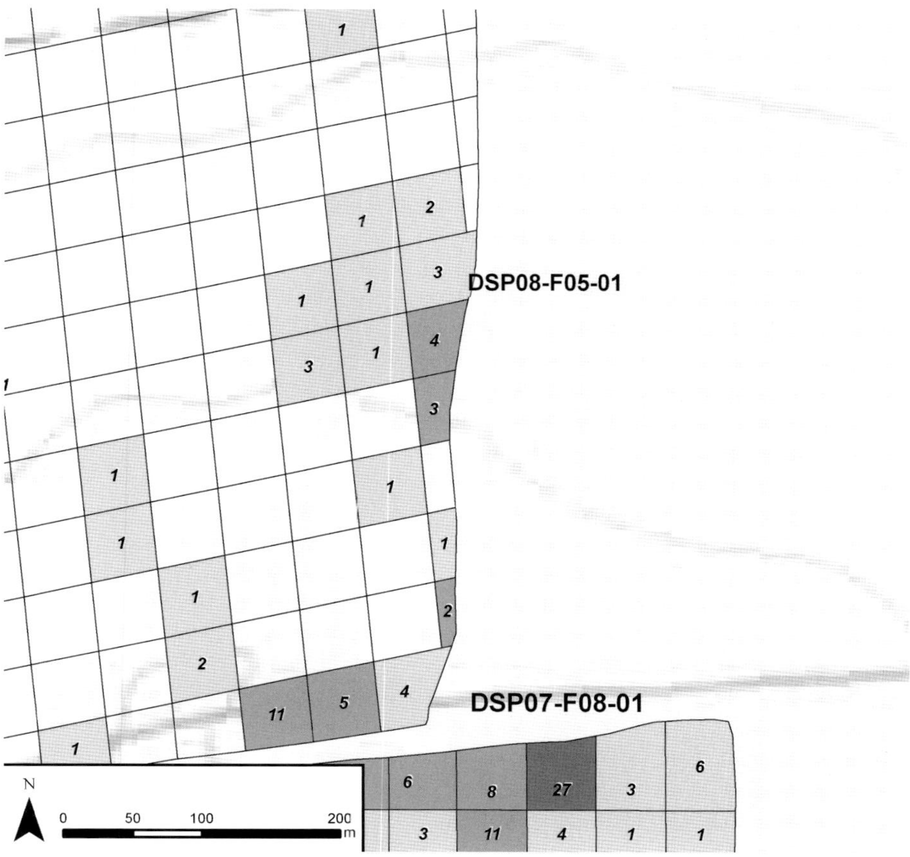

Fig. 4.131 • Distribution map of DSP08-F05-01 (LCEH).

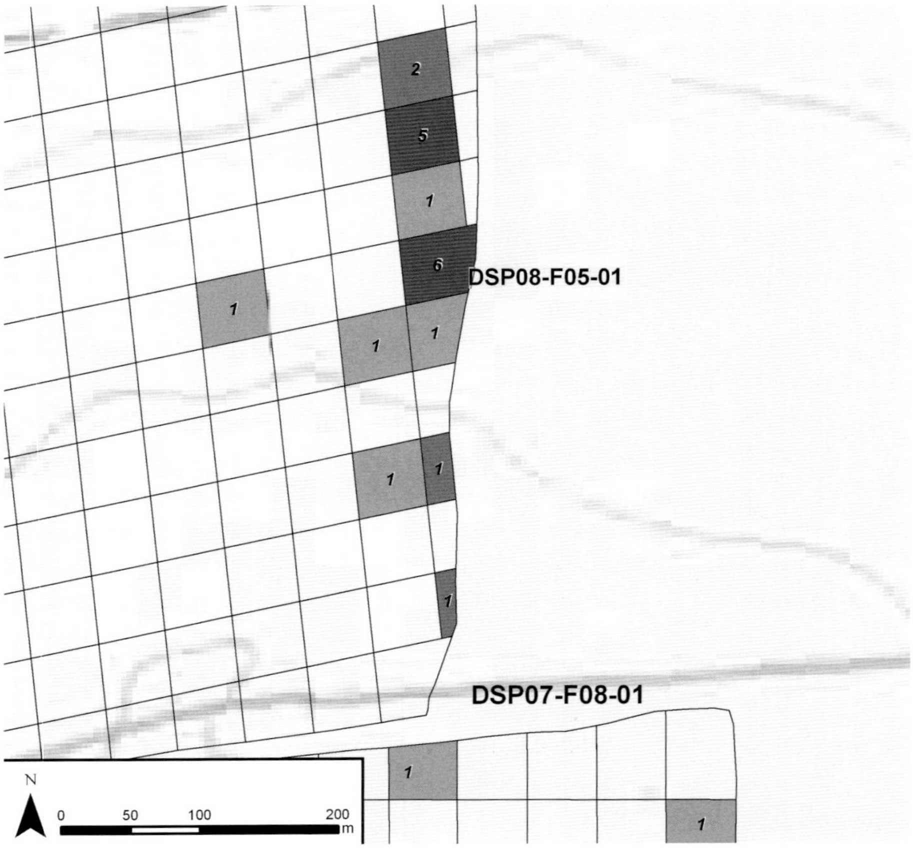

Fig. 4.132 • Distribution map of DSP08-F05-01 (EM).

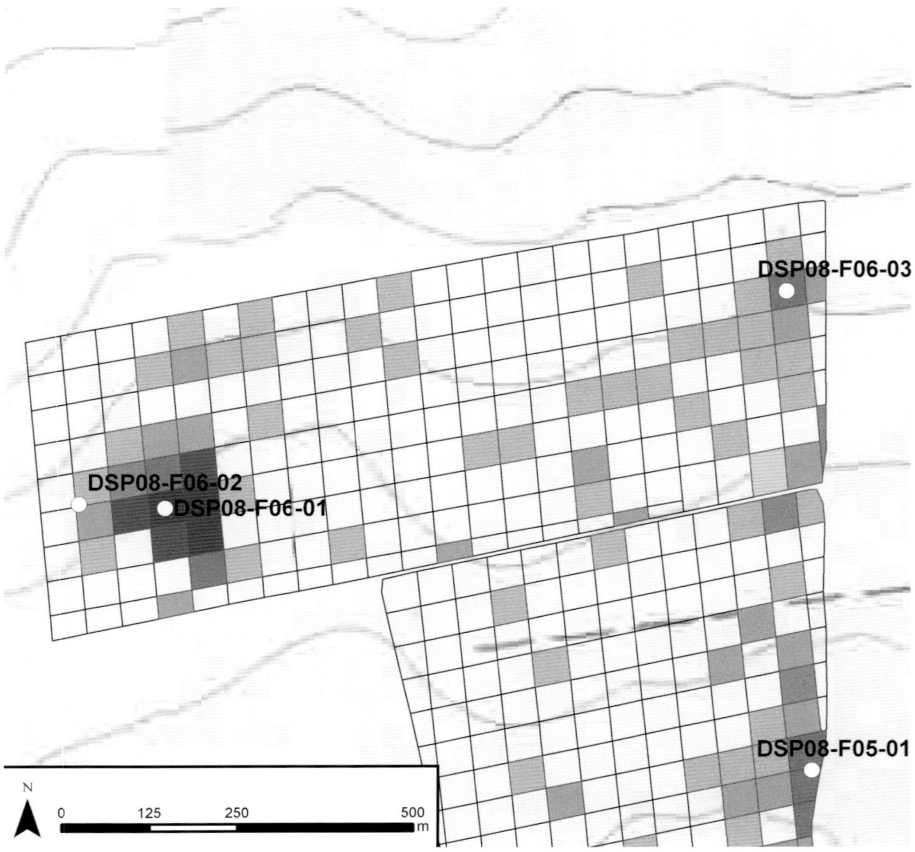

Fig. 4.133 • Distribution map of all finds from DSP08-F06.

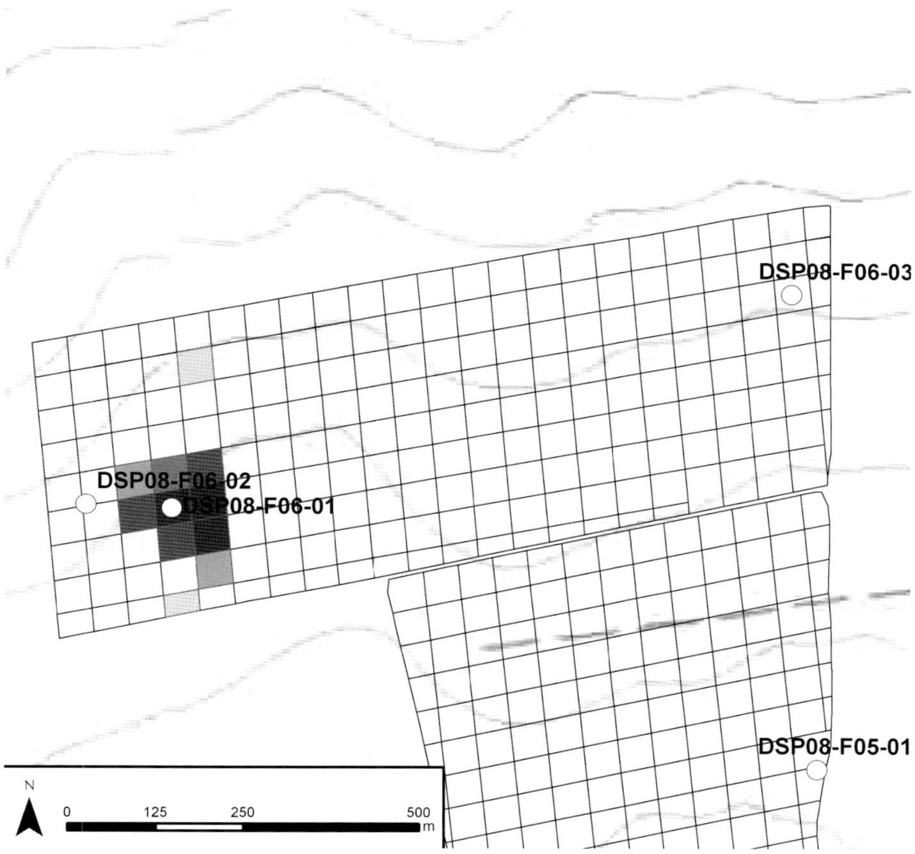

Fig. 4.134 • Distribution map of BA finds from DSP08-F06.

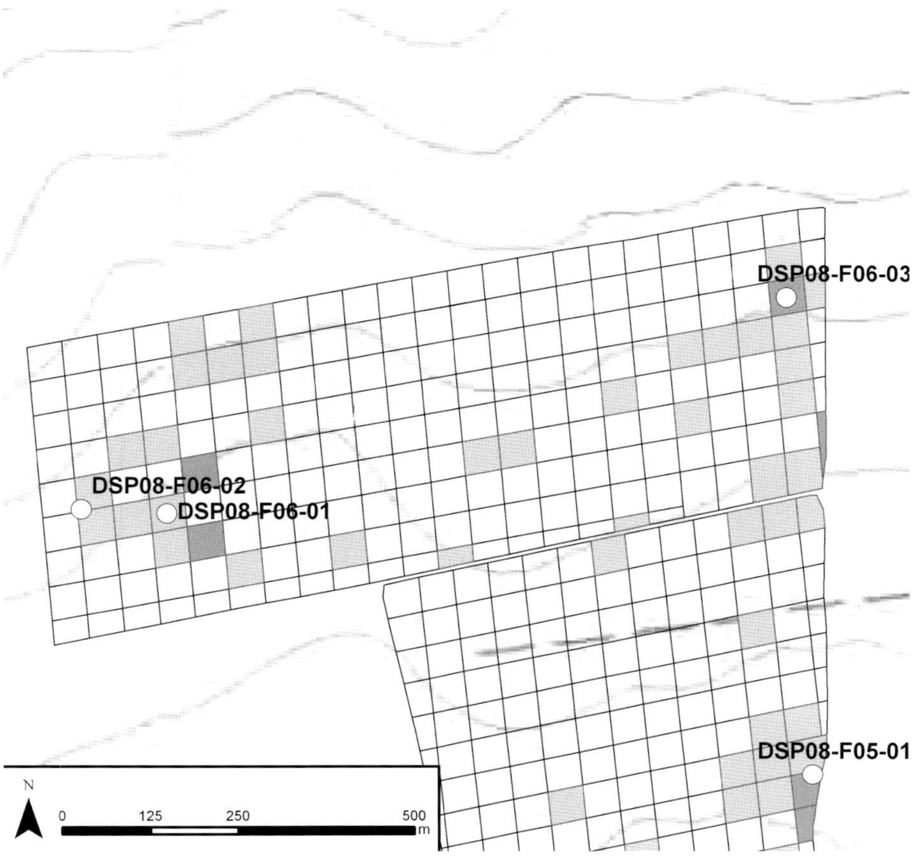

Fig. 4.135 • Distribution map of LCEH finds from DSP08-F06.

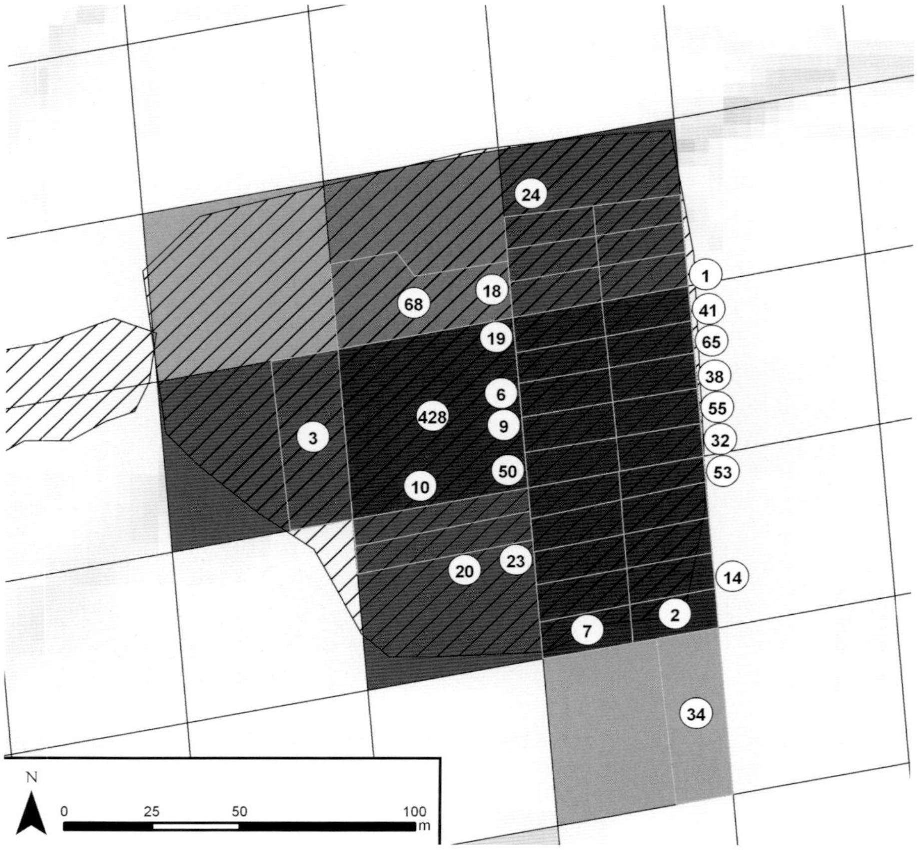

Fig. 4.136 • Distribution map of DSP08-F06-01-02 (BA). DSP08-F06-02 is located on the far left.

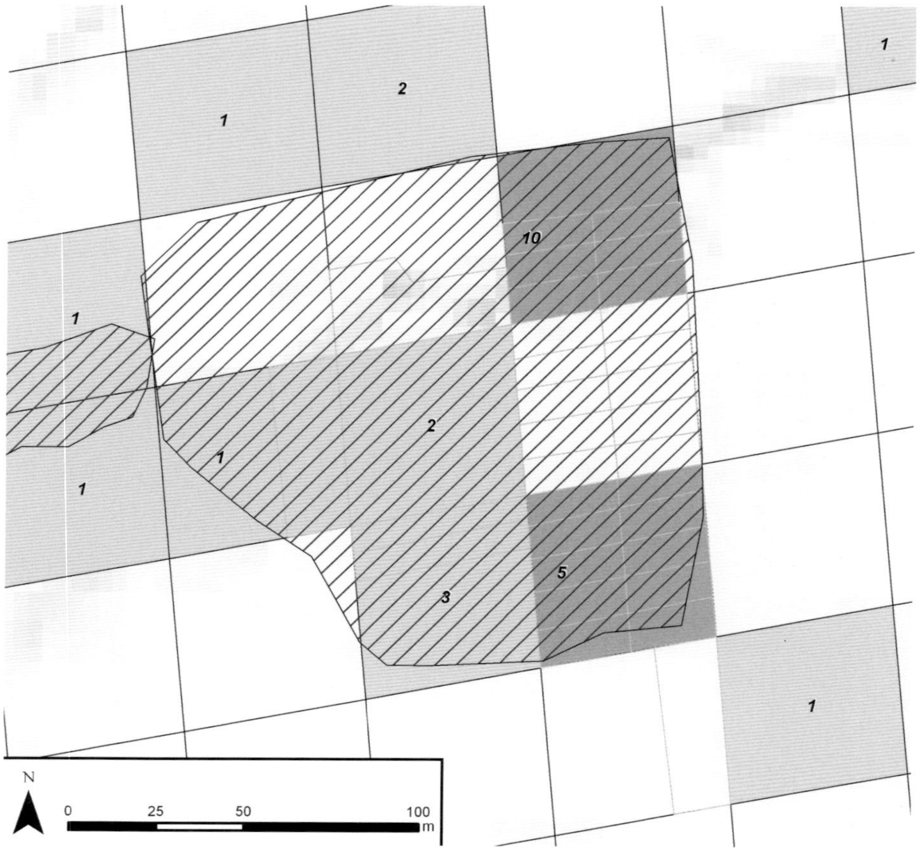

Fig. 4.137 • Distribution map of DSP08-F06-01-02 (LCEH). DSP08-F06-02 is located on the far left.

Fig. 4.138 • DSP08-F06-01 and 02 from the southeast. DSP08-F06-02 is located on the far left horizon.

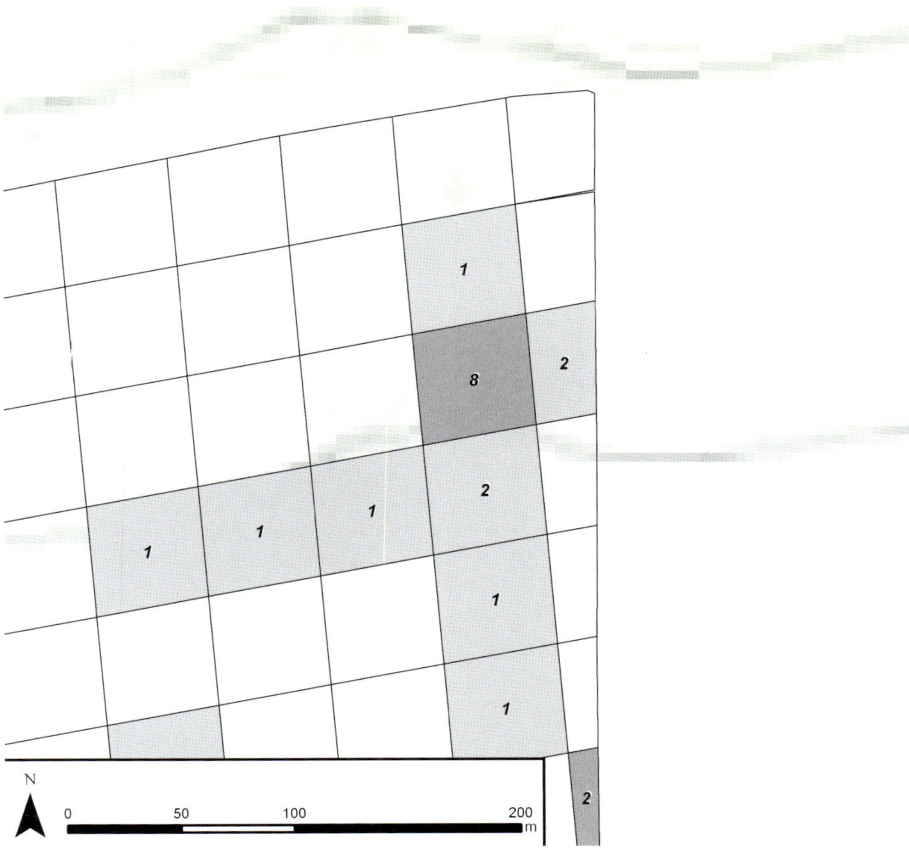

Fig. 4.139 • Distribution map of LCEH finds from DSP08-F06-03.

Fig. 4.140 • DSP08-H02-03: dammed ravine east of Hill 2.

Fig. 4.141 • Shallow soils with many tiny rock fragments.

Fig. 4.142 • Dark "deep" soils with no or little exposed bedrock.

Fig. 4.143 • Core at Foundation 47.

Fig. 4.144 • Red loess built-up at the tip of promontory on the western side of the ravine.

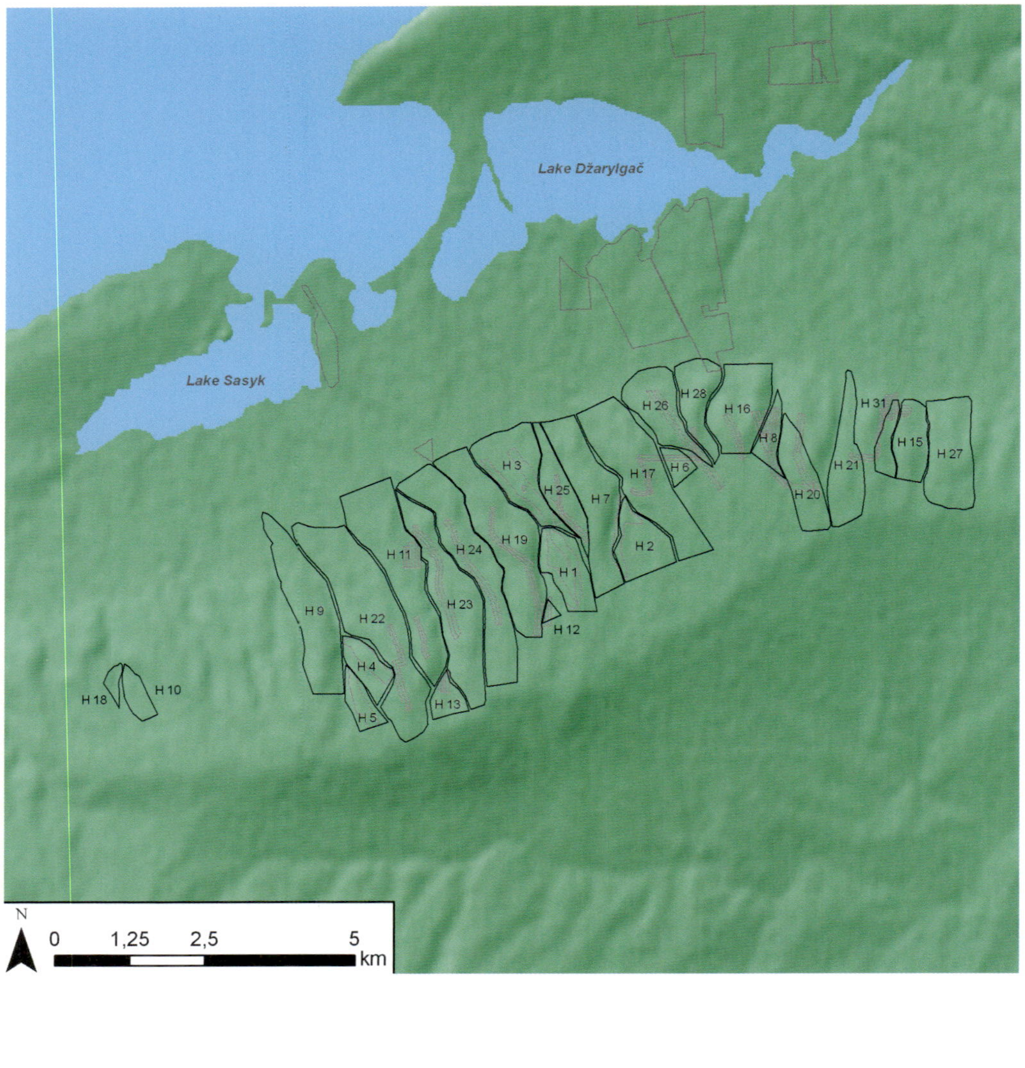

Fig. 4.145 • Map of the slope units in relation to hills.

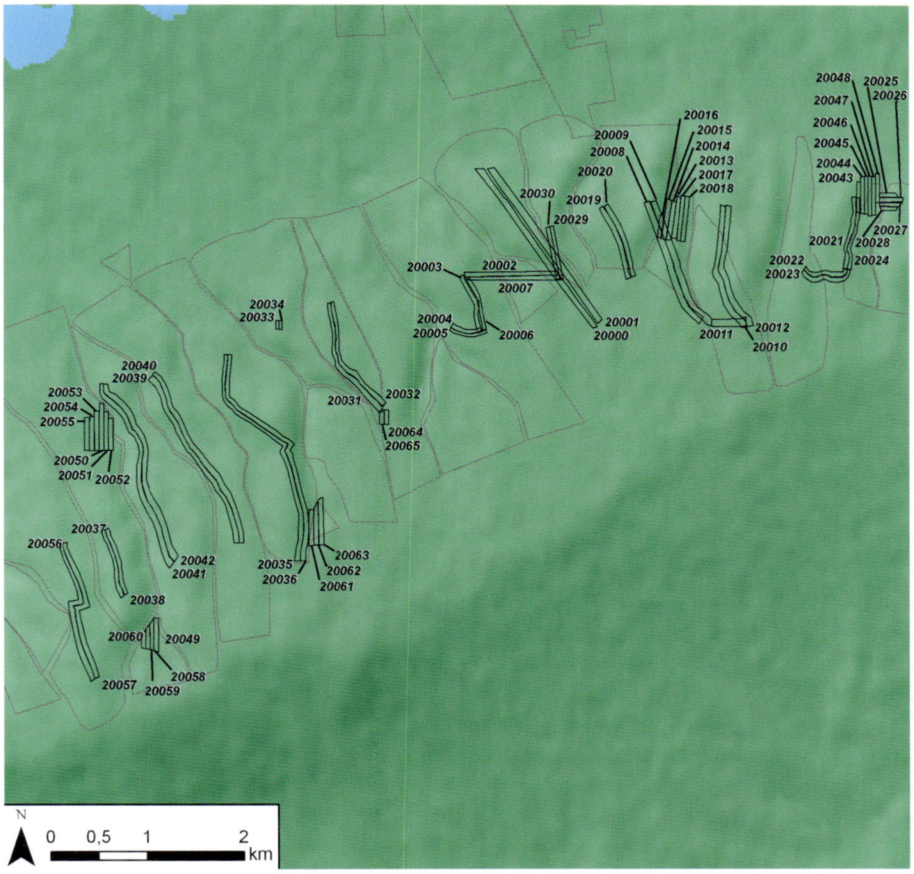

Fig. 4.146 • Map of the slope units.

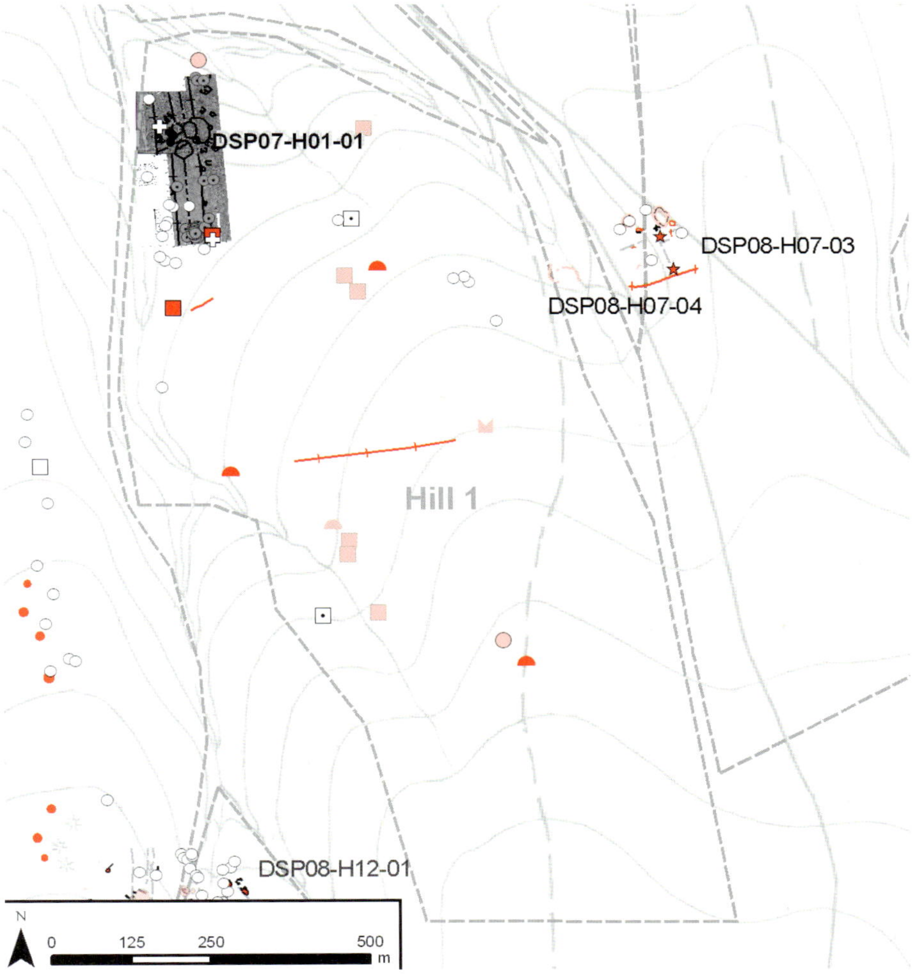

Fig. 4.147 • Distribution map of Hill 1. For legends see Fig. 4.40.

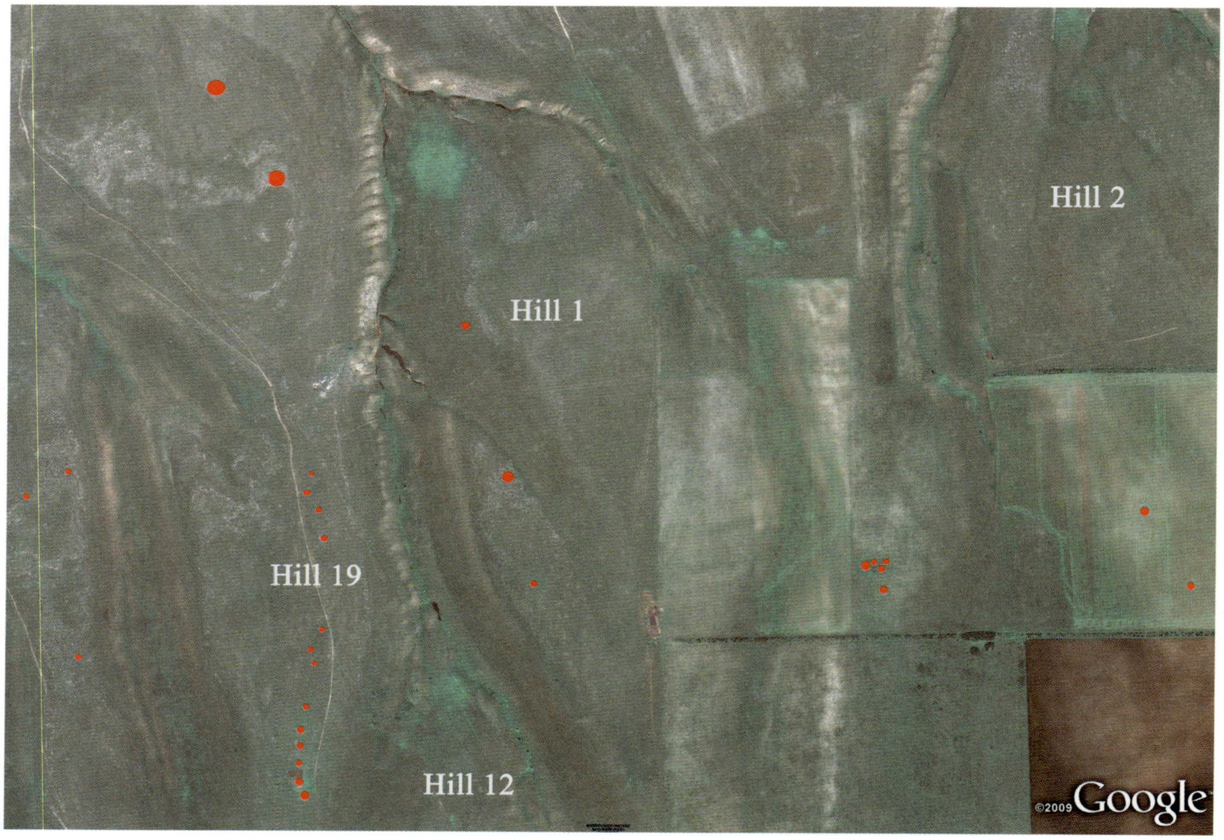

Fig. 4.148 • Hills 1, 12, and 19, satellite image (Google Earth); red dots mark kurgans.

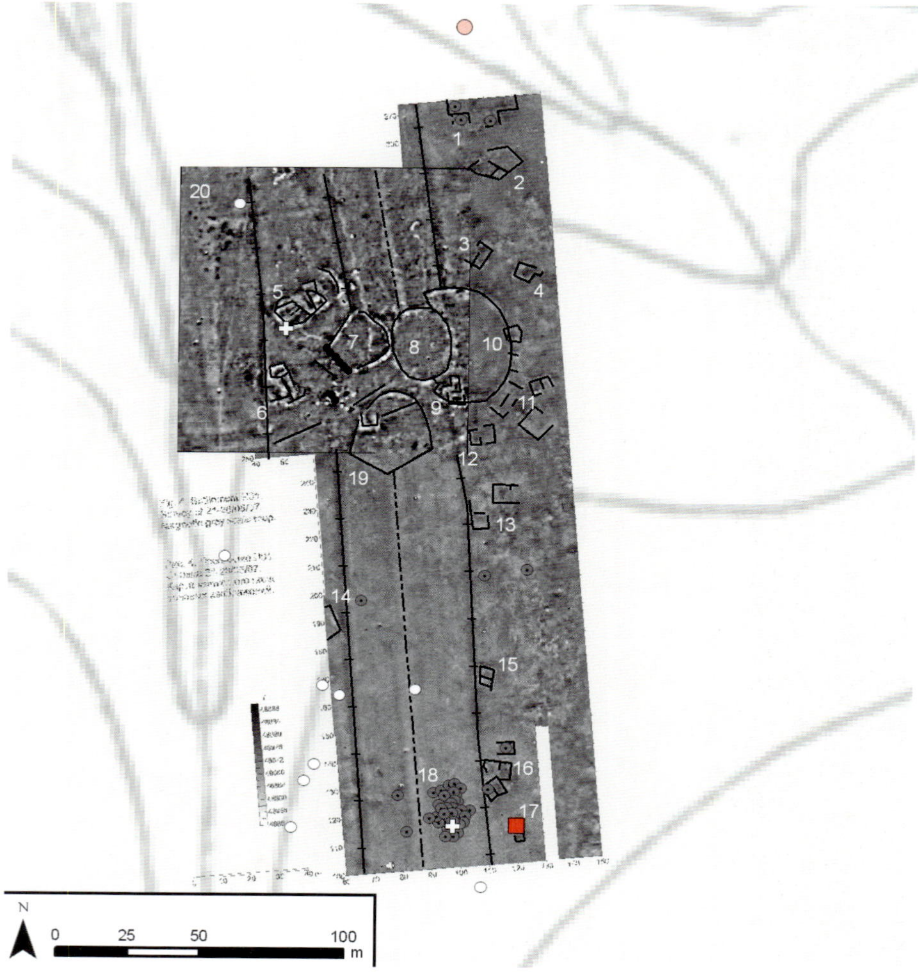

Fig. 4.149 • Distribution map of DSP07-H01-01.

Fig. 4.150 • View across DSP07-H01-01 from the south toward the sea.

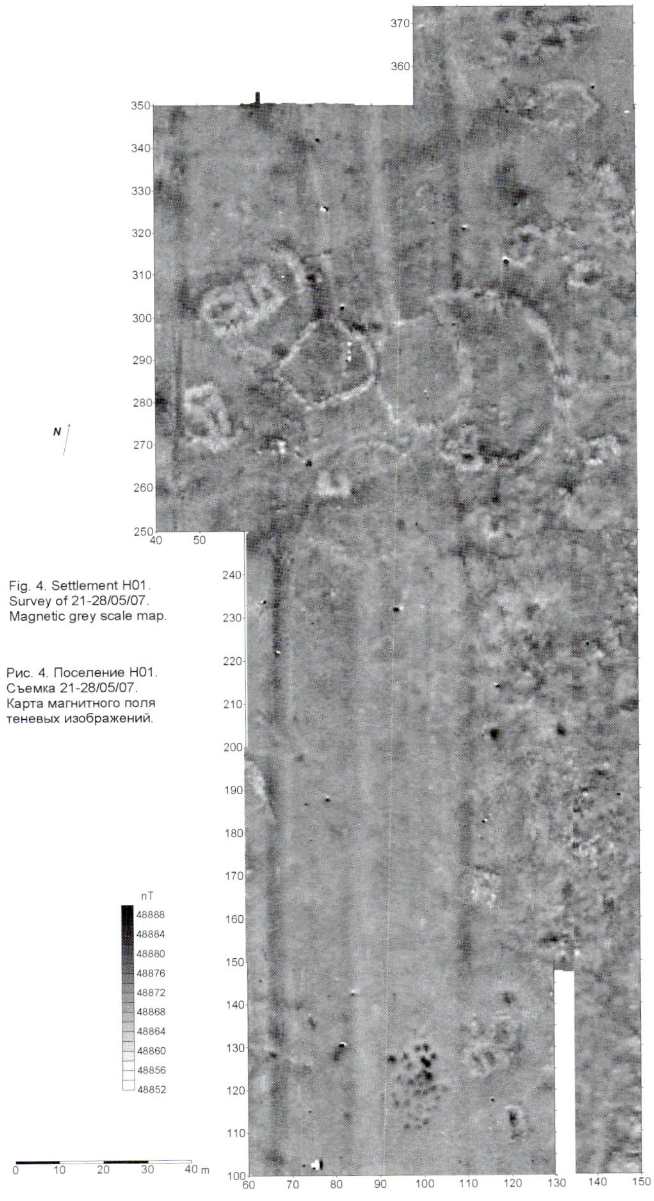

Fig. 4.151 • Geomagnetic map of DSP07-H01-01.

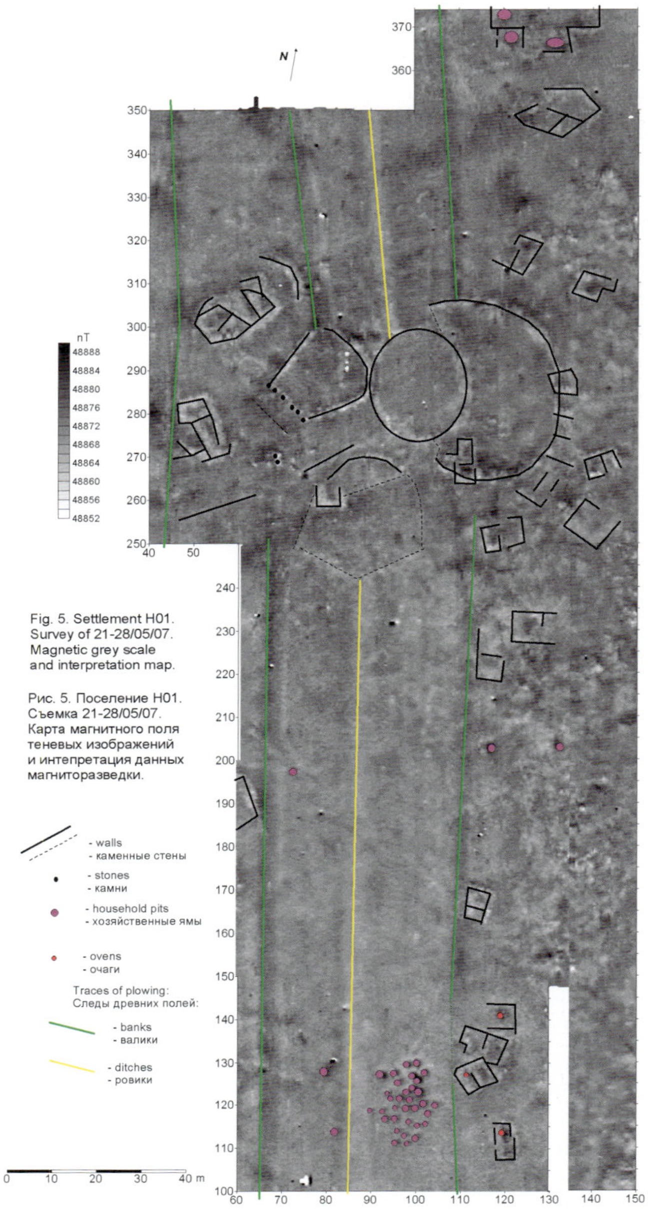

Fig. 4.152 • Geomagnetic map of DSP07-H01-01, interpretation.

Fig. 4.153 • Aerial photo of DSP07-H01-01.

Fig. 4.154 • Trial trench DSP07-H01-01 Feature 18.

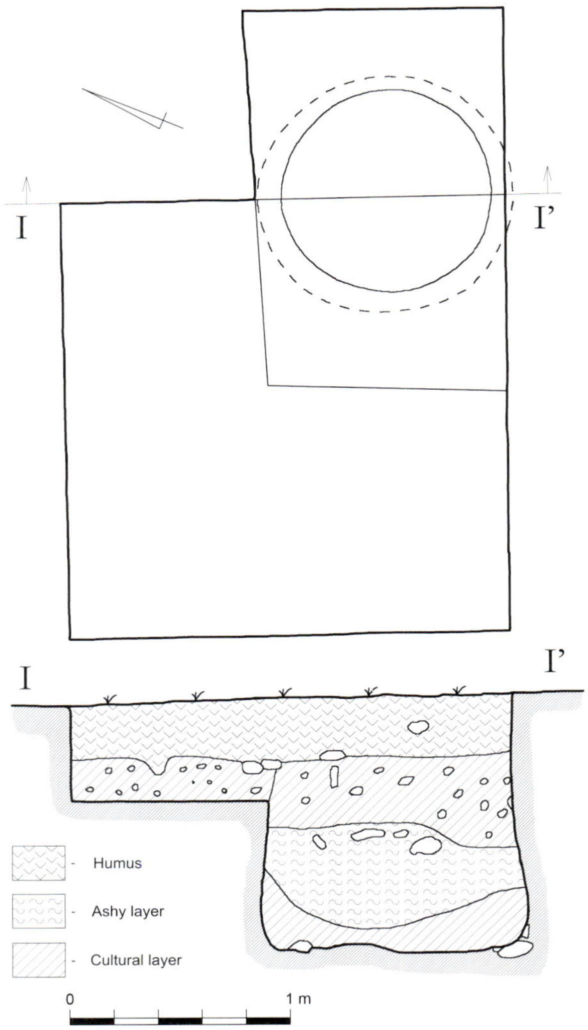

Fig. 4.155 • Drawings of trial trench DSP07-H01-01 Feature 18.

Fig. 4.156 • Trial trench DSP07-H01-01 Feature 5.

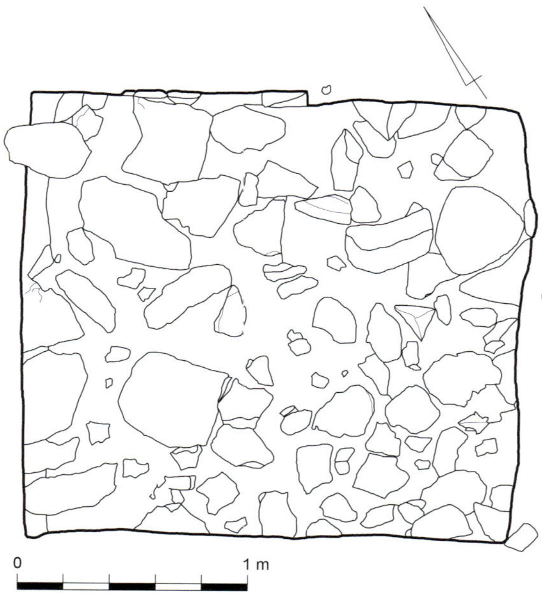

Fig. 4.157 • Drawing of trial trench DSP07-H01-01 Feature 5.

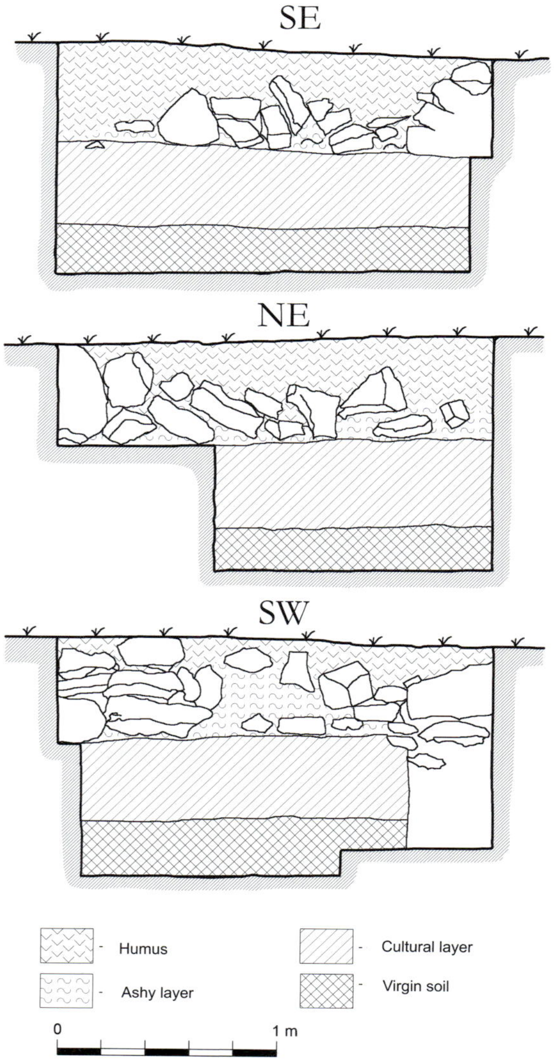

Fig. 4.158 • Drawing of profile of trial trench DSP07-H01-01 Feature 5.

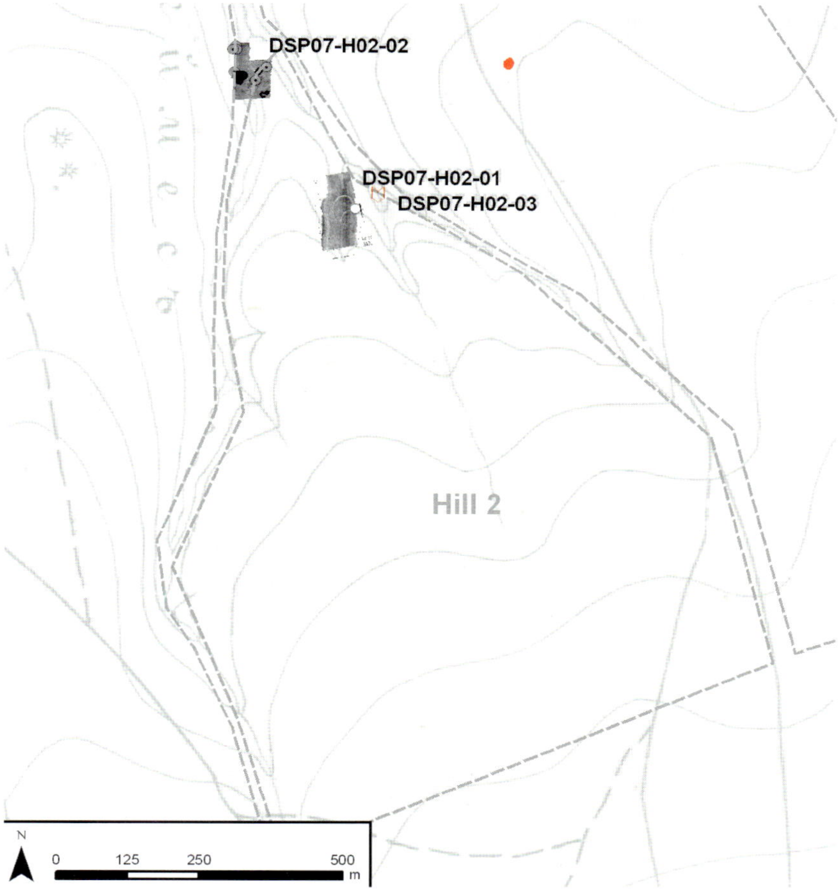

Fig. 4.159 • Distribution map of Hill 2.

Fig. 4.160 • Hill 2, satellite image (Google Earth); red dots mark kurgans.

Fig. 4.161 • Hill 2 from the north. Dammed ravine on the eastern side (DSP08-H02-03).

Fig. 4.162 • The upper tip of Hill 2 from the south. DSP08-H02-01 on the right.

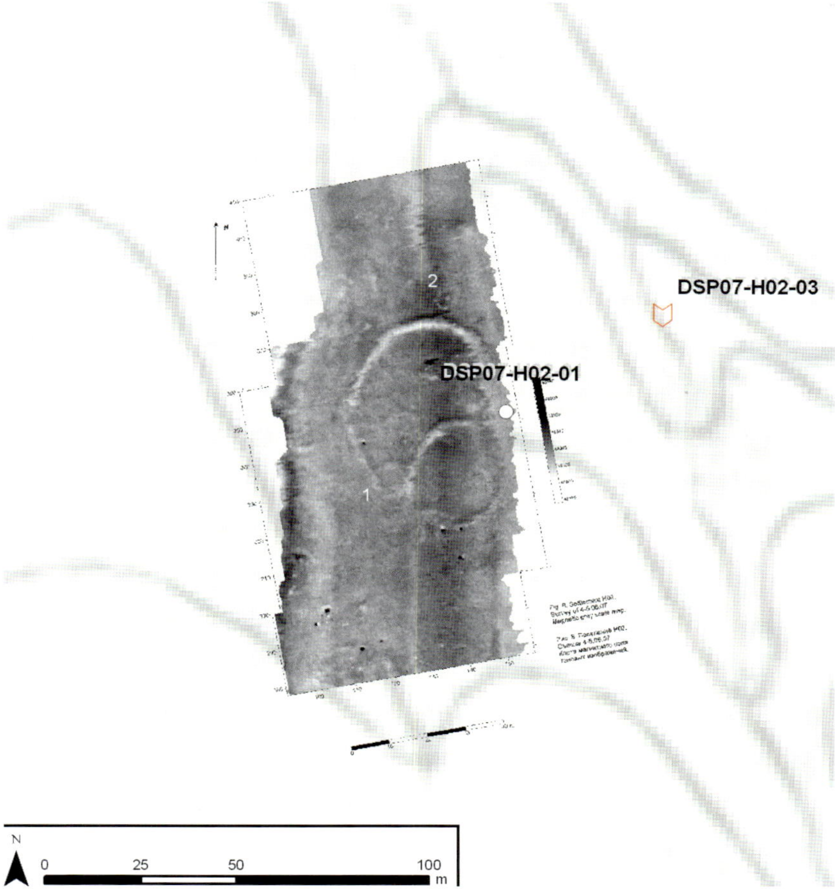

Fig. 4.163 • Distribution map of DSP07-H02-01 and 03.

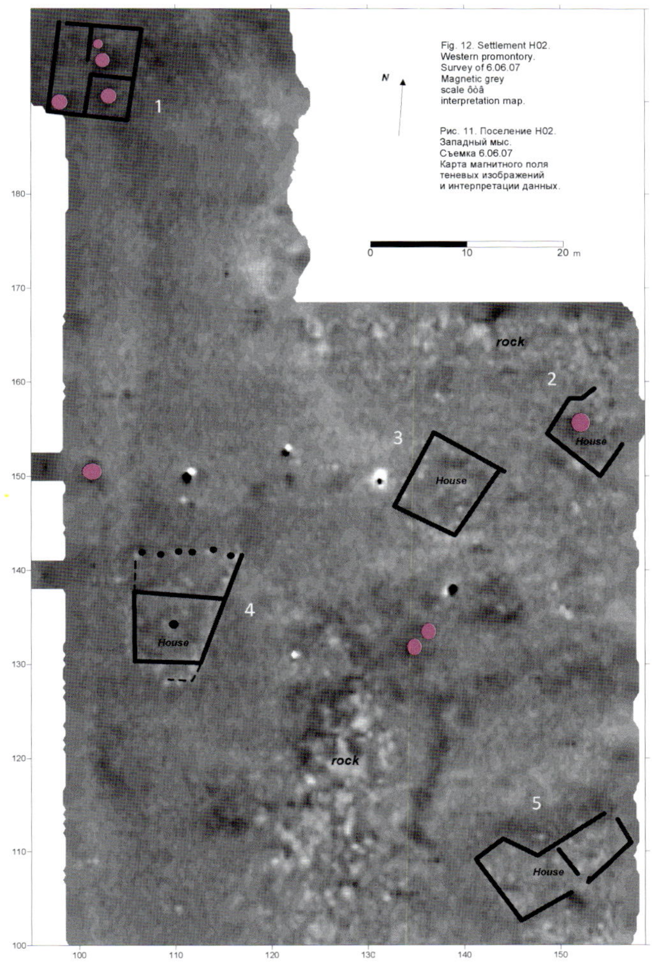

Fig. 4.164 • Geomagnetic map of DSP07-H02-02, interpretation.

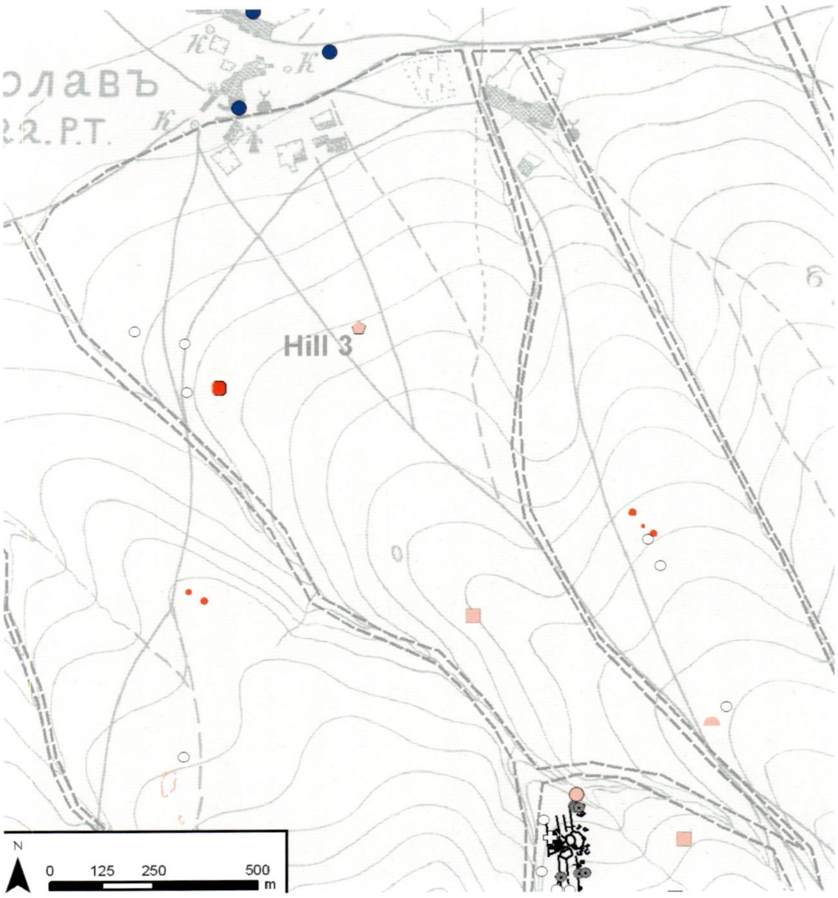

Fig. 4.165 • Distribution map of Hill 3.

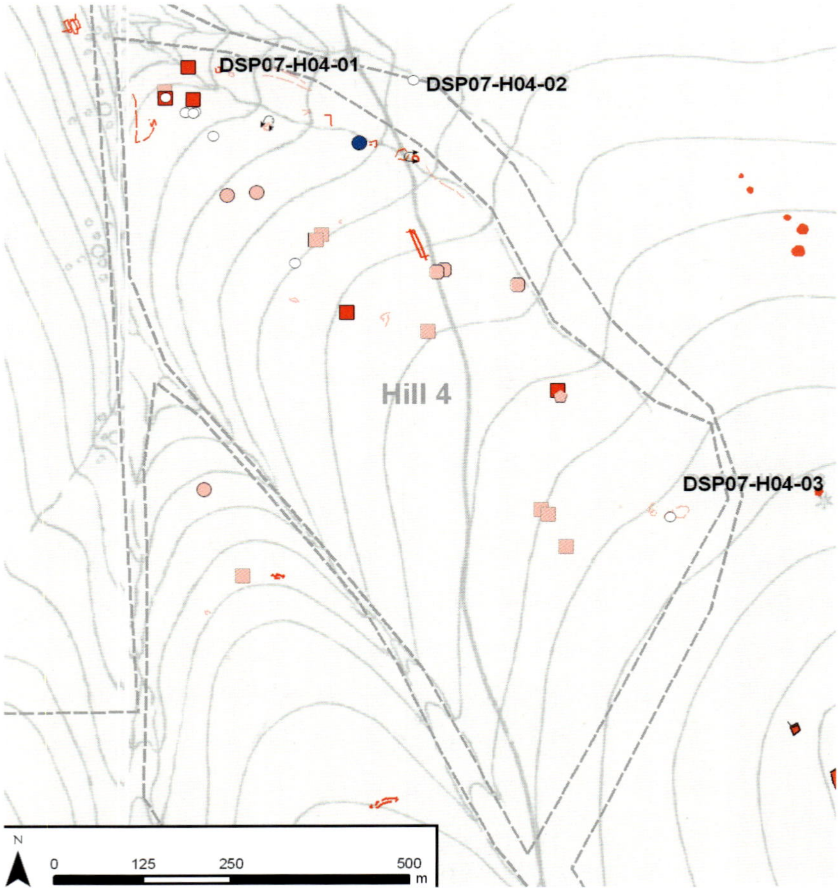

Fig. 4.166 • Distribution map of Hill 4.

Fig. 4.167 • Hill 04 and 05, satellite image (Google Earth); red dots mark kurgans.

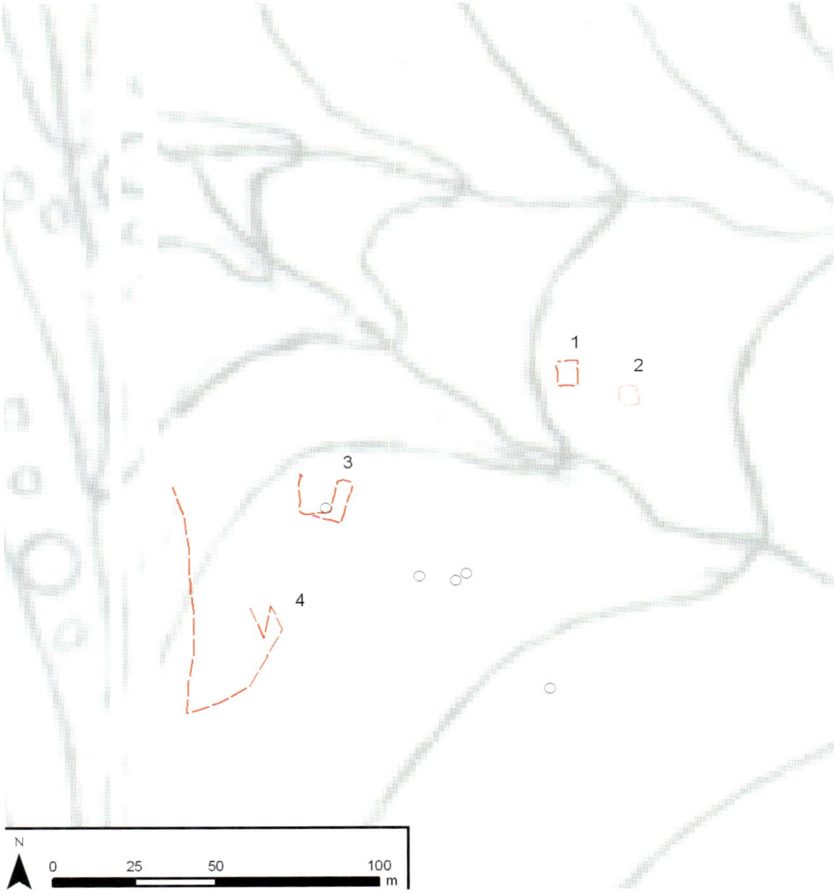

Fig. 4.168 • Distribution map of DSP07-H04-01.

Fig. 4.169 • DSP07-H04-01 from the west.

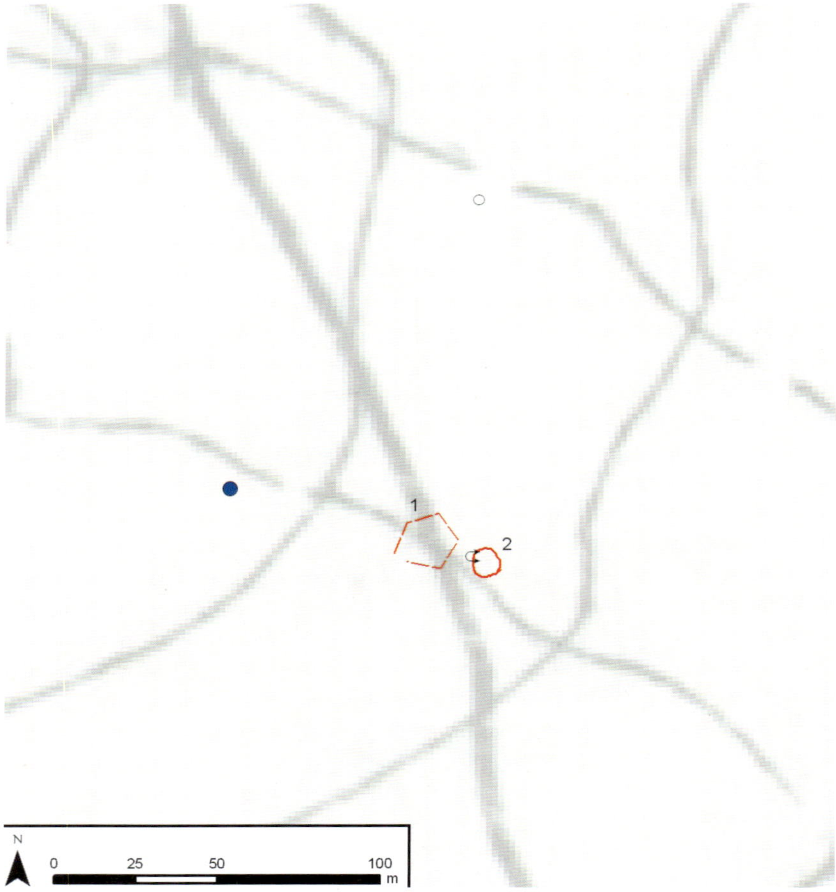

Fig. 4.170 • Distribution map of DSP07-H04-02.

Fig. 4.171 • DSP07-H04-02 from the southeast.

Fig. 4.172 • Distribution map of DSP07-H04-03.

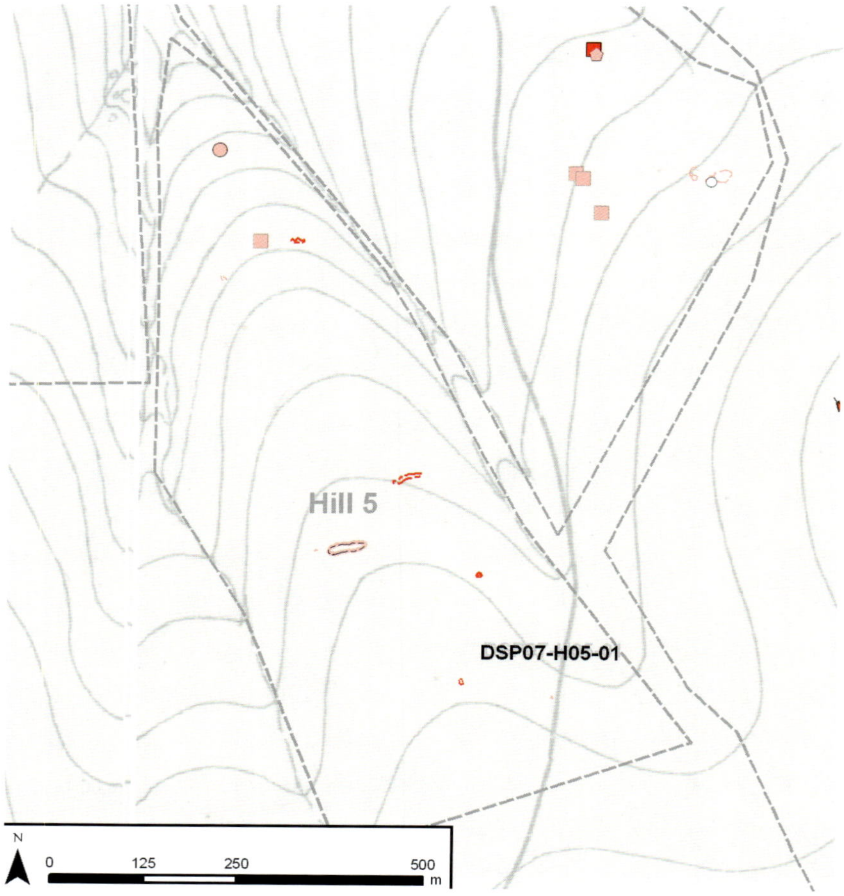

Fig. 4.173 • Distribution map of Hill 5.

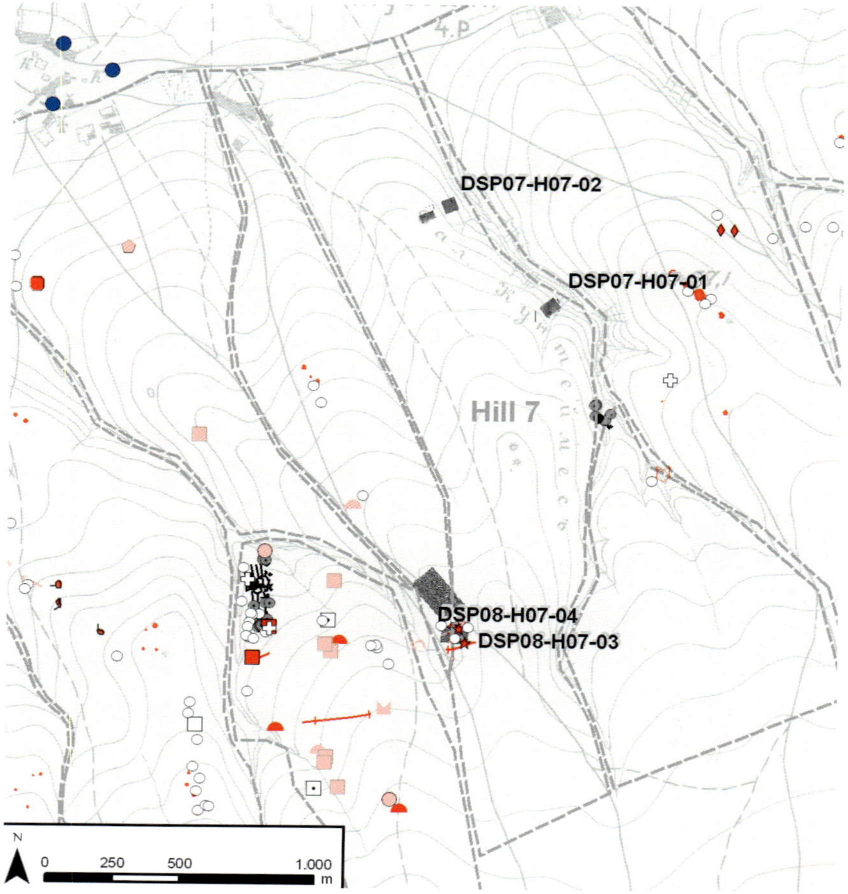

Fig. 4.174 • Distribution map of Hill 7.

Fig. 4.175 • DSP07-H07-01 Feature 2.

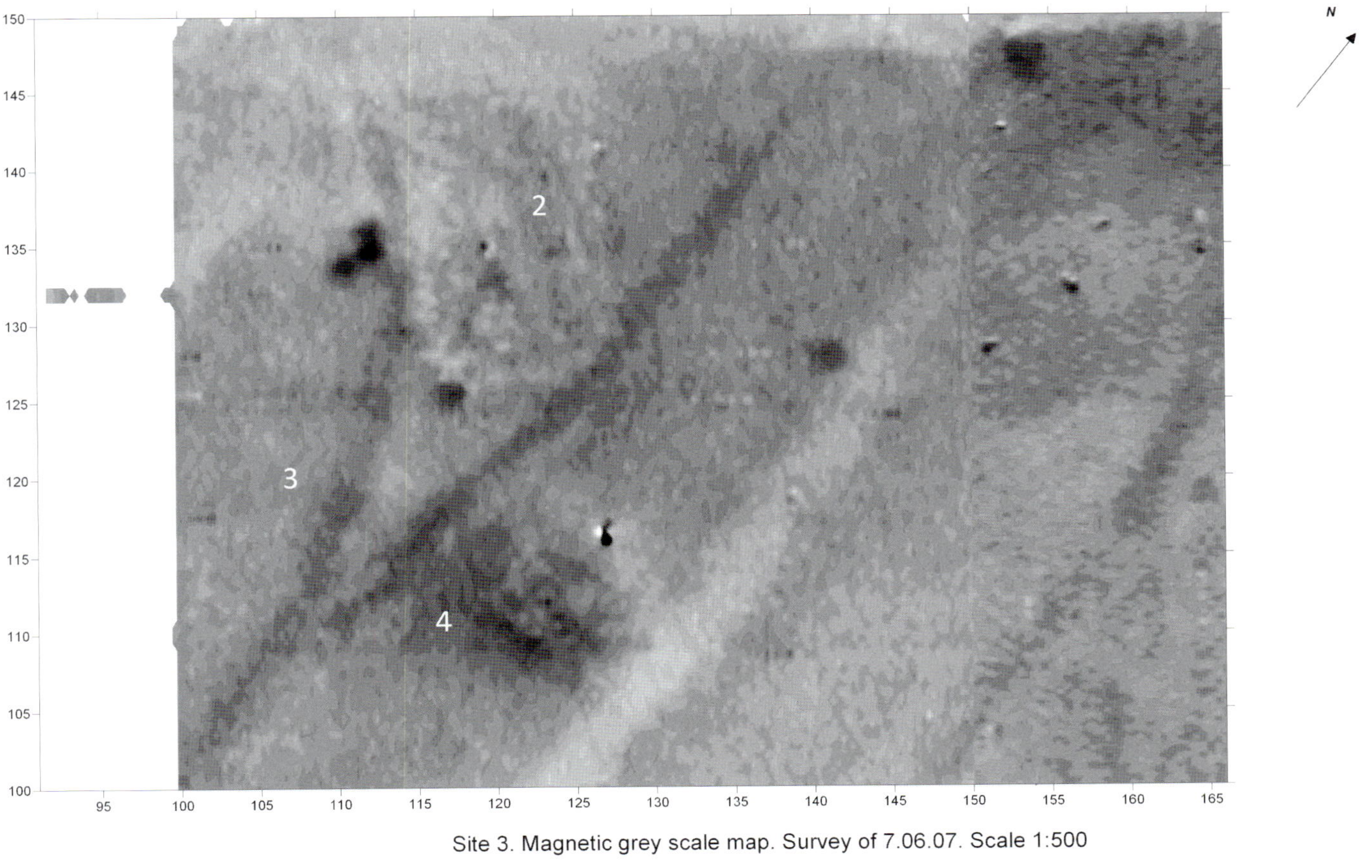

Fig. 4.176 • Geomagnetic map of DSP07-H07-01.

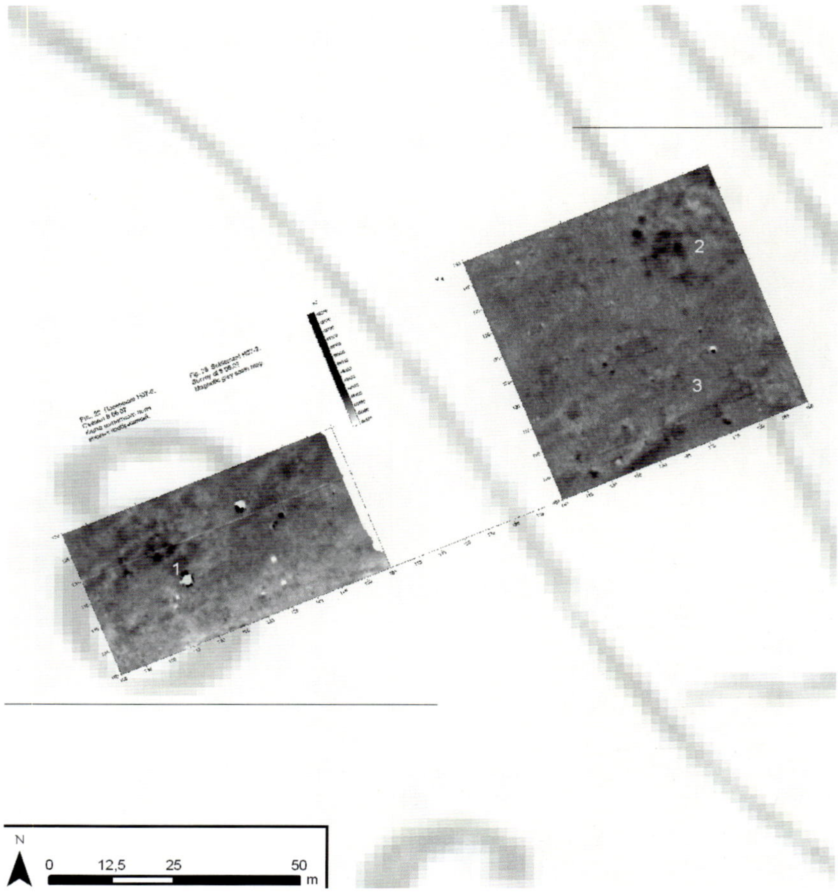

Fig. 4.177 • Distribution map of DSP07-H07-02.

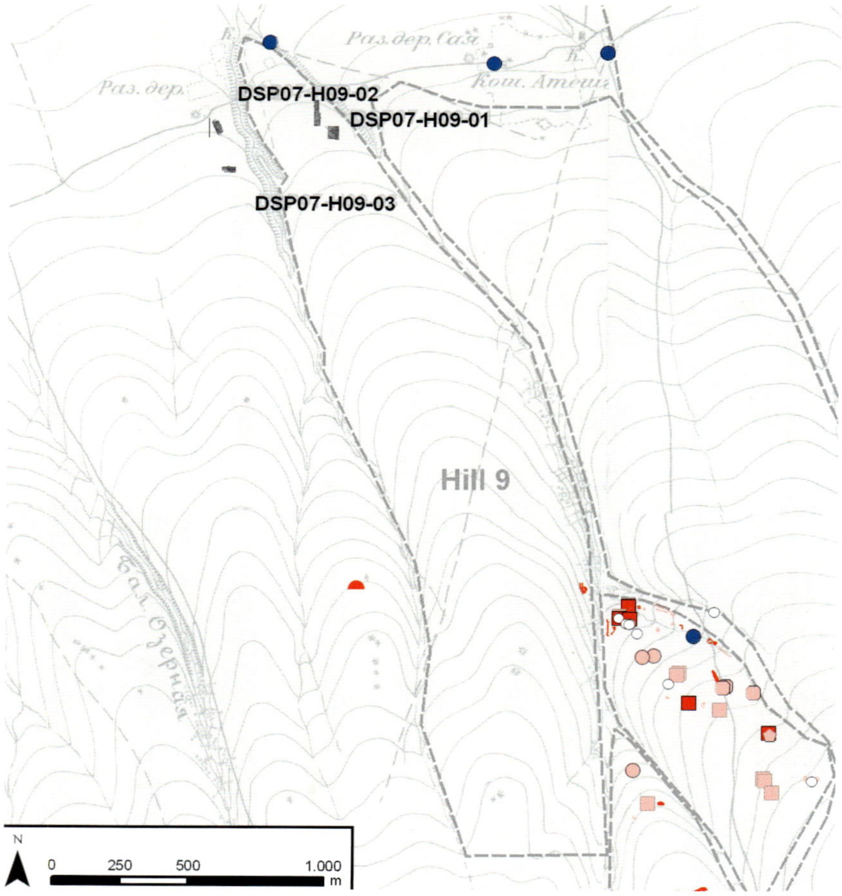

Fig. 4.178 • Map of Hill 9.

Fig. 4.179 • DSP07-H09-02 Feature (number unknown).

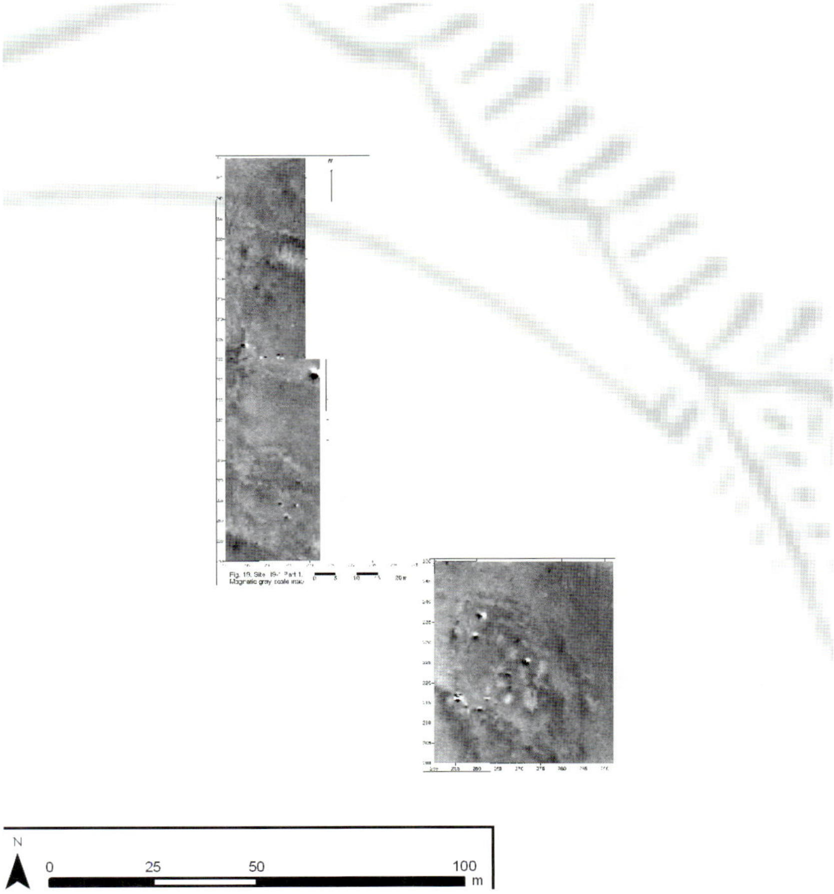

Fig. 4.180 • Distribution map of DSP07-H09-01.

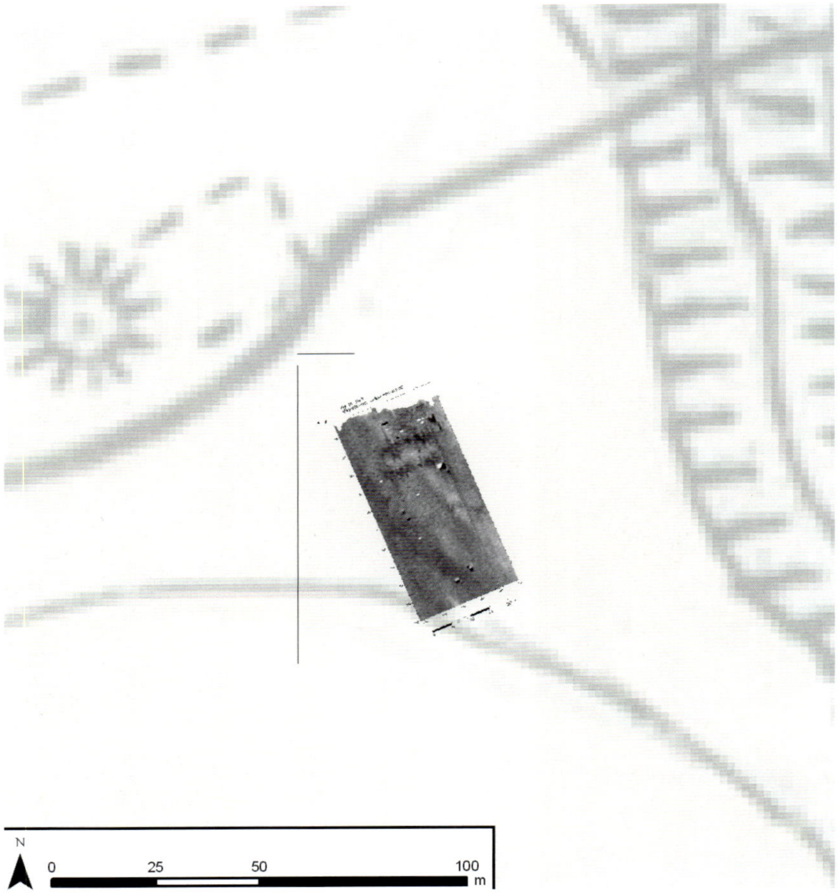

Fig. 4.181 • Distribution map of DSP07-H09-02.

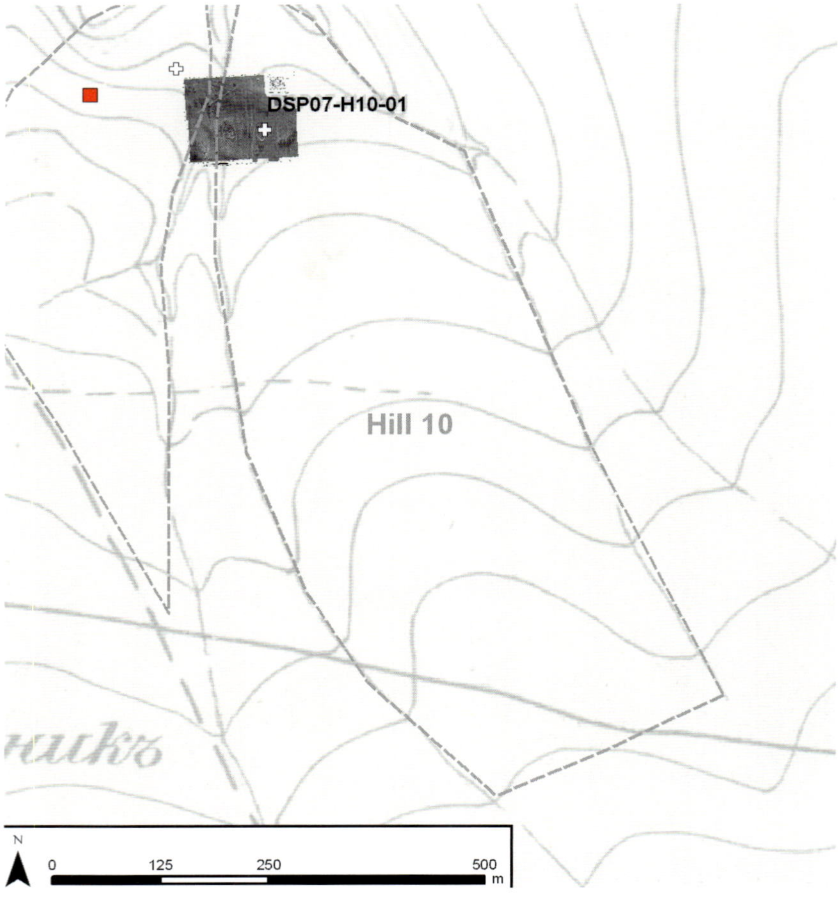

Fig. 4.182 • Map of Hill 10.

Fig. 4.183 • View of Hill 10 from the northwest.

Fig. 4.184 • DSP07-H10-01 Feature 2 from the southeast. Hill 18 in the left back ground.

Fig. 4.185 • Google map of DSP07-H10-01 in the centre and H18-01-01 to the northwest. DSP07-H10-01 Feature 2 is seen very clearly towards the eastern ravine.

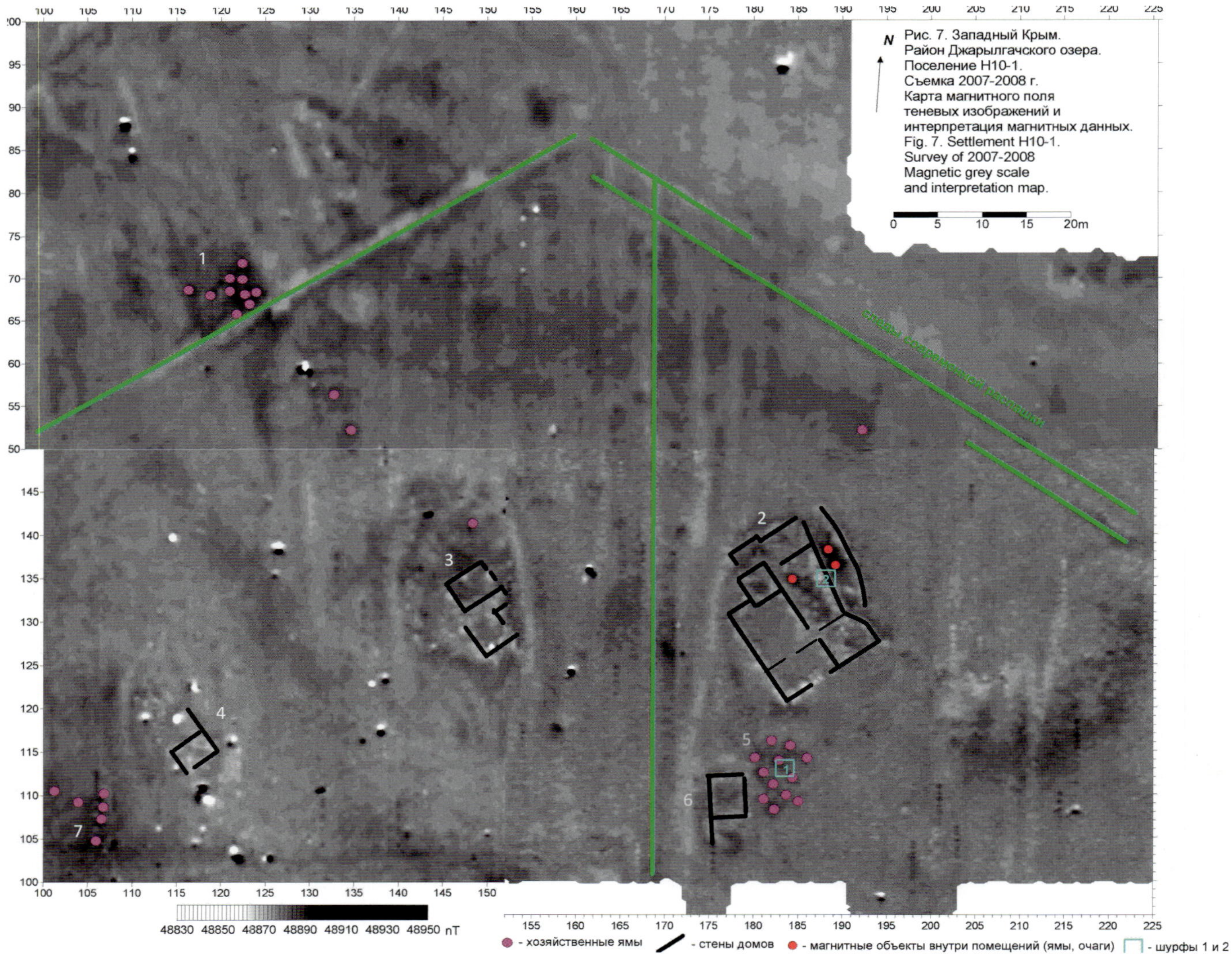

Fig. 4.186 • Geomagnetic map of DSP07-H10-01, interpretation.

Fig. 4.187 • Trial trench on the north wall of DSP07-H10-01 Feature 2.

Fig. 4.188 • Trial trench on the north wall of DSP07-H10-01 Feature 2.

Fig. 4.189 • Drawing of trial trench on the north wall of DSP07-H10-01 Feature 2: 0-20 cm below the surface.

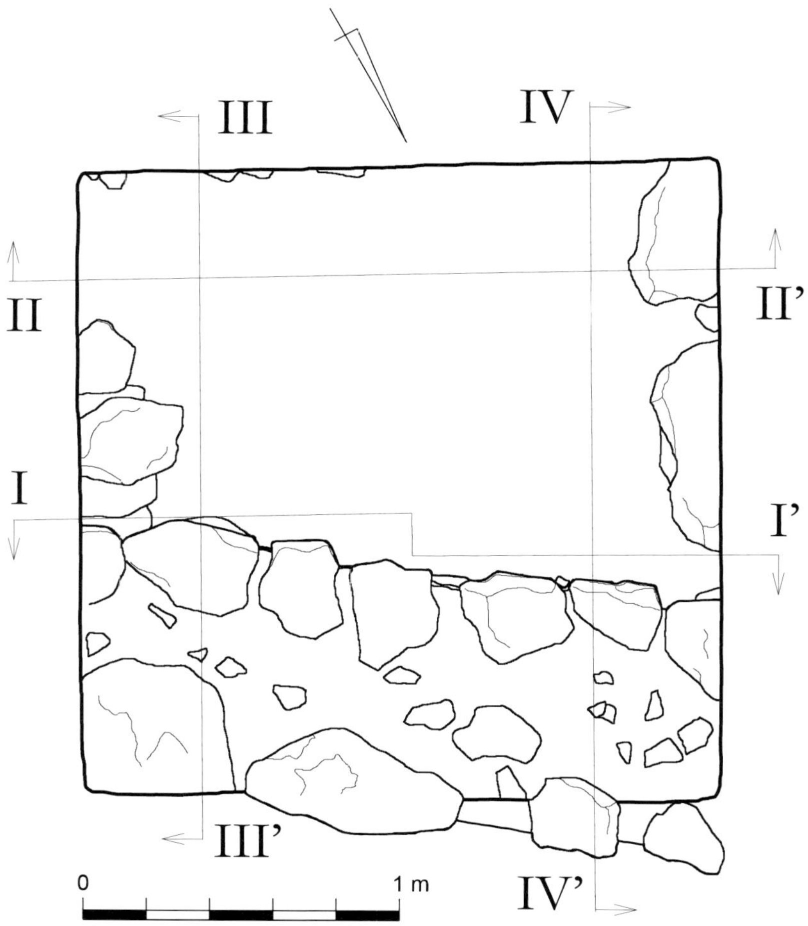

Fig. 4.190 • Drawing of trial trench on the north wall of DSP07-H10-01 Feature 2.

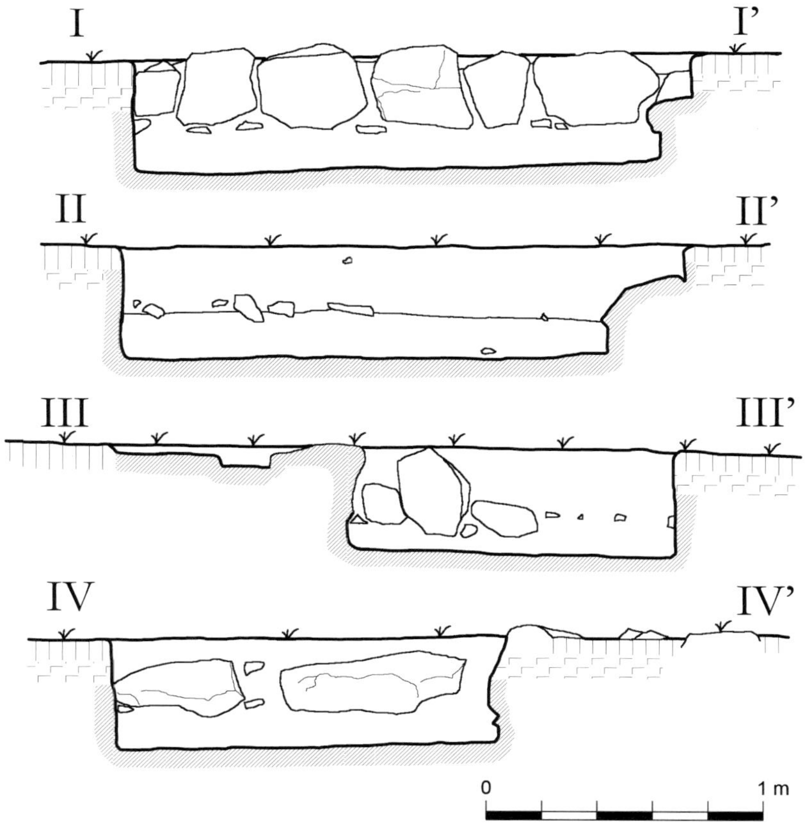

Fig. 4.191 • Profile drawings of trial trench on the north wall of DSP07-H10-01 Feature 2.

Fig. 4.192 • Trial trench DSP07-H10-01 Feature 5.

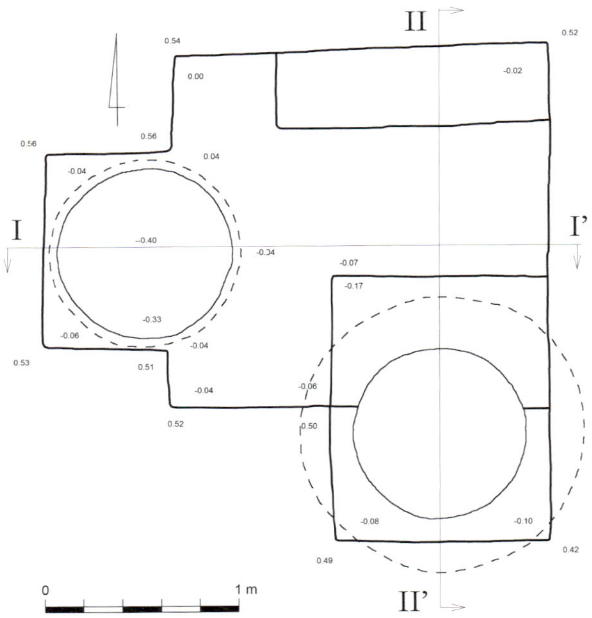

Fig. 4.193 • Drawing of trial trench DSP07-H10-01 Feature 5.

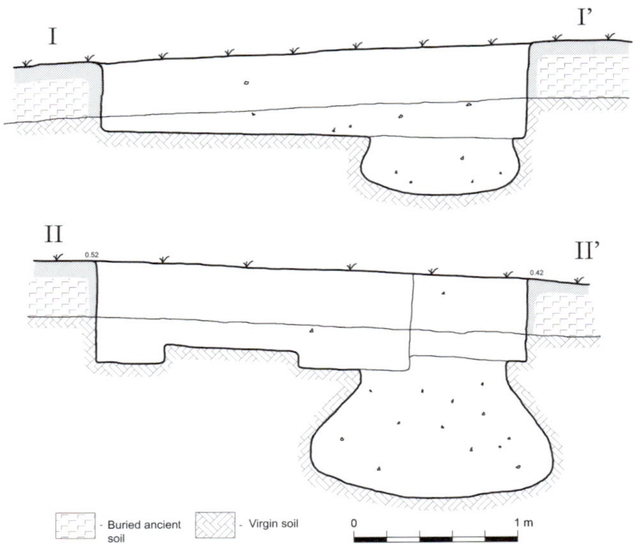

Fig. 4.194 • Profile drawings of trial trench DSP07-H10-01 Feature 5.

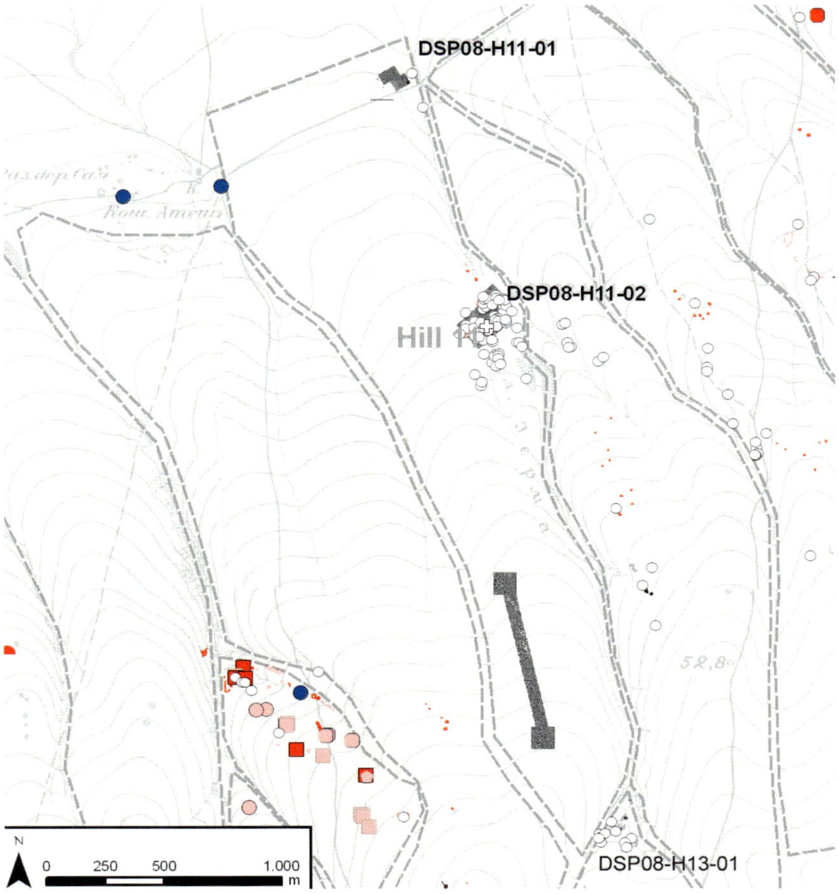

Fig. 4.195 • Map of Hill 11.

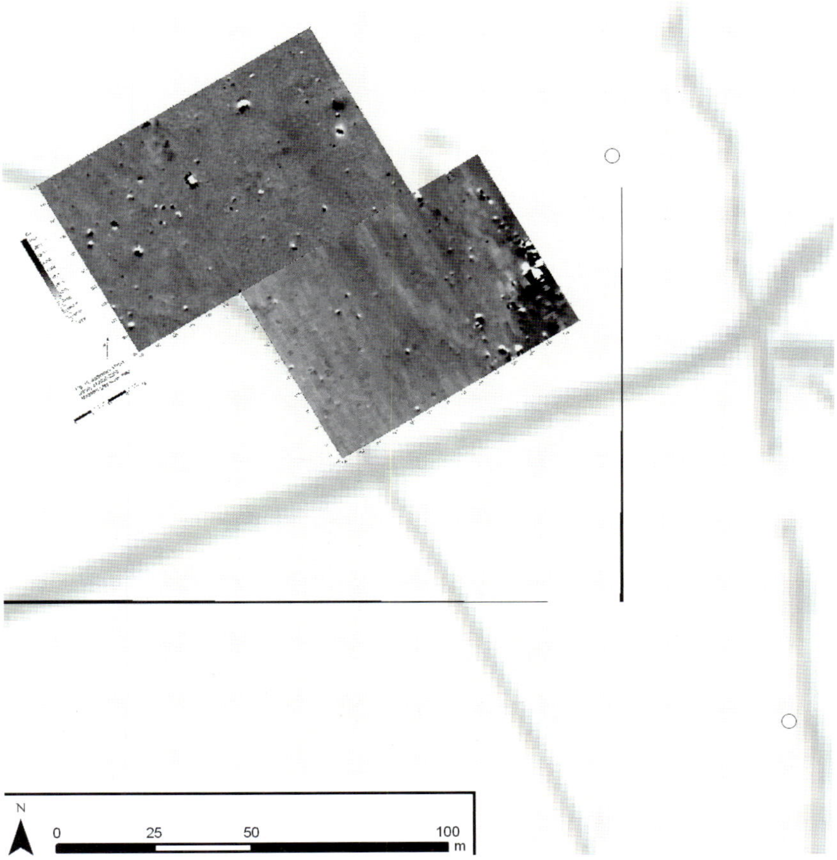

Fig. 4.196 • Distribution map of DSP07-H11-01.

Fig. 4.197 • DSP07-H11-01.

Fig. 4.198 • Line of stones on the eastern bank of the shallow ravine on the eastern flank of DSP07-H11-01.

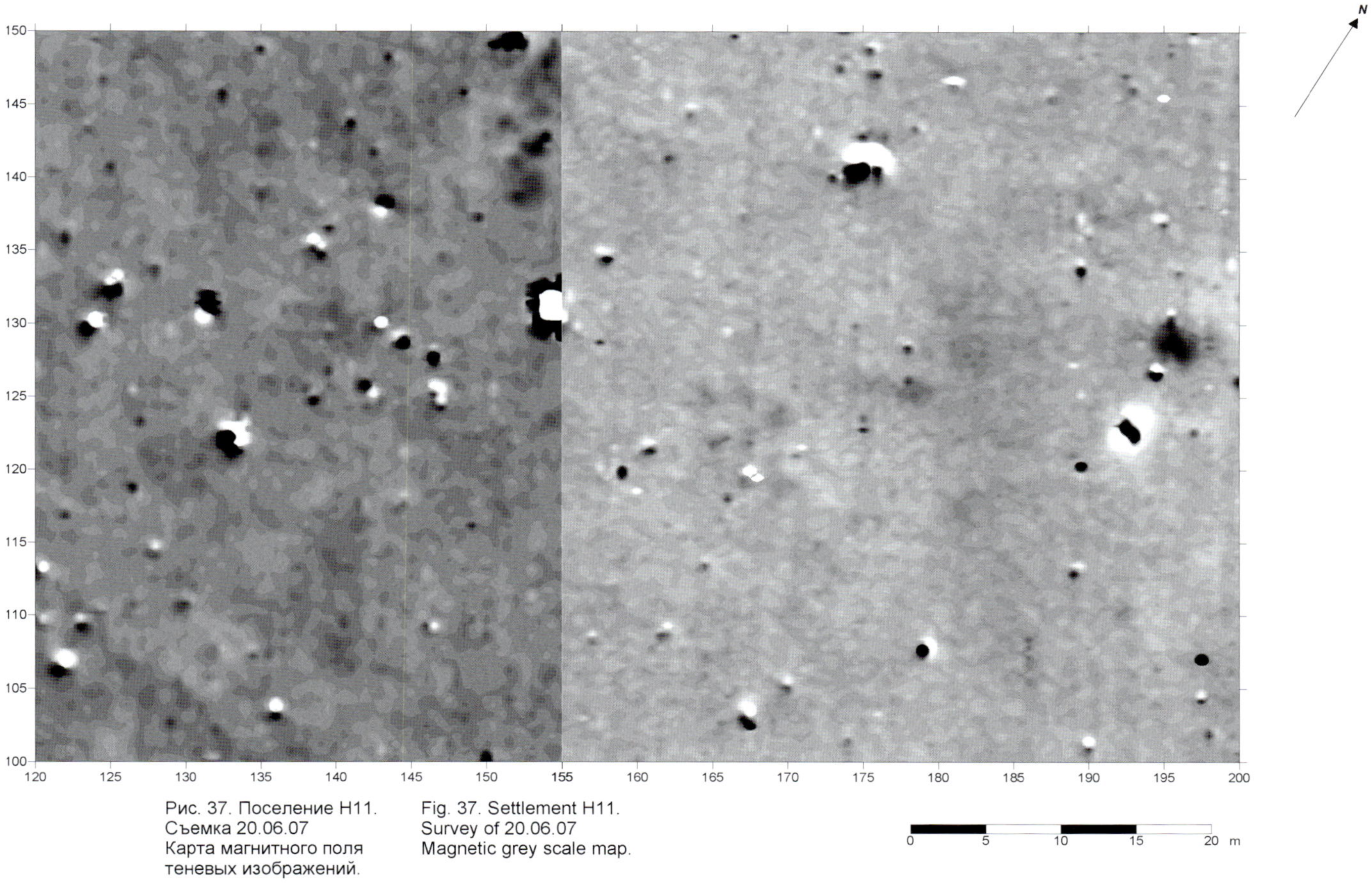

Fig. 4.199 • Geomagnetic map of DSP07-H11-01.

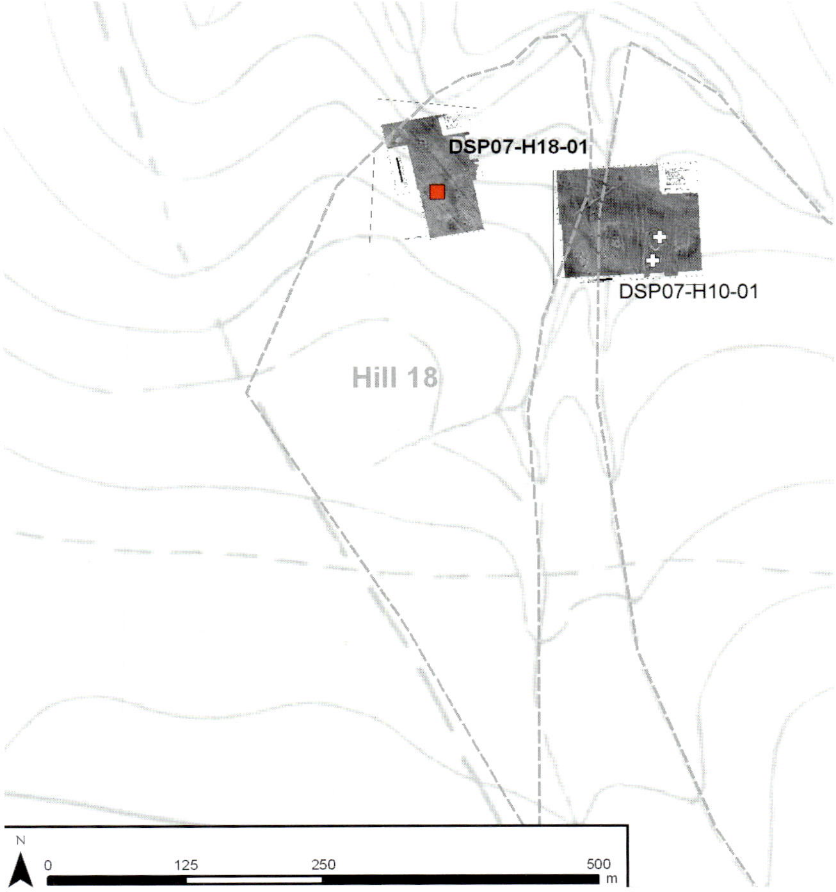

Fig. 4.200 • Map of Hill 18.

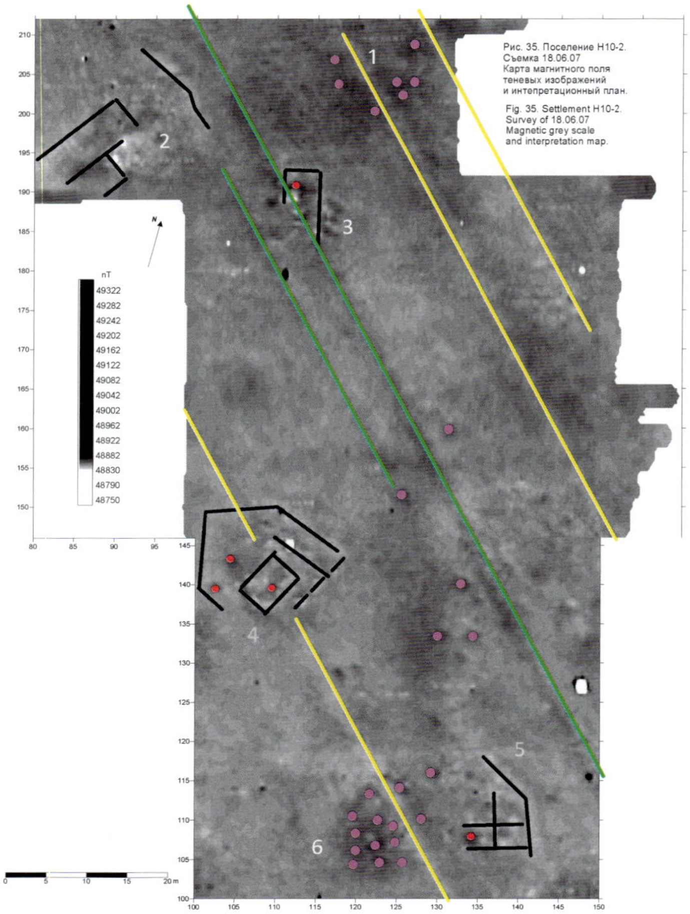

Fig. 4.201 • Geomagnetic map of DSP07-H18-01, interpretation.

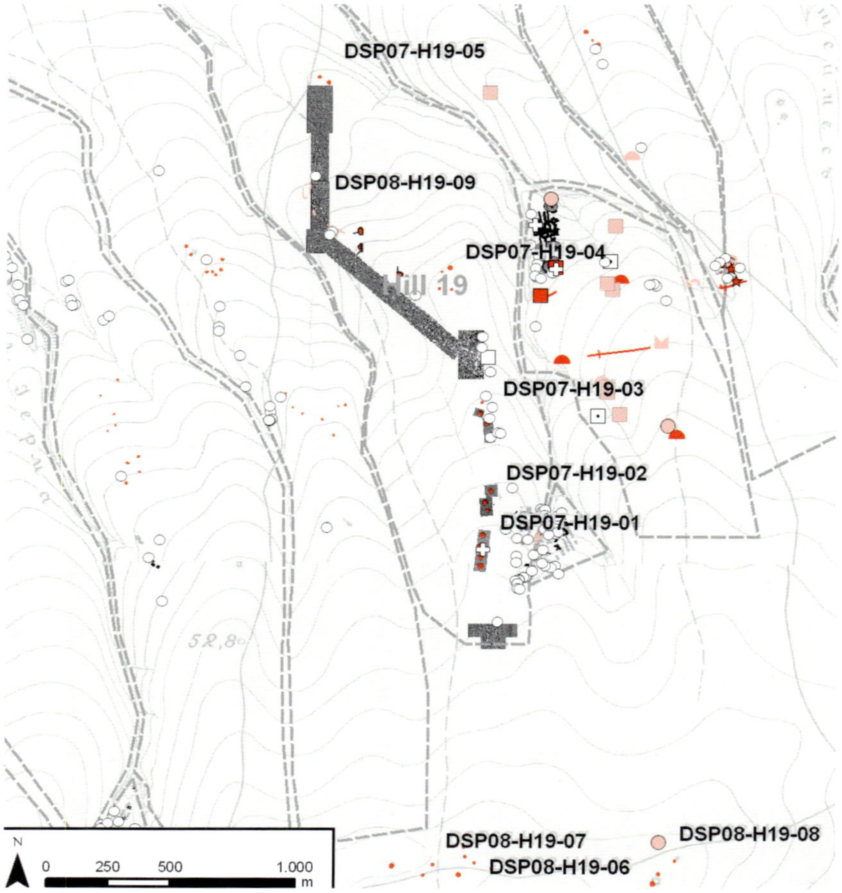

Fig. 4.202 • Distribution map of Hill 19.

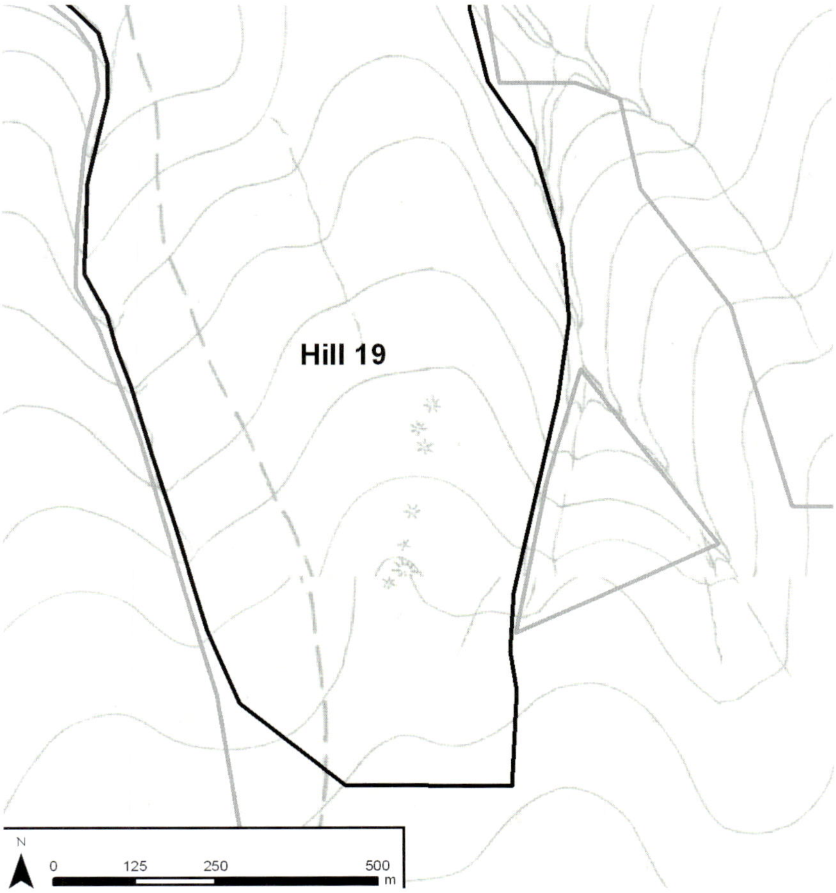

Fig. 4.203 • Half-verst map showing kurgans located on Hill 19.

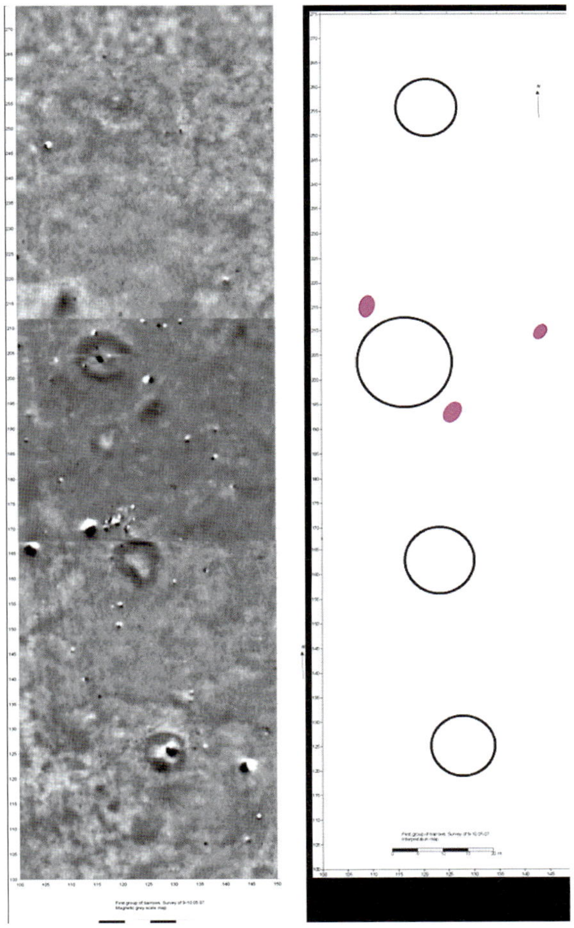

Fig. 4.204 • Distribution map of DSP07-H19-01.

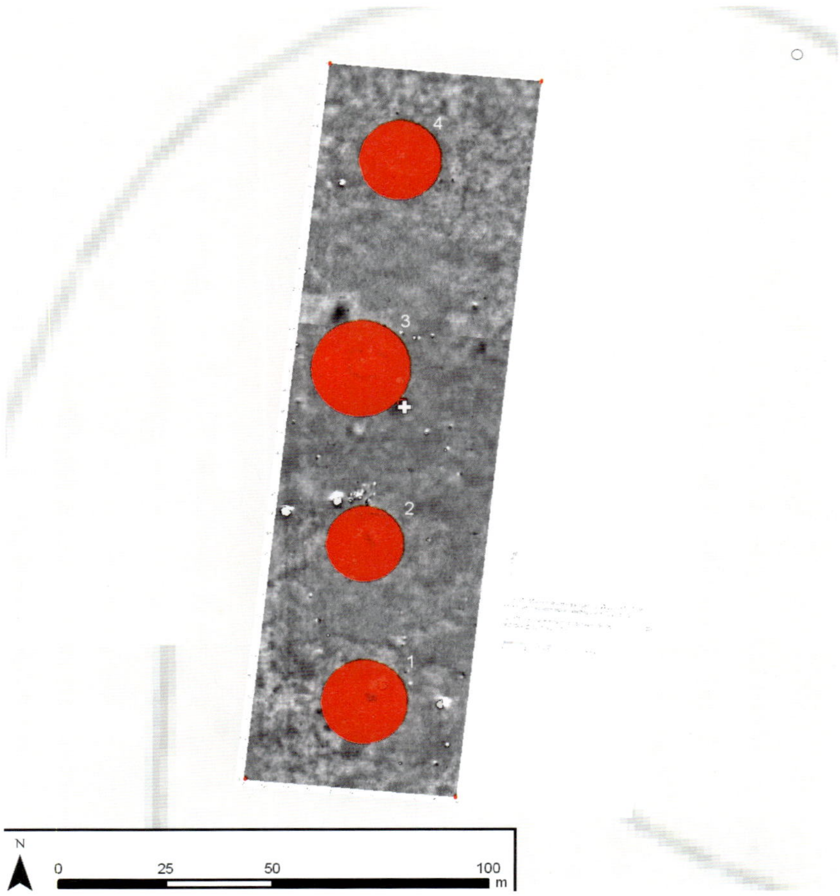

Fig. 4.205 • Distribution map of DSP07-H19-01.

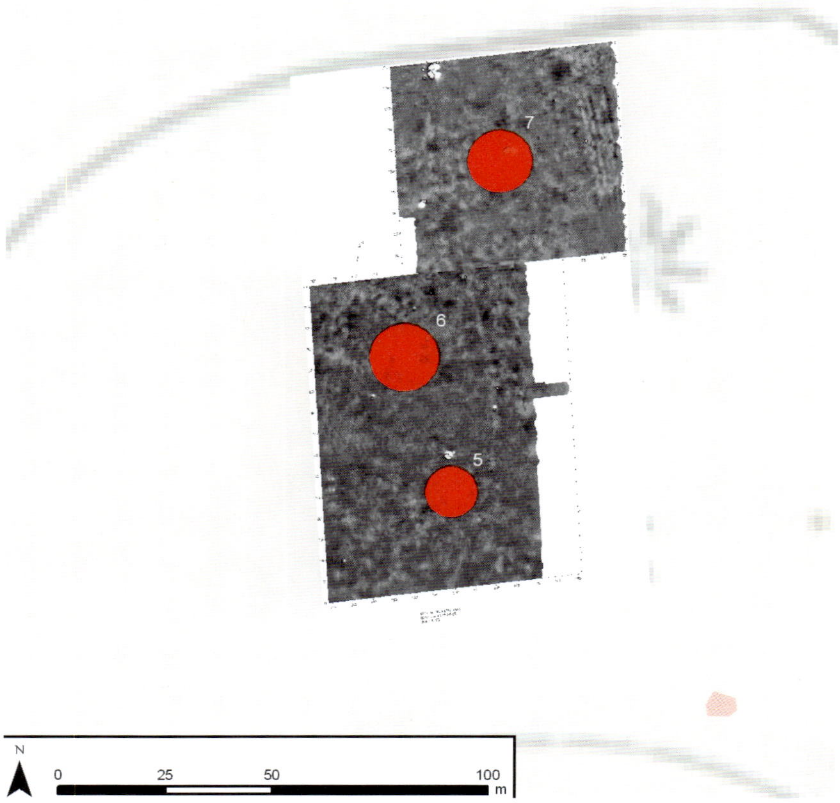

Fig. 4.206 • Distribution map of DSP07-H19-02.

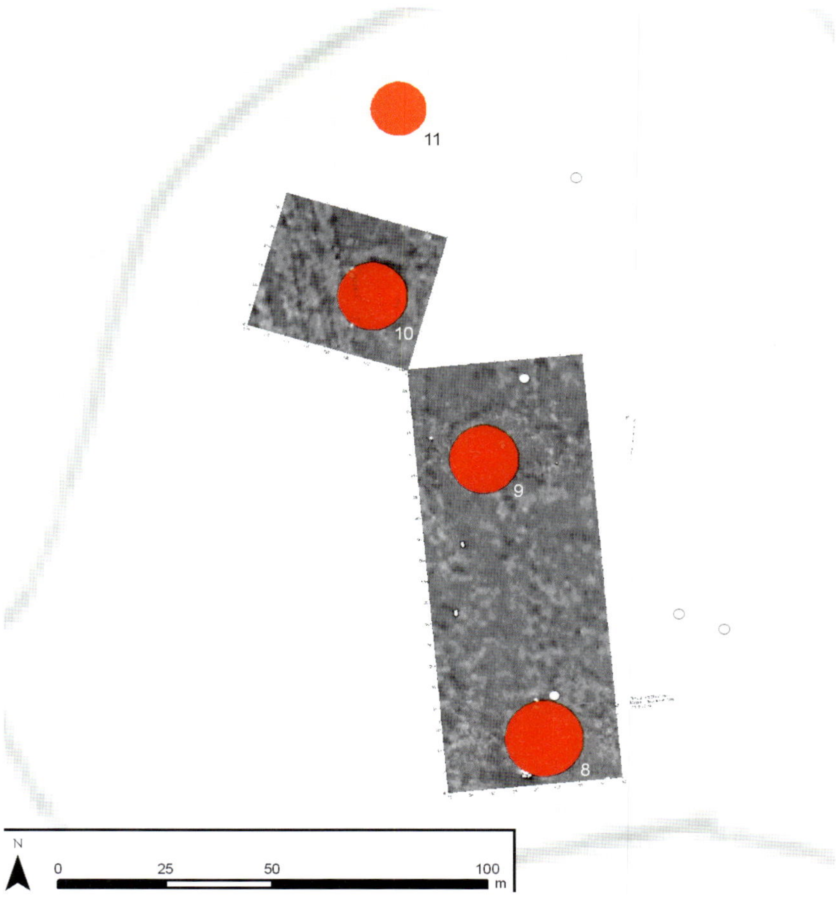

Fig. 4.207 • Distribution map of DSP07-H19-03.

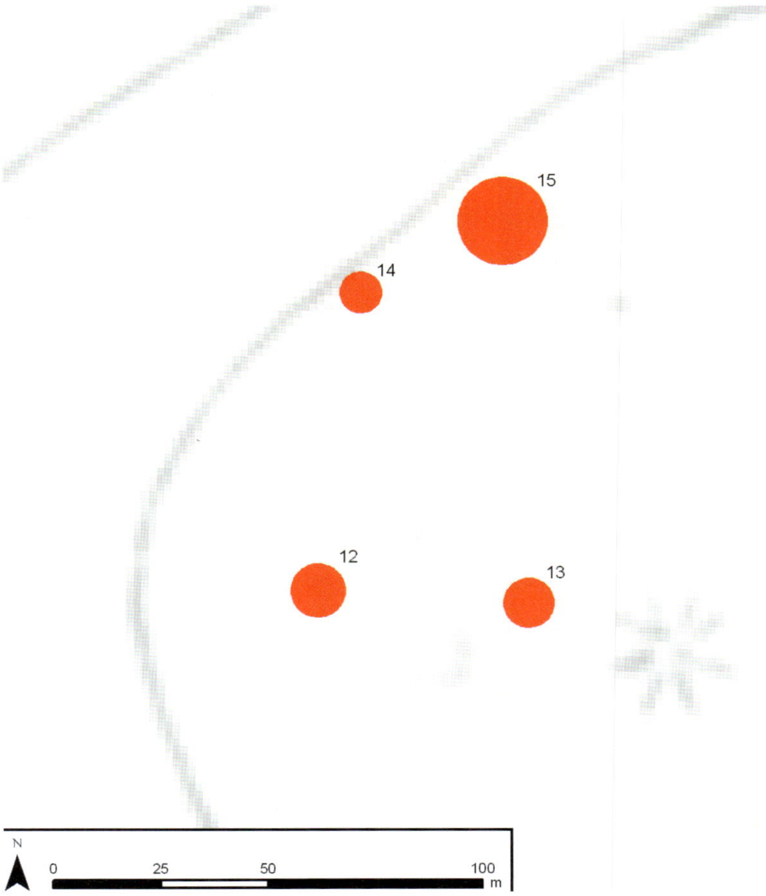

Fig. 4.208 • Distribution map of DSP07-H19-04.

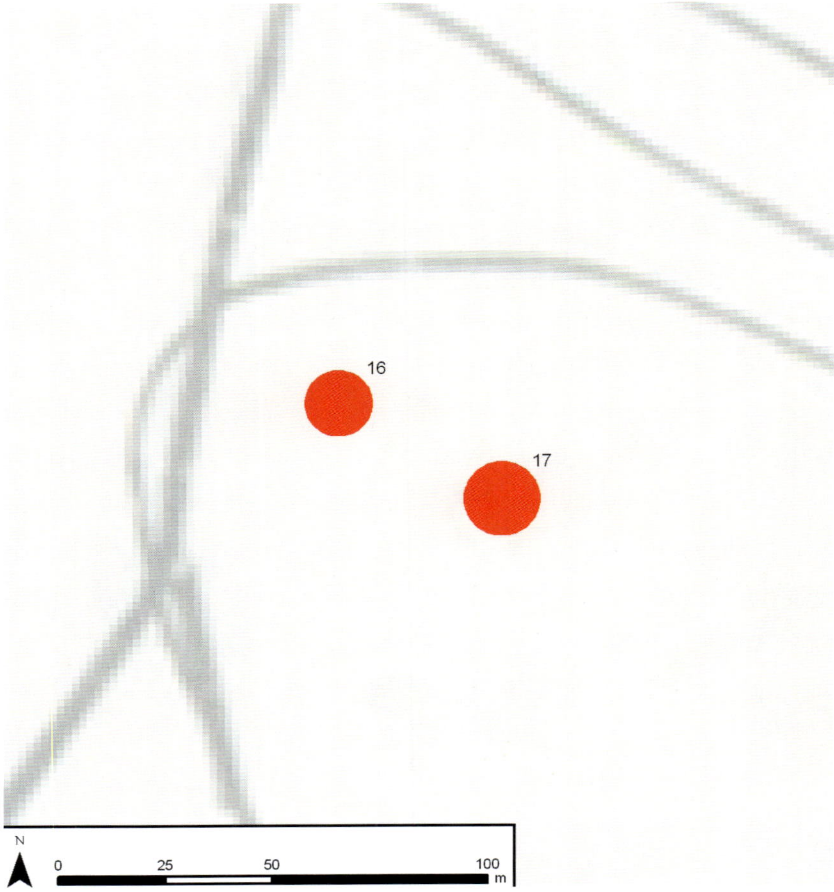

Fig. 4.209 • Distribution map of DSP07-H19-05.

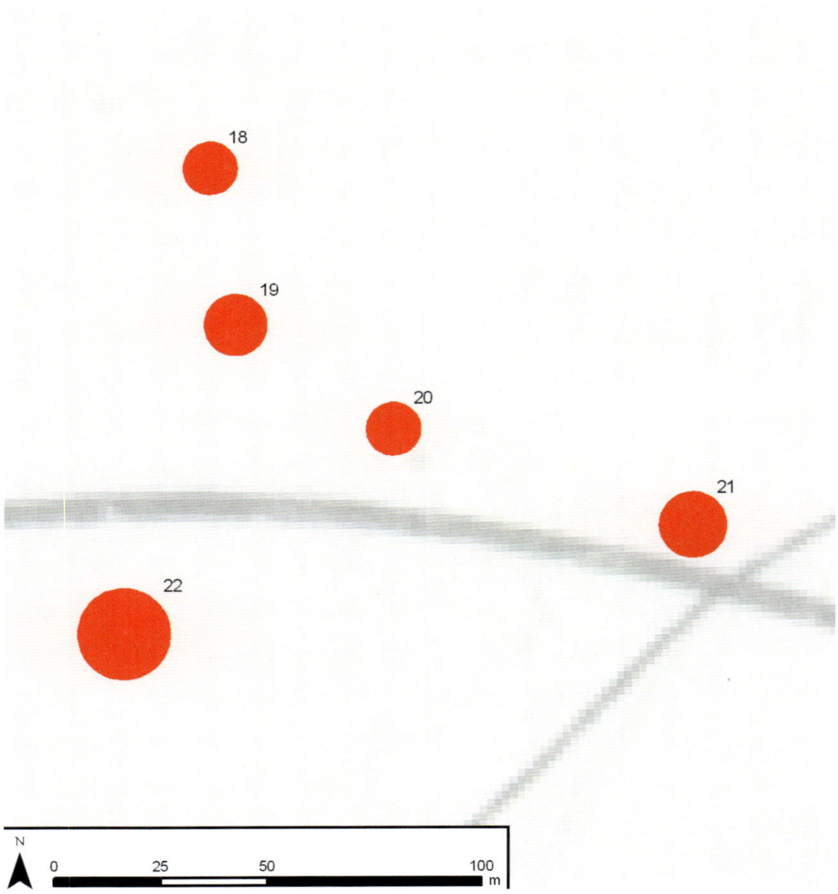

Fig. 4.210 • Distribution map of DSP07-H19-06.

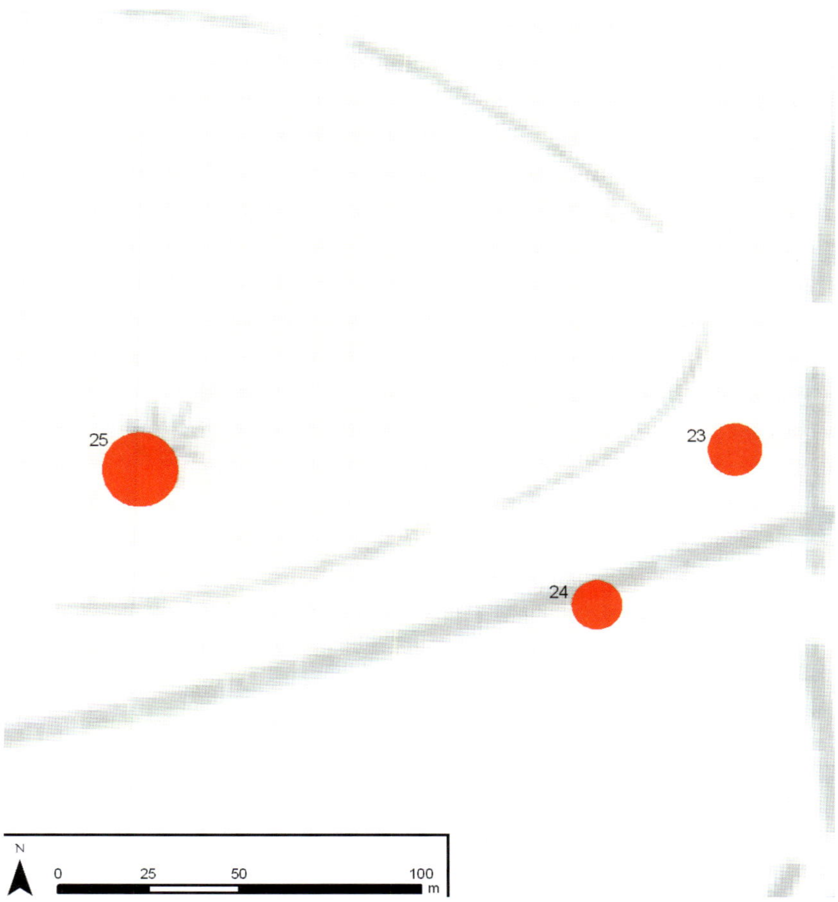

Fig. 4.211 • Distribution map of DSP08-H19-07.

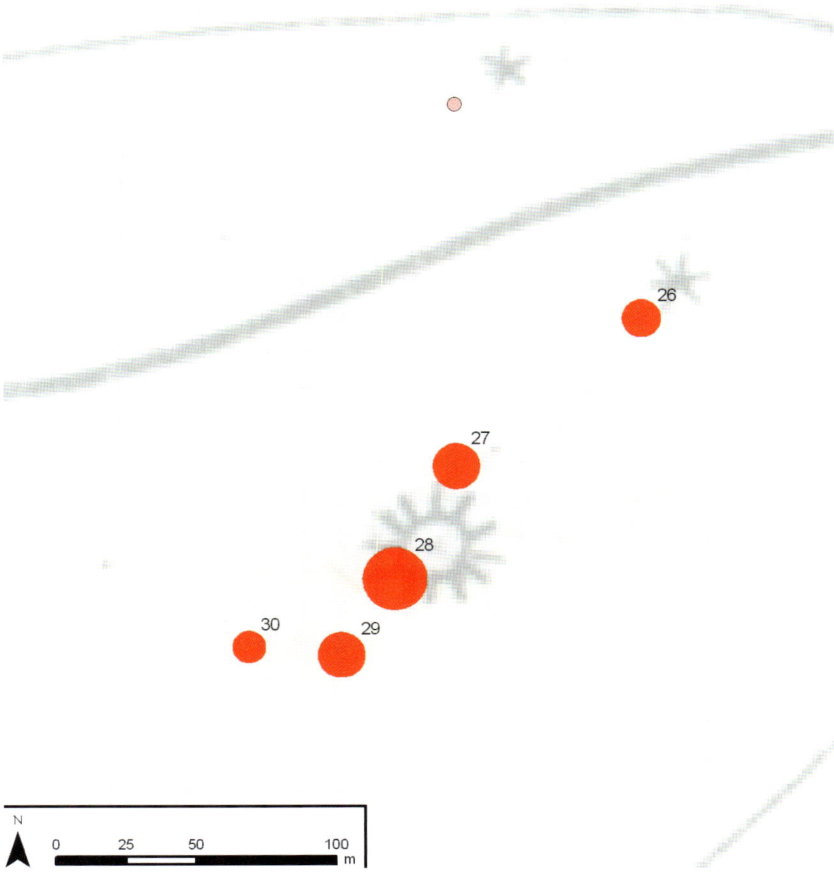

Fig. 4.212 • Distribution map of DSP08-H19-08.

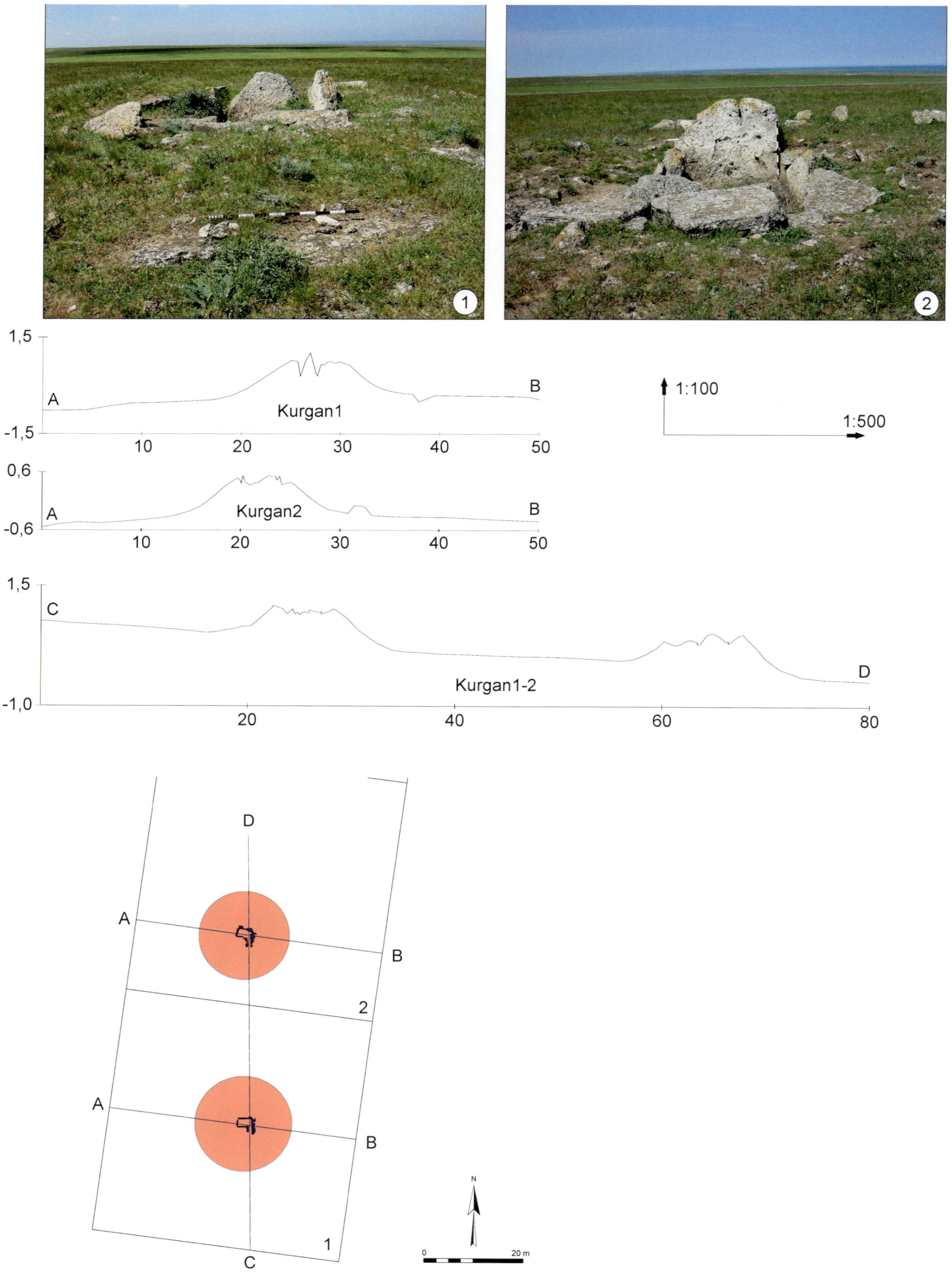

Fig. 4.213 • Photos and cross-sections of Kurgans 1 and 2.

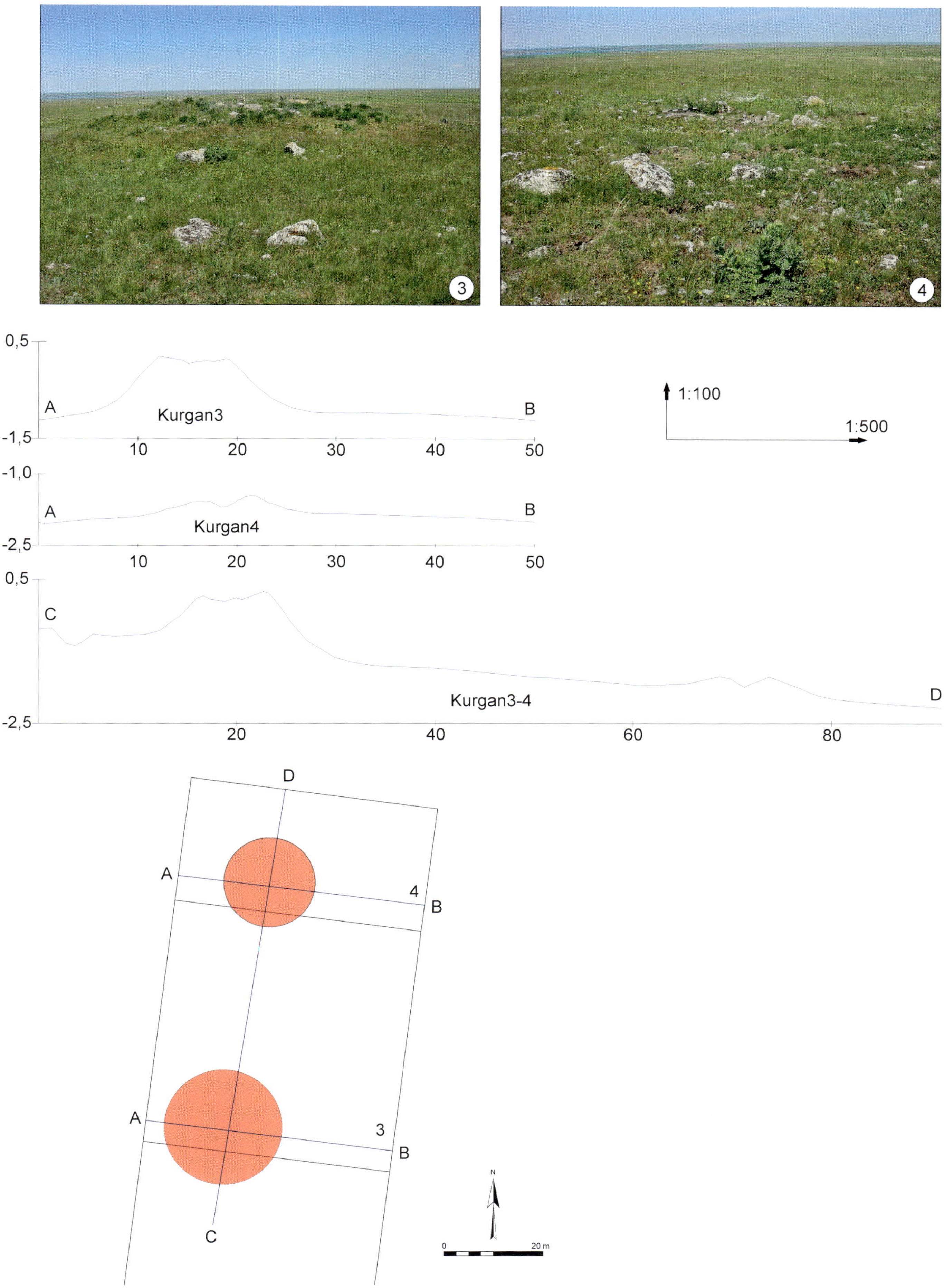

Fig. 4.214 • Photos and cross-sections of Kurgans 3 and 4.

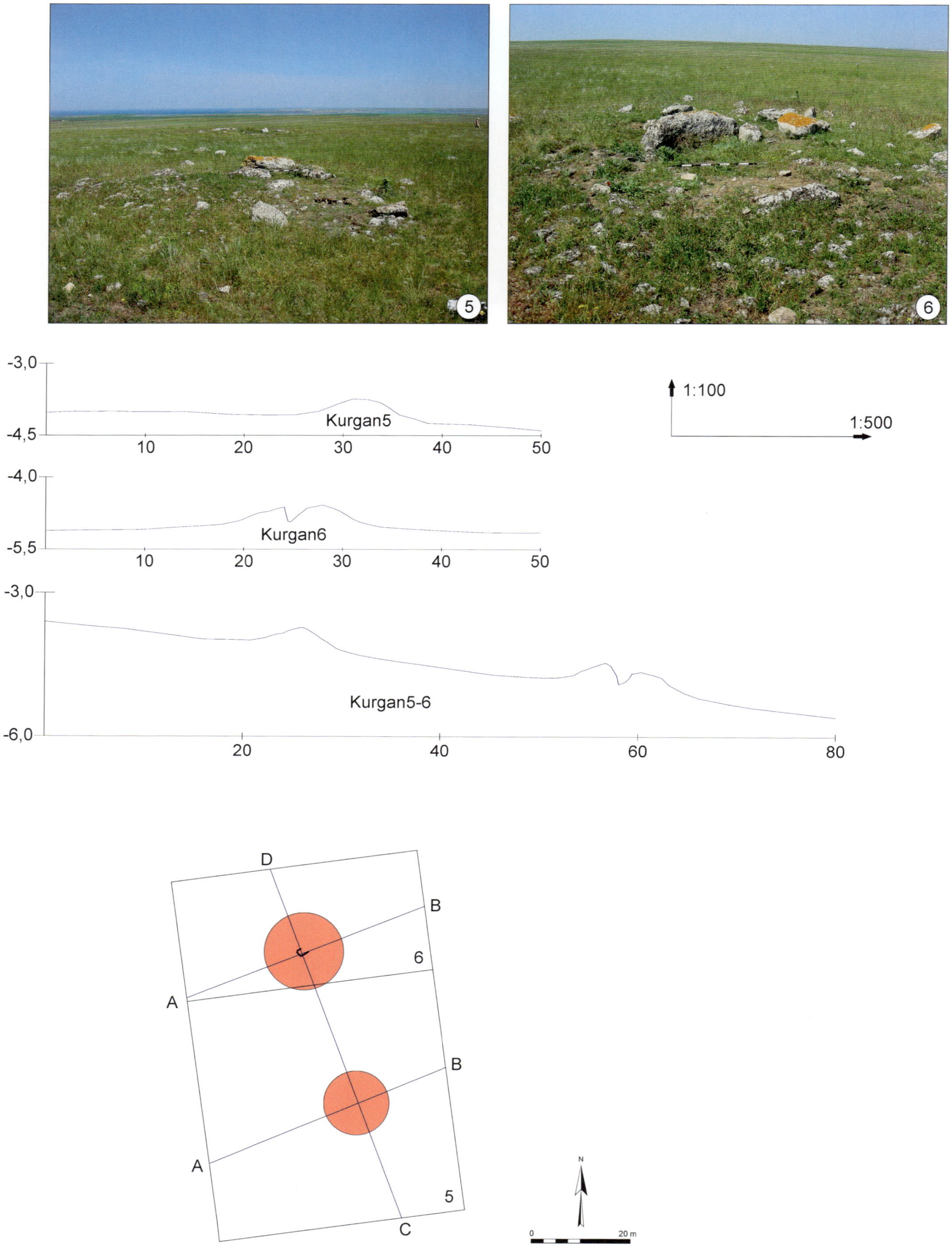

Fig. 4.215 • Photos and cross-sections of Kurgans 5 and 6.

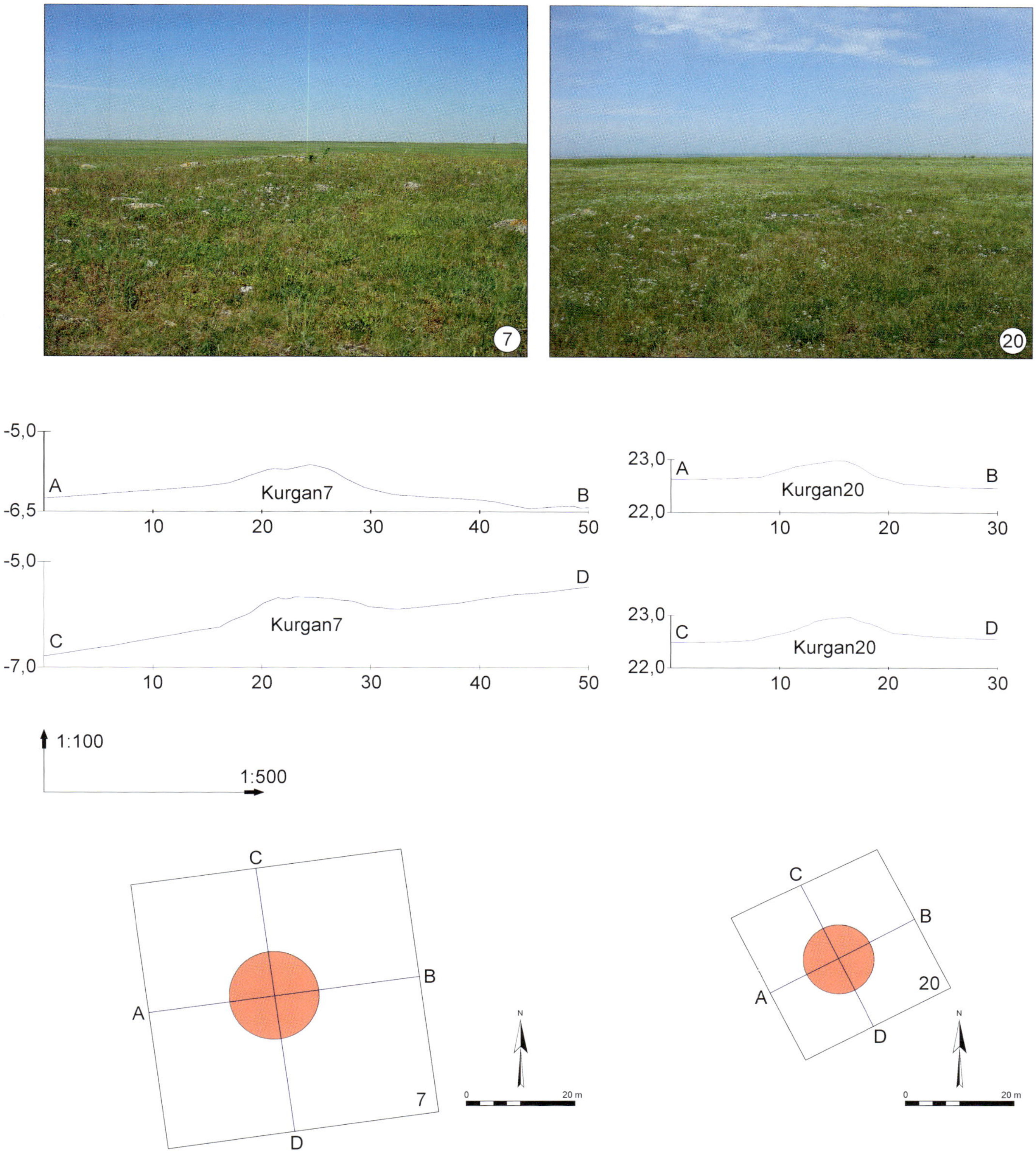

Fig. 4.216 • Photos and cross-sections of Kurgans 7 and 20.

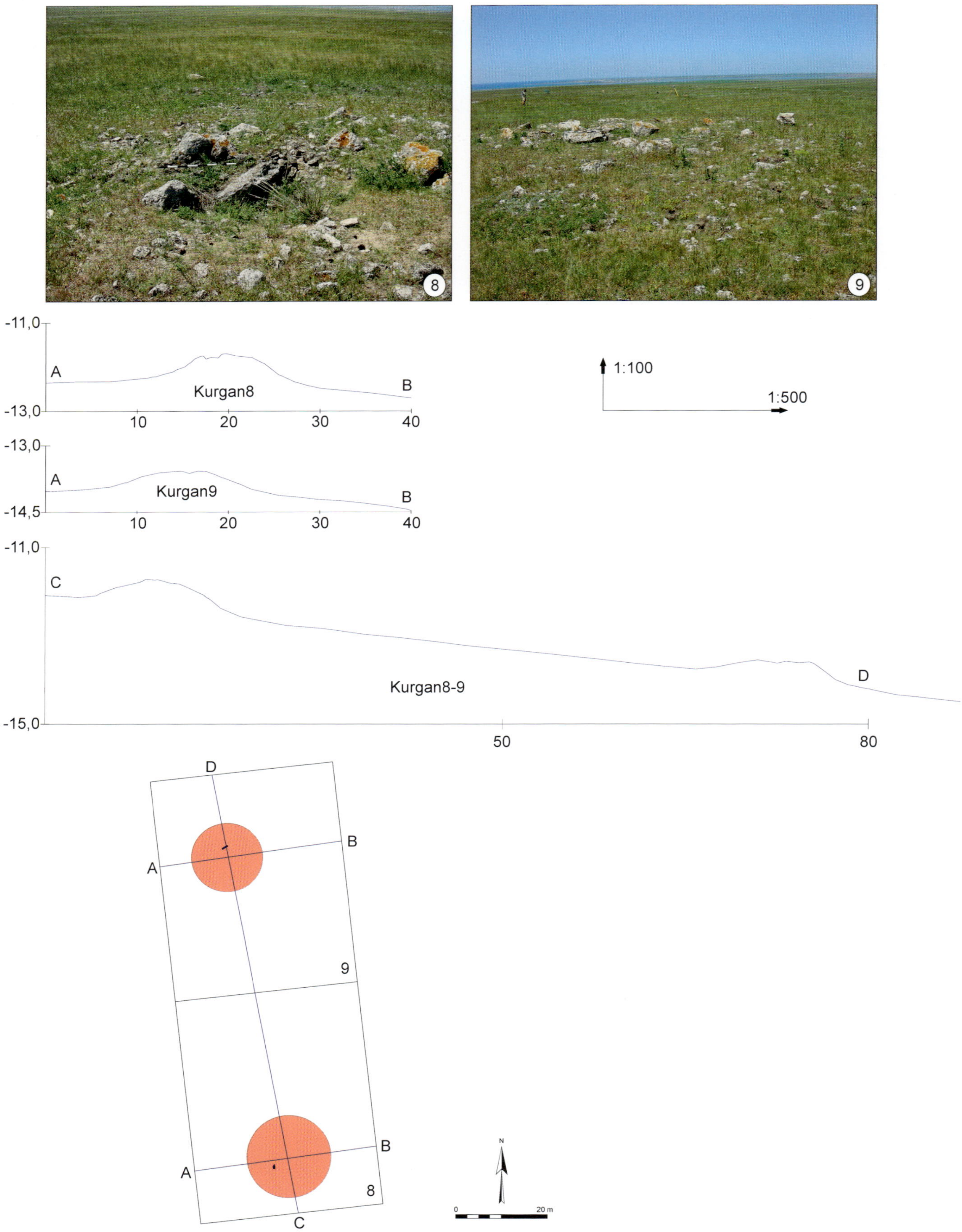

Fig. 4.217 • Photos and cross-sections of Kurgans 8 and 9.

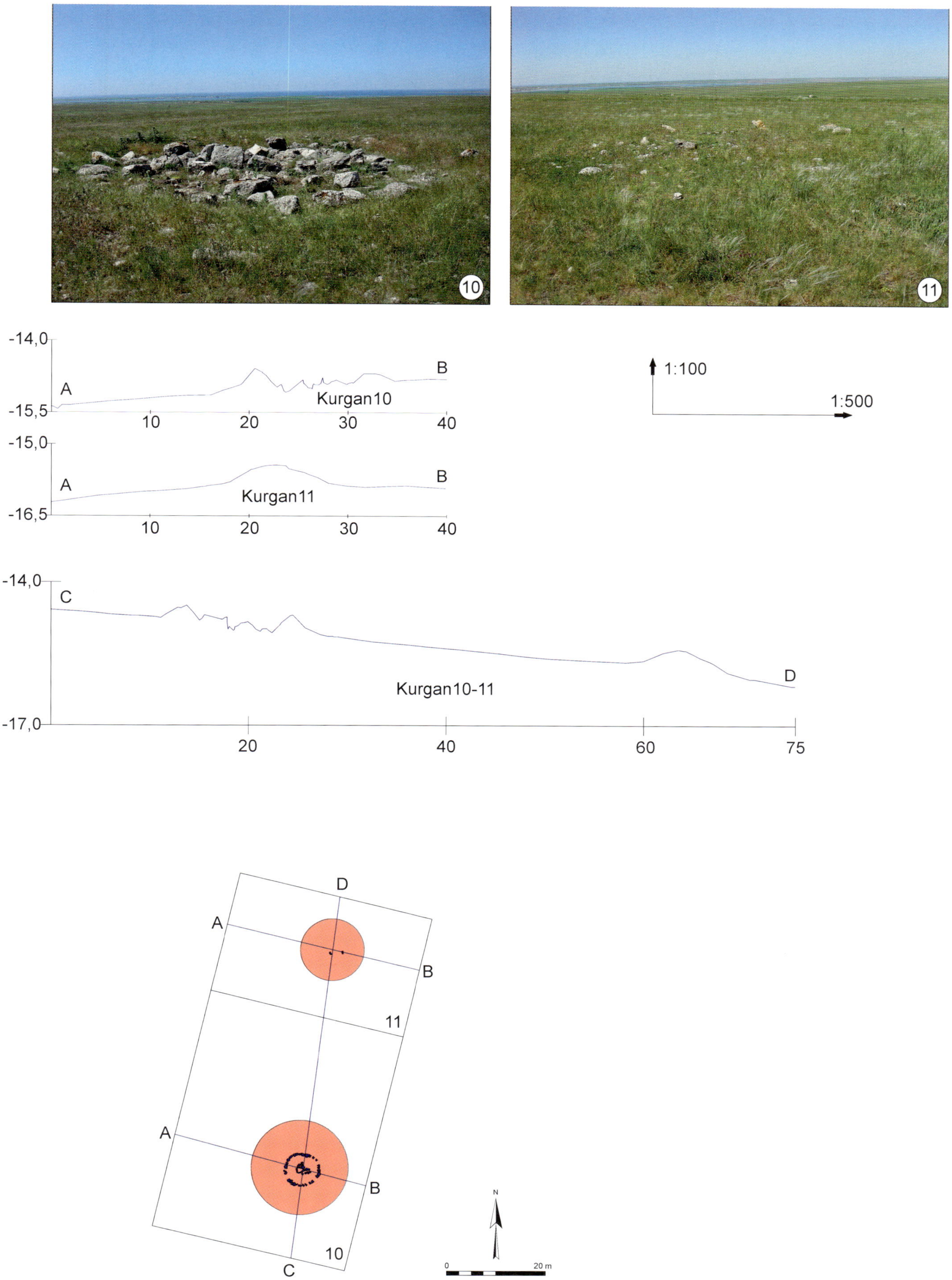

Fig. 4.218 • Photos and cross-sections of Kurgans 10 and 11.

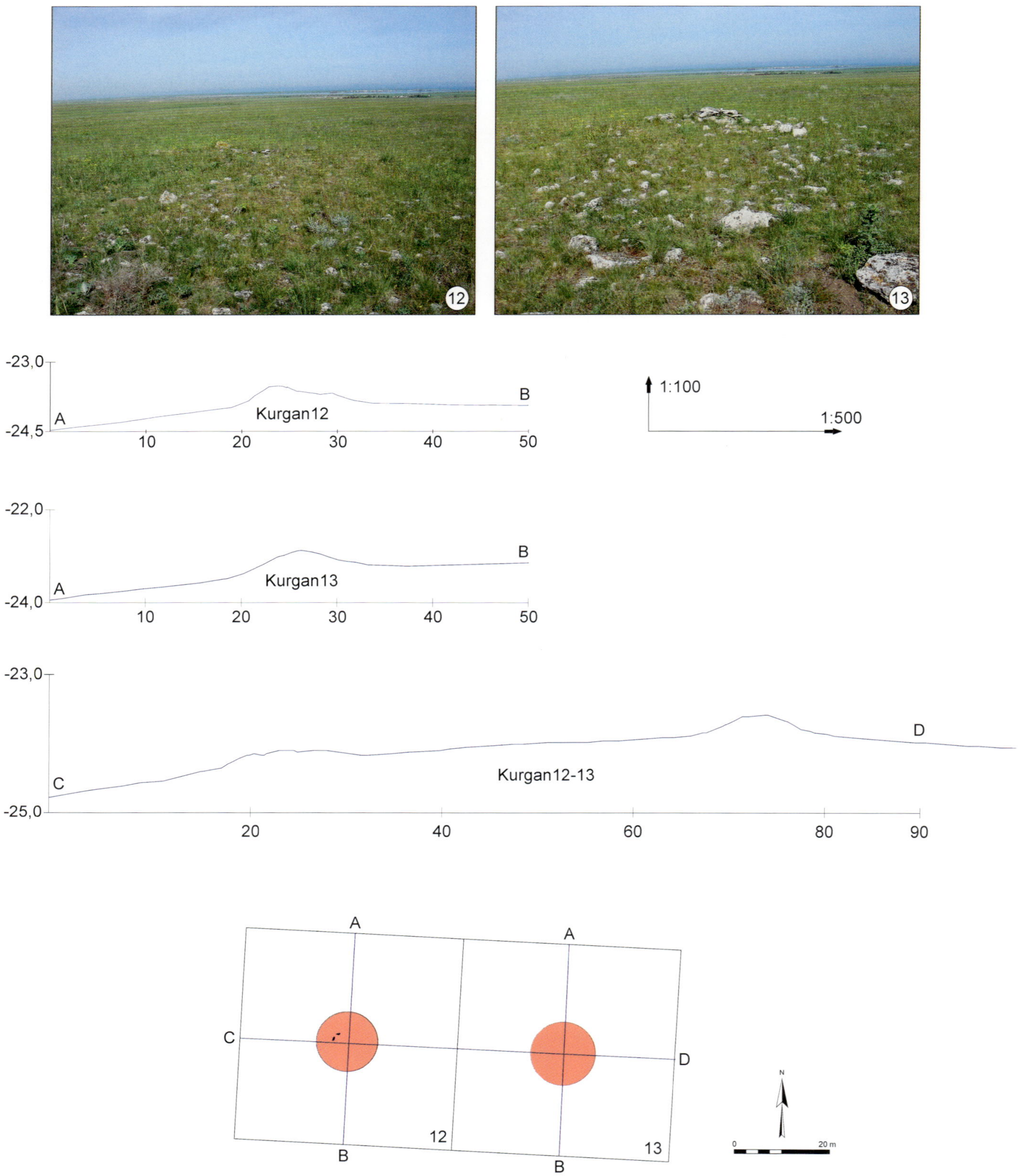

Fig. 4.219 • Photos and cross-sections of Kurgans 12 and 13.

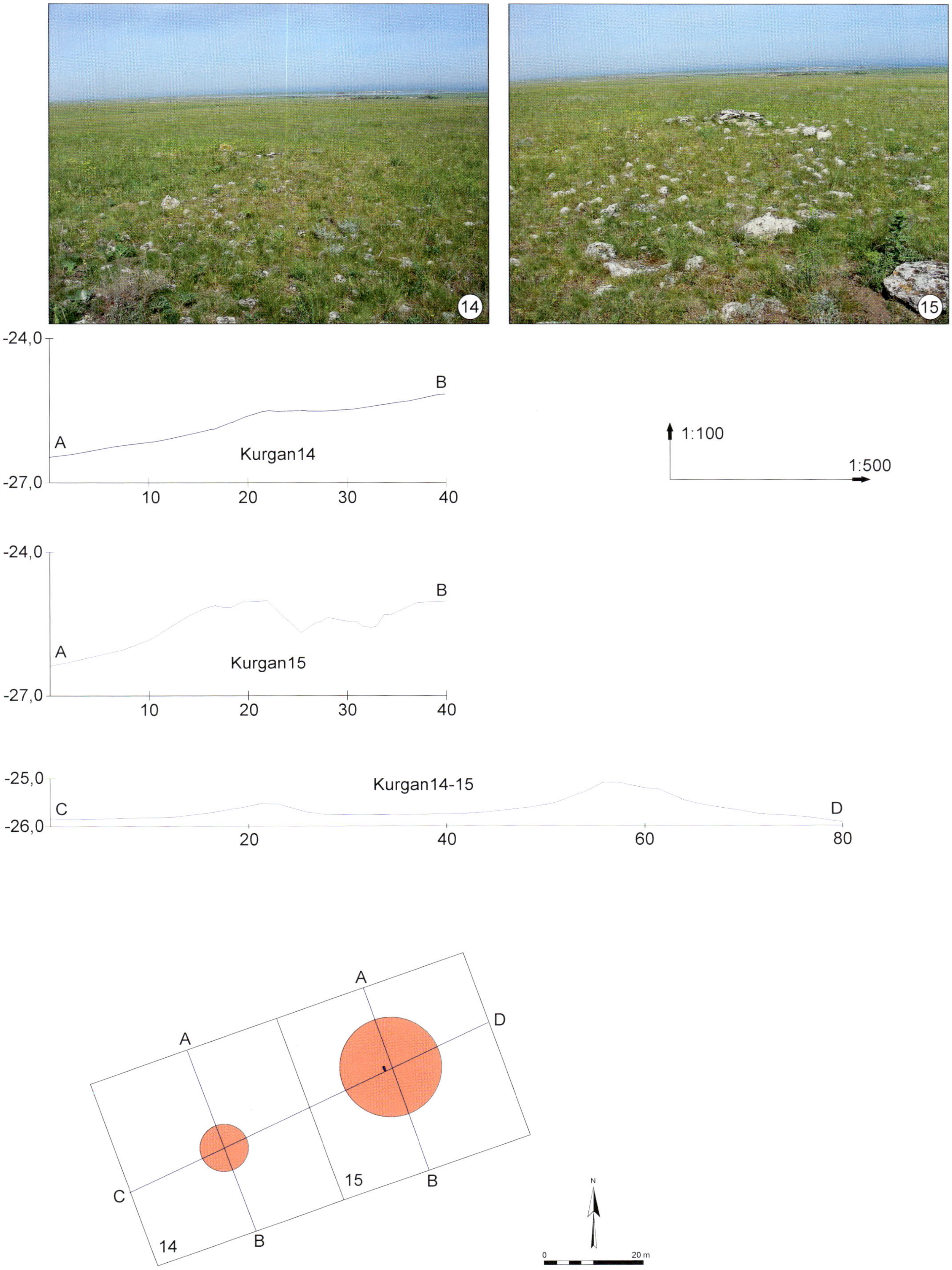

Fig. 4.220 • Photos and cross-sections of Kurgans 14 and 15.

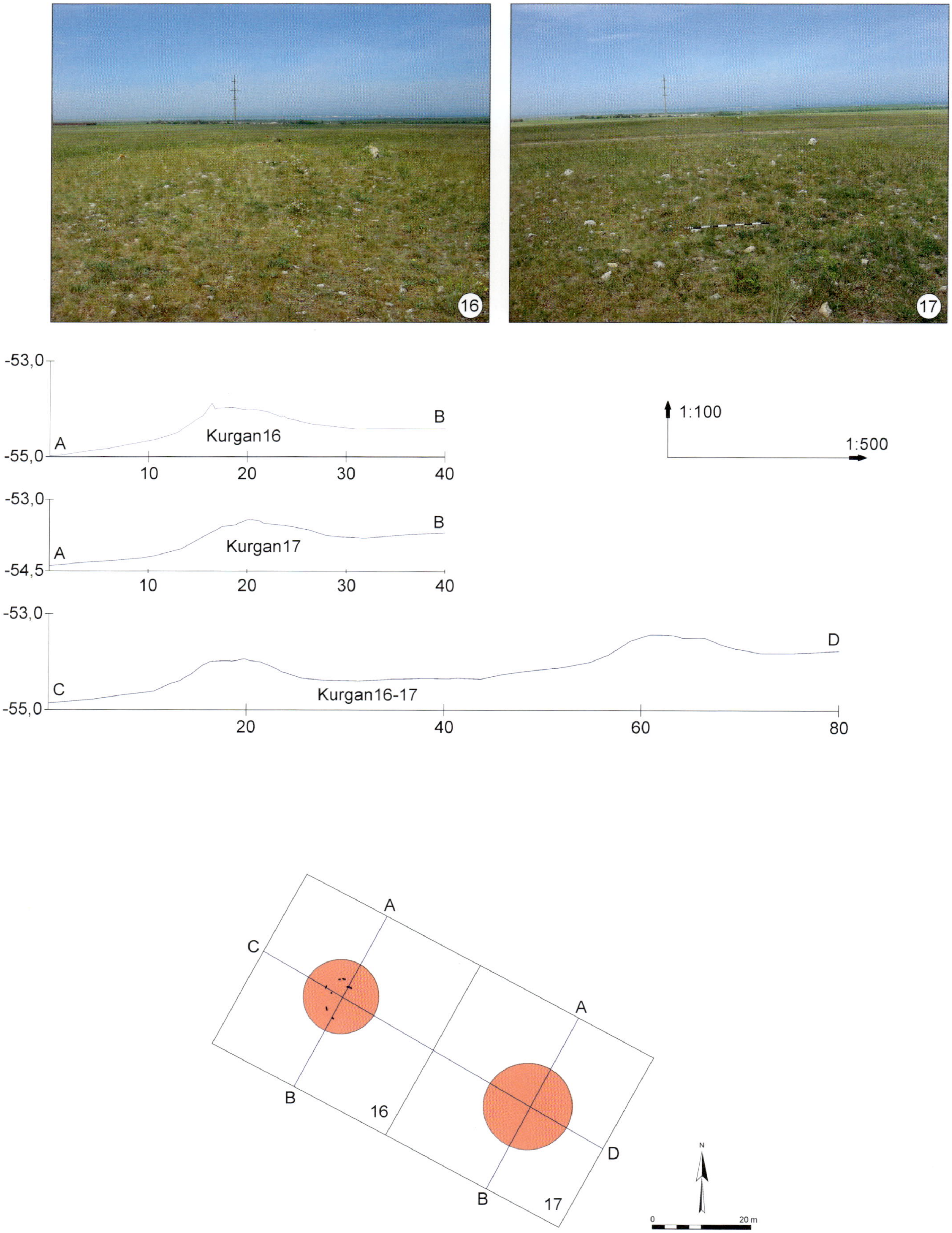

Fig. 4.221 • Photos and cross-sections of Kurgans 16 and 17.

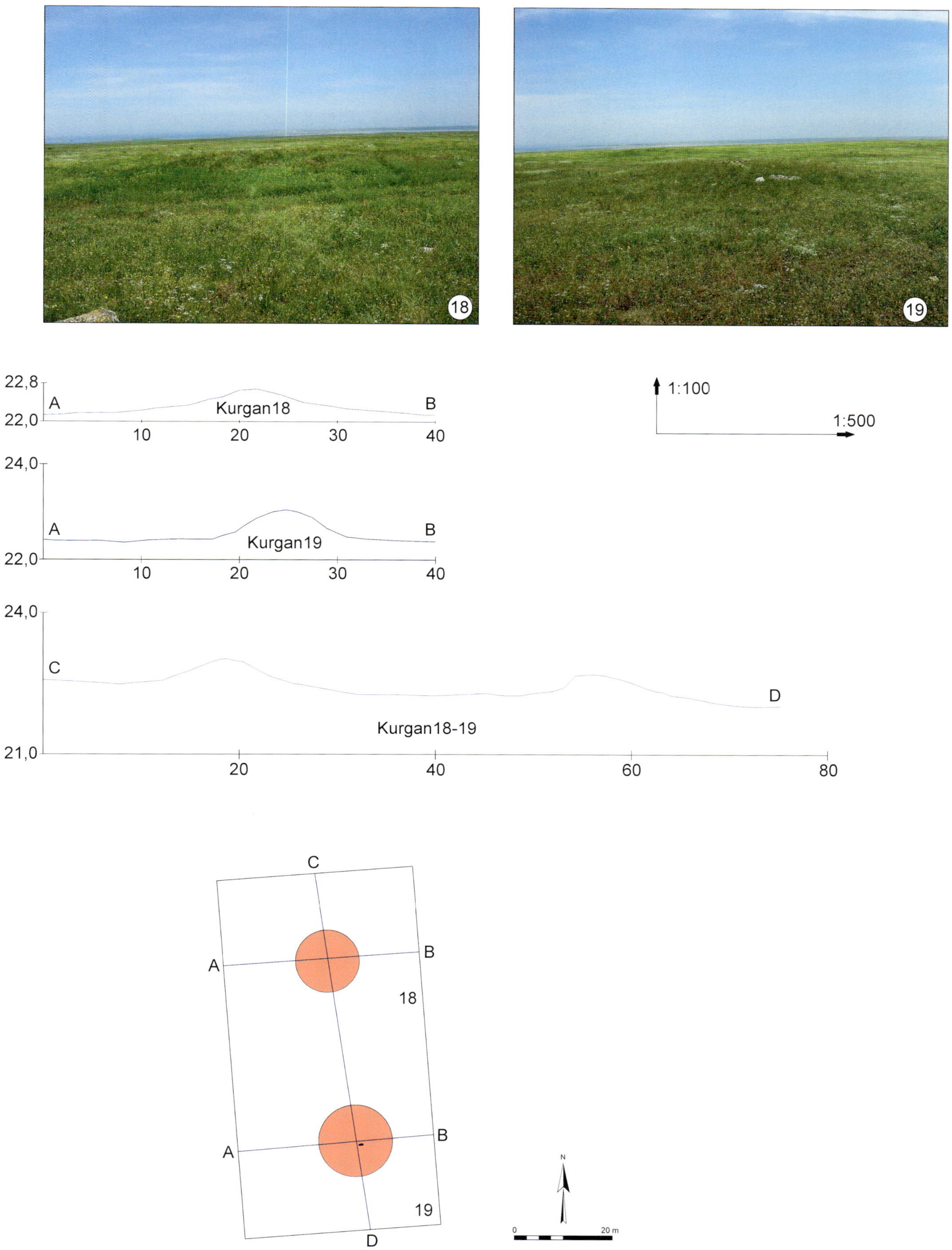

Fig. 4.222 • Photos and cross-sections of Kurgans 18 and 19.

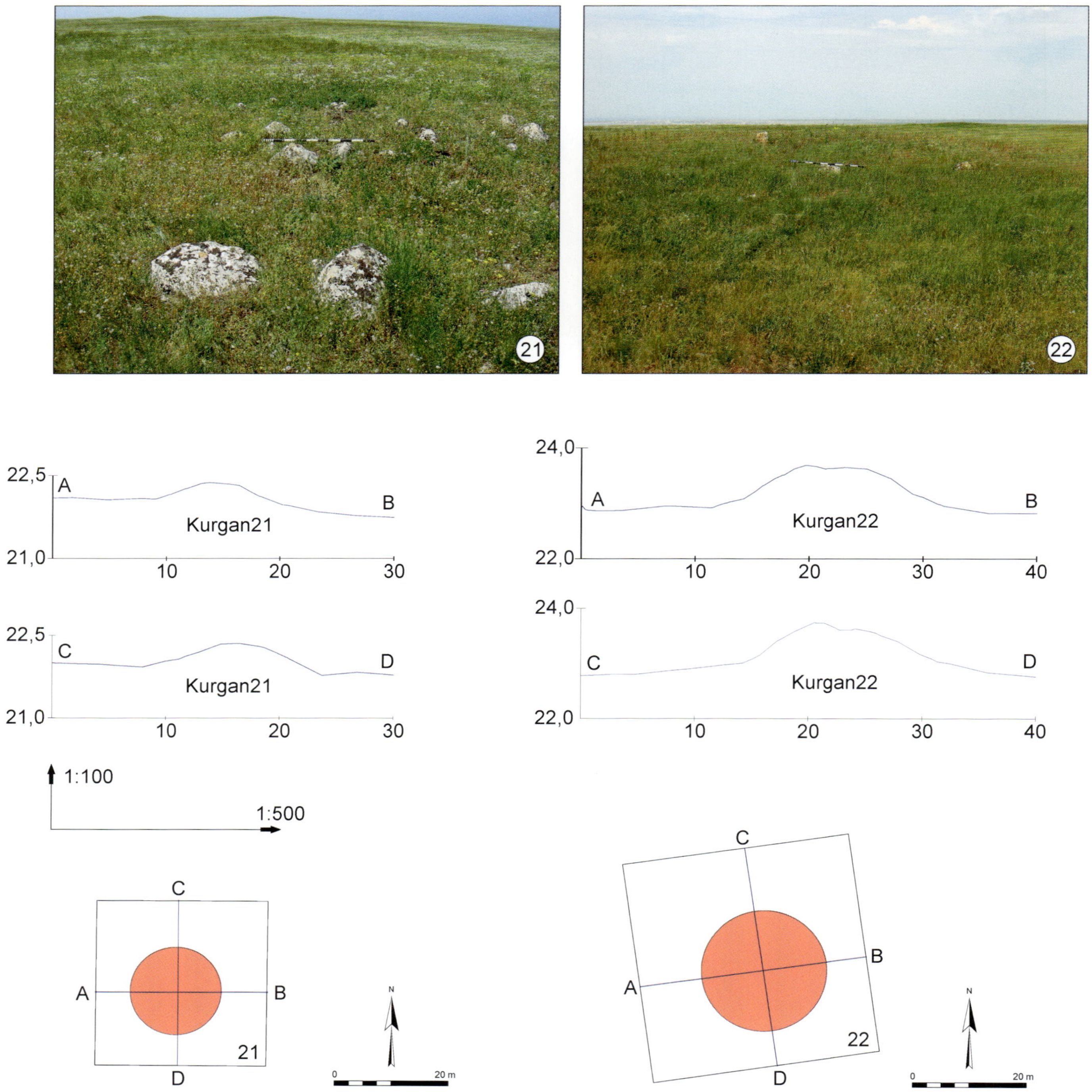

Fig. 4.223 • Photos and cross-sections of Kurgans 21 and 22.

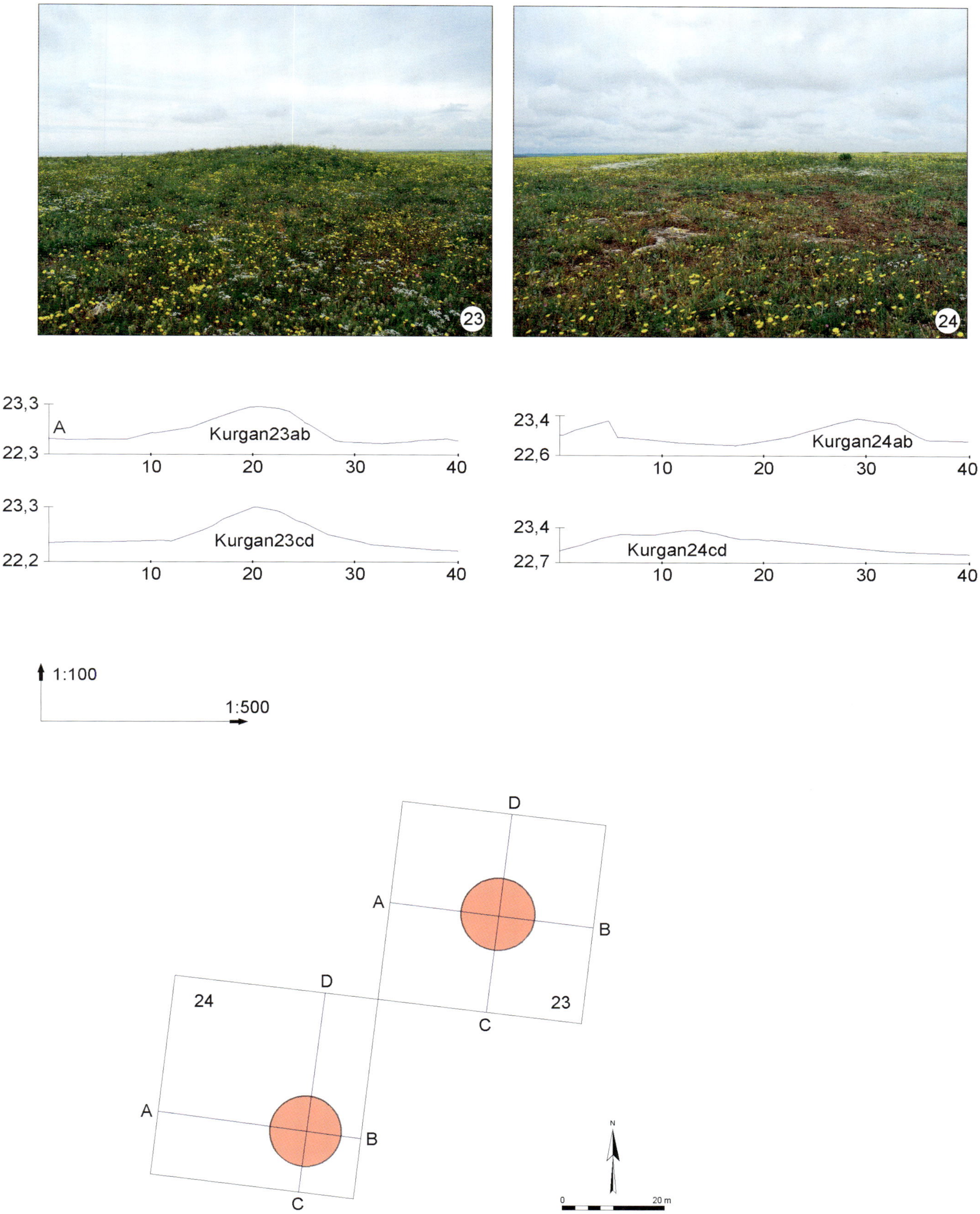

Fig. 4.224 • Photos and cross-sections of Kurgans 23 and 24.

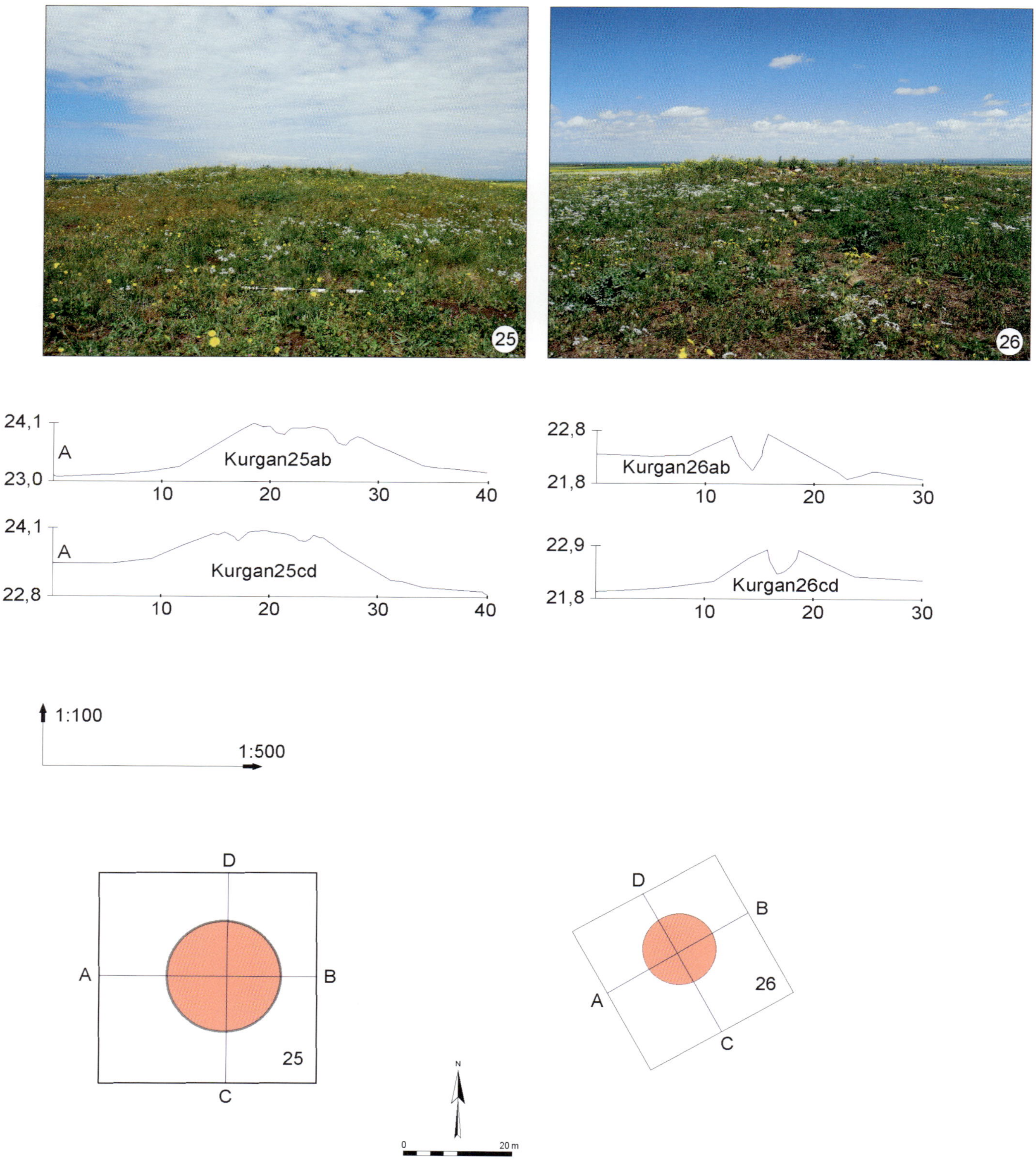

Fig. 4.225 • Photos and cross-sections of Kurgans 25 and 26.

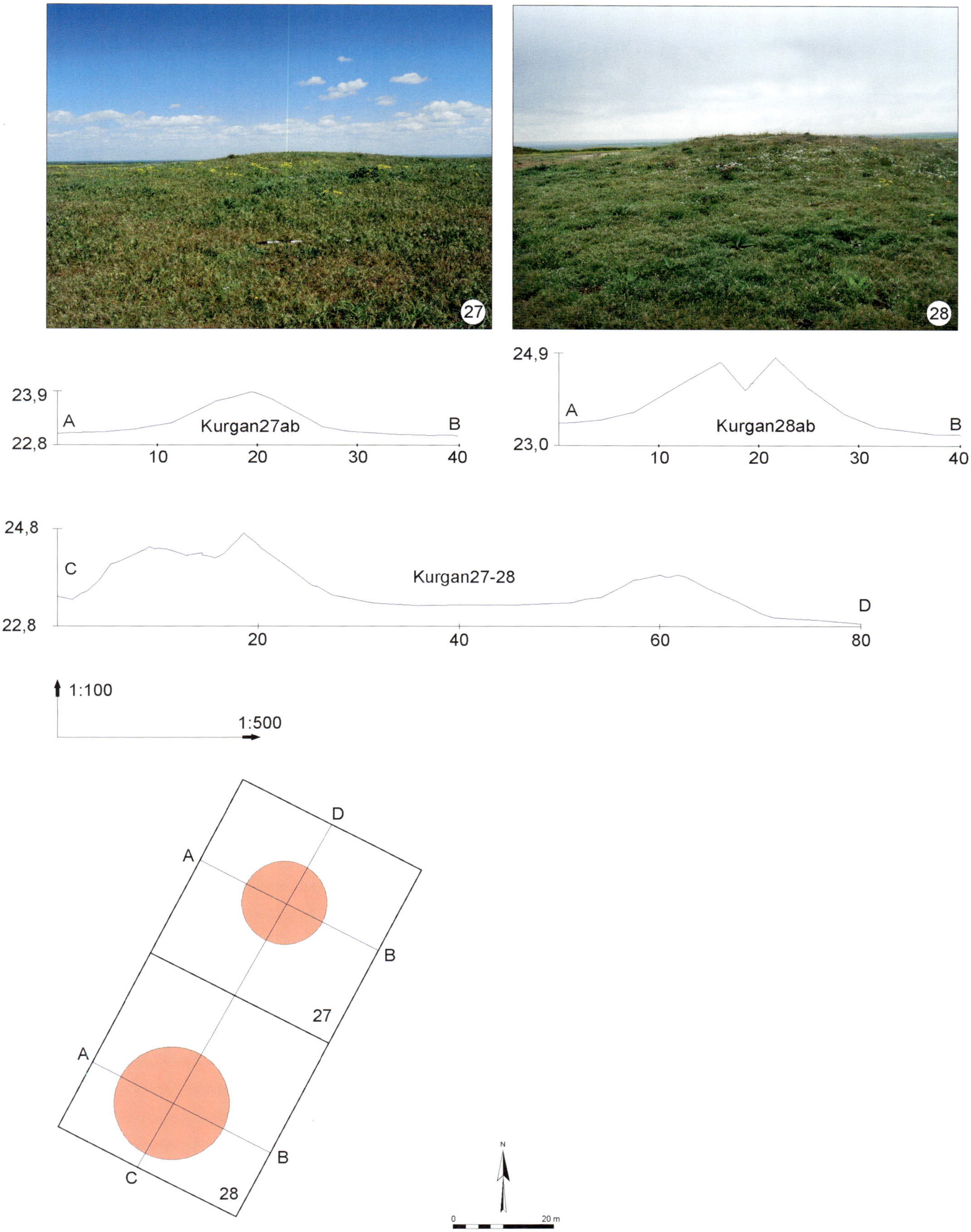

Fig. 4.226 • Photos and cross-sections of Kurgans 27 and 28.

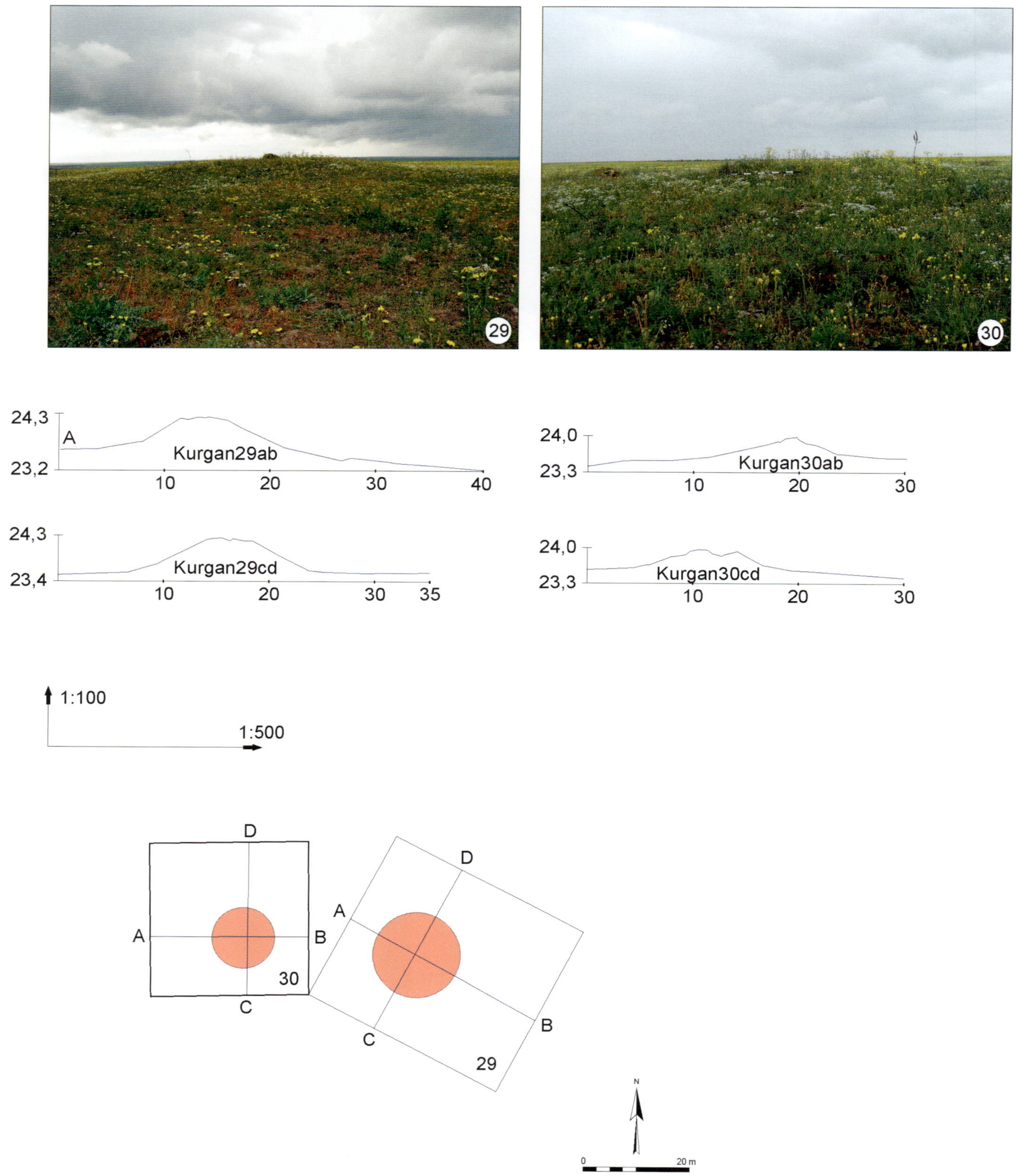

Fig. 4.227 • Photos and cross-sections of Kurgans 29 and 30.

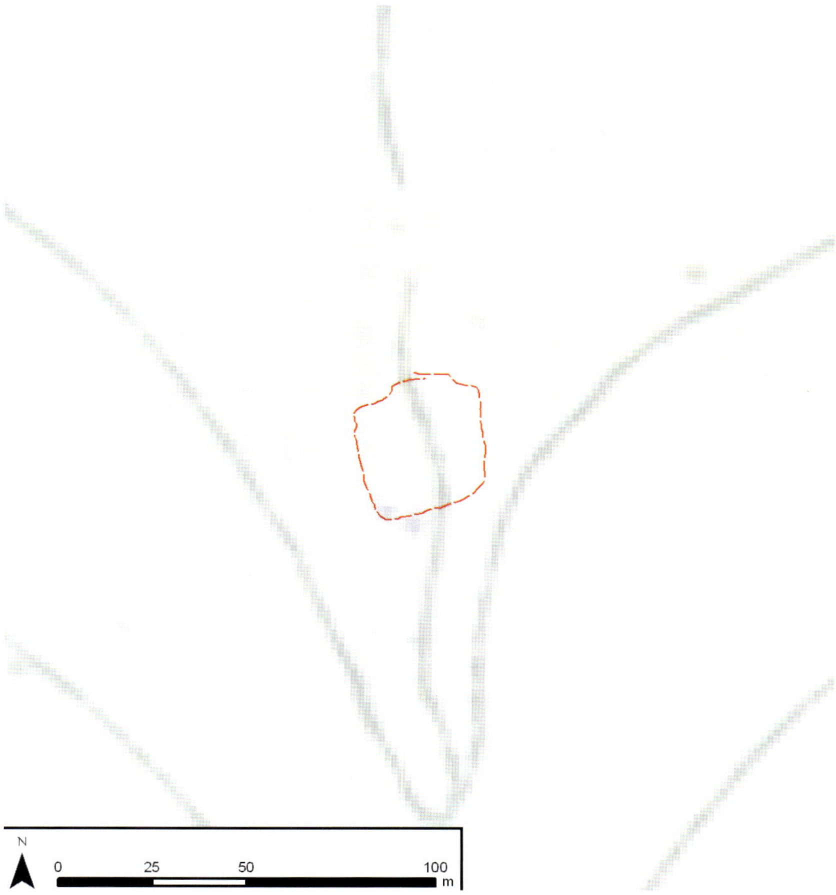

Fig. 4.228 • Distribution map of DSP08-H27-01.

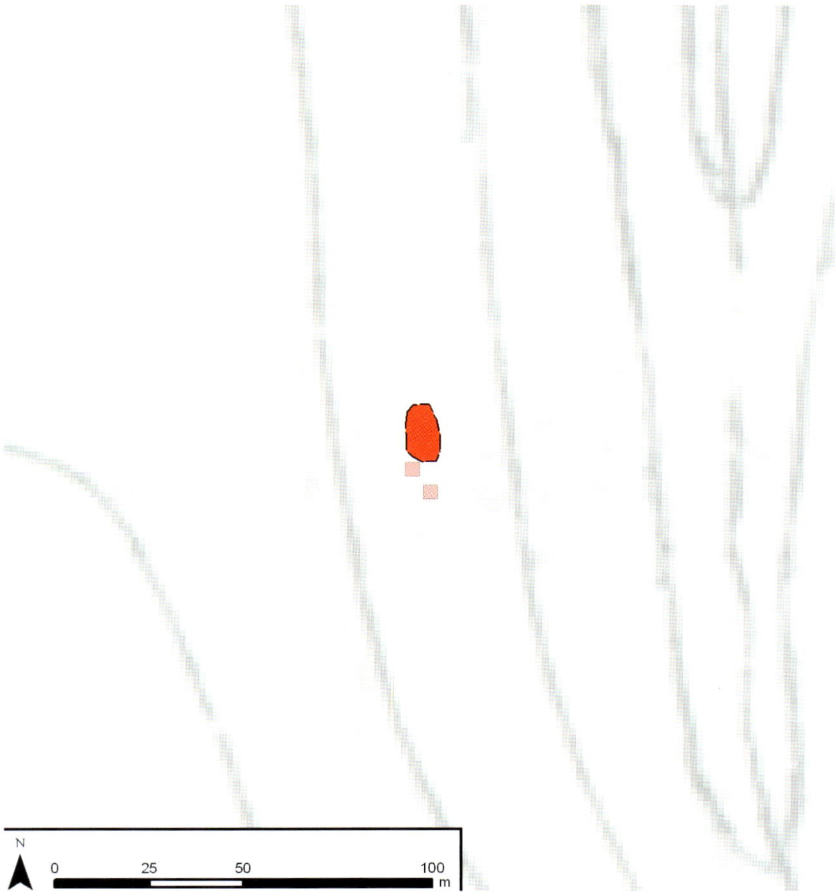

Fig. 4.229 • Distribution map of DSP08-H28-01.

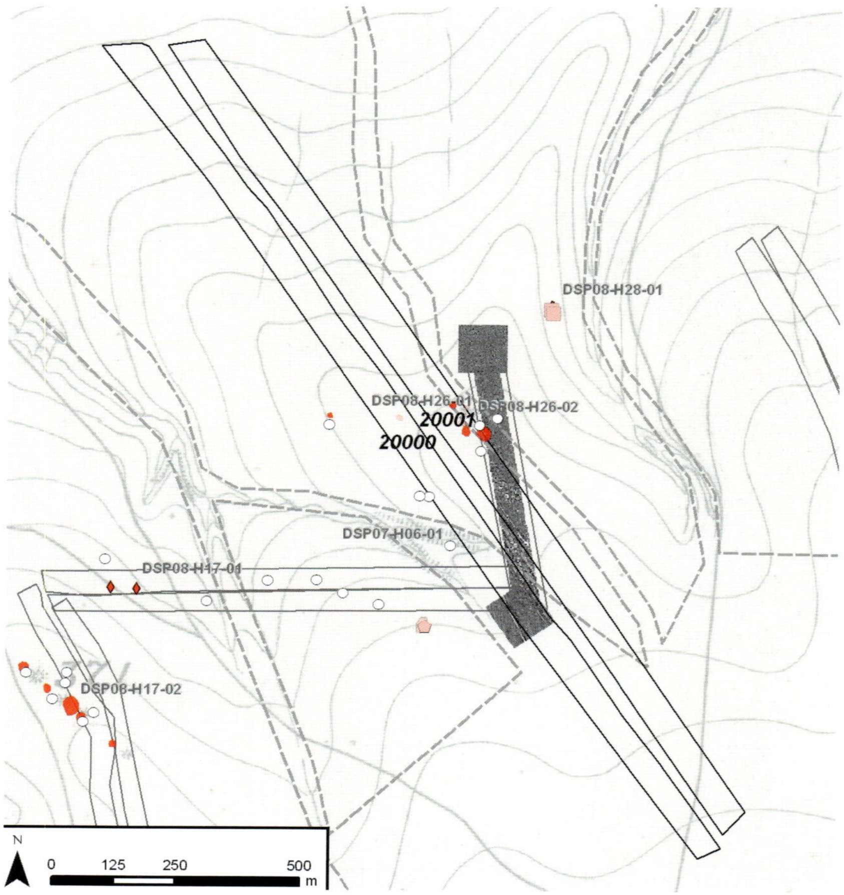

Fig. 4.230 • Distribution map of Units 20000-20001.

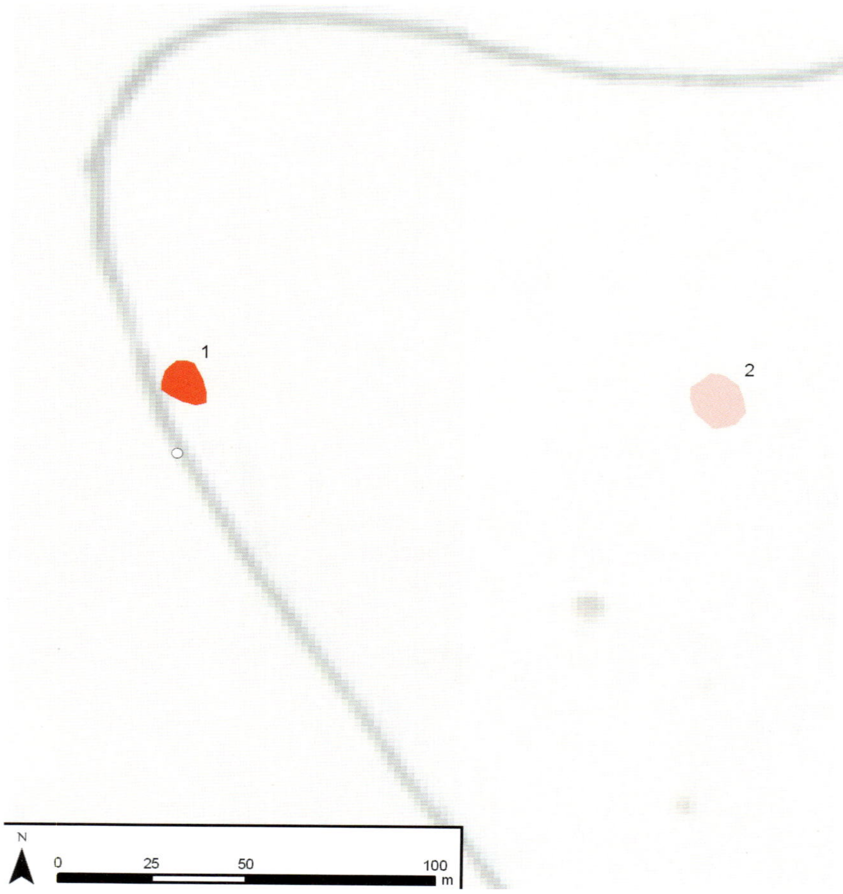

Fig. 4.231 • Distribution map of DSP08-H26-01.

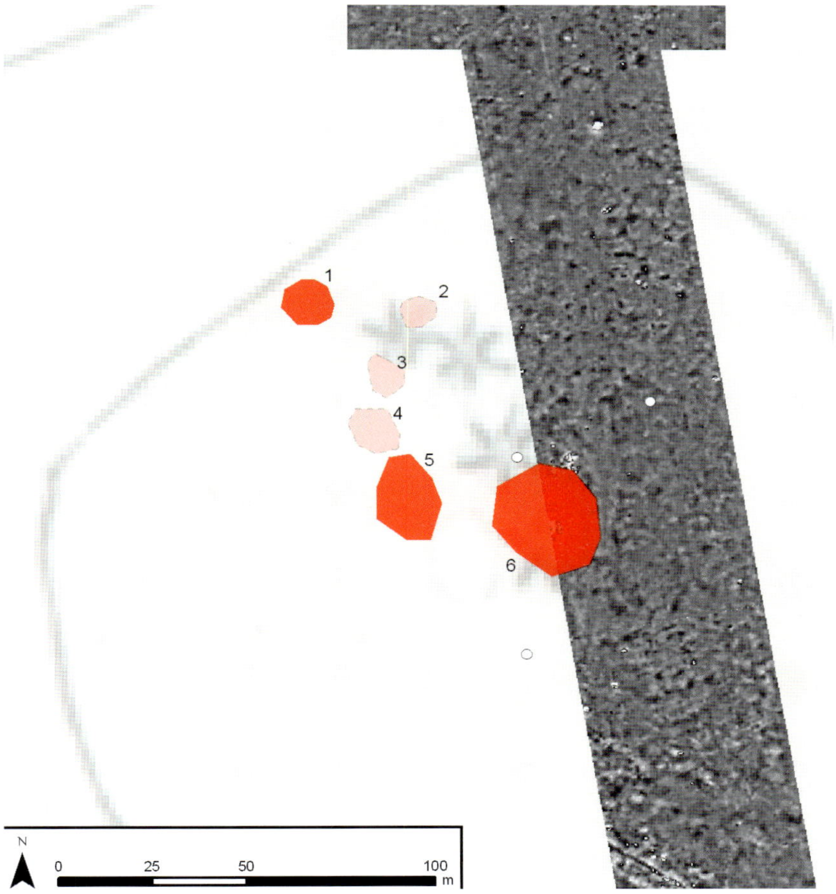

Fig. 4.232 • Distribution map of DSP08-H26-02.

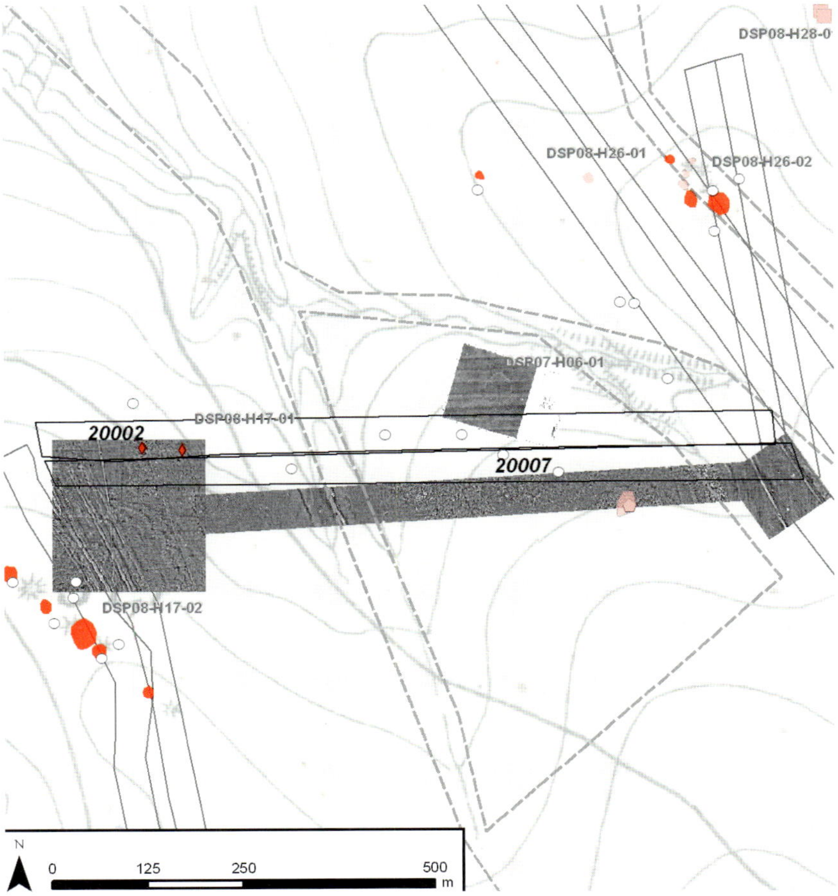

Fig. 4.233 • Distribution map of Units 20002 and 20007.

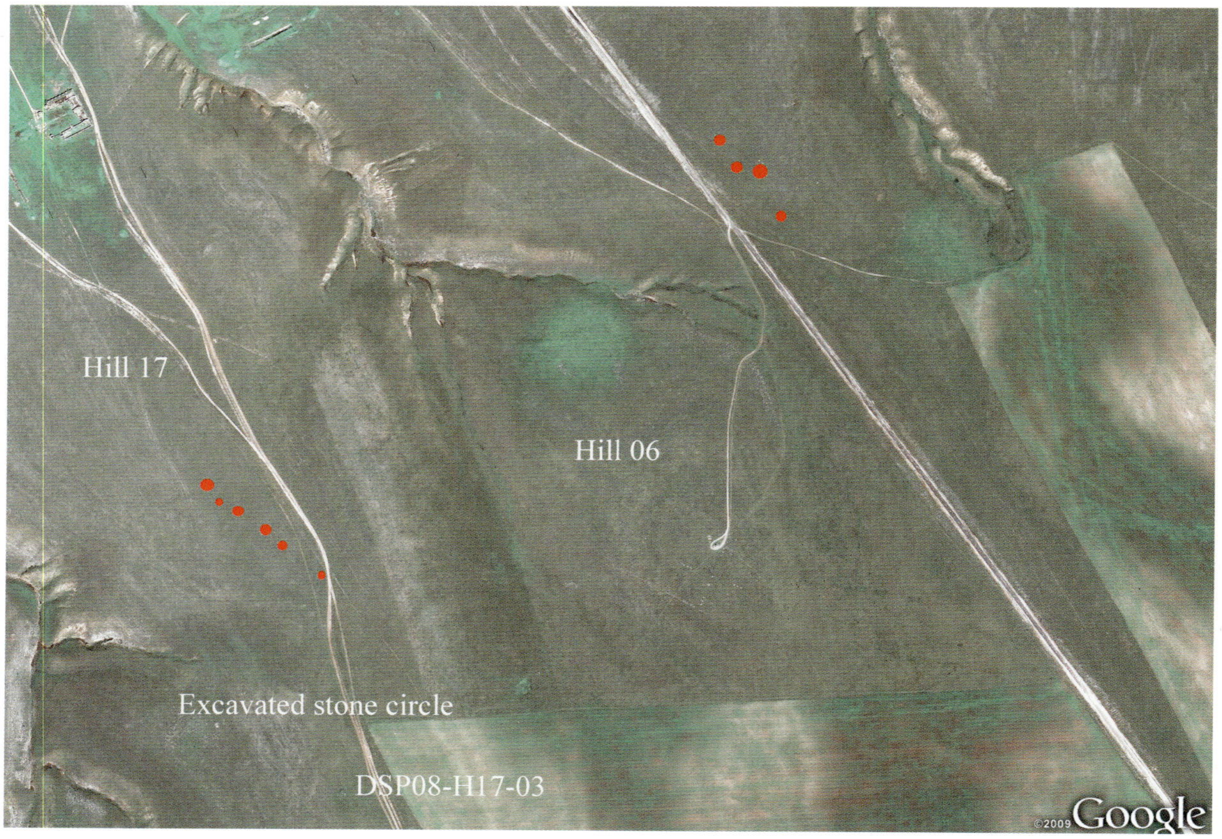

Fig. 4.234 • Hills 6 and 17, satellite image (Google Earth); red dots mark kurgans.

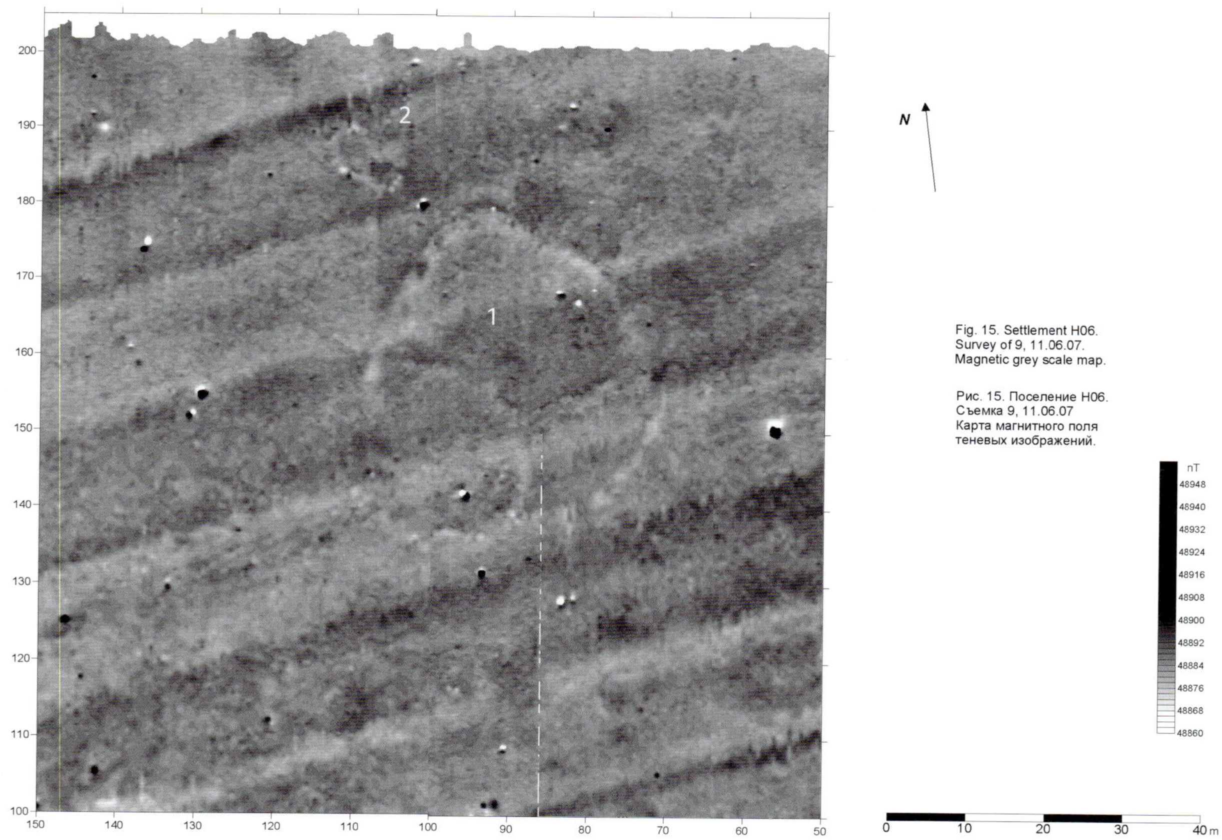

Fig. 4.235 • Geomagnetic map of DSP07-H06-01.

Fig. 4.236 • Stone with a circular depression.

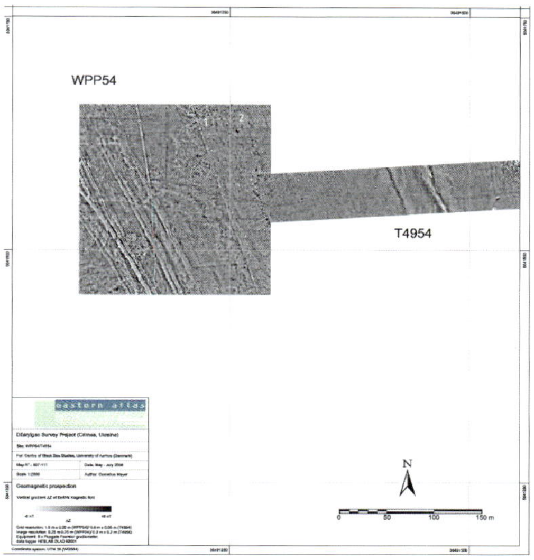

Fig. 4.237 • Distribution map of DSP08-H17-01.

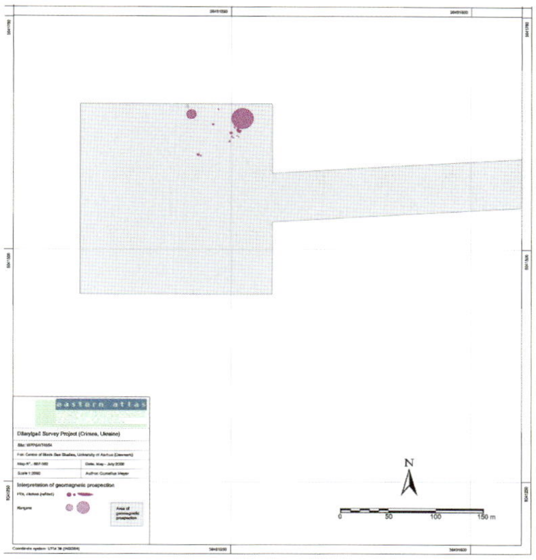

Fig. 4.238 • Geomagnetic map of DSP08-H17-01, interpretation.

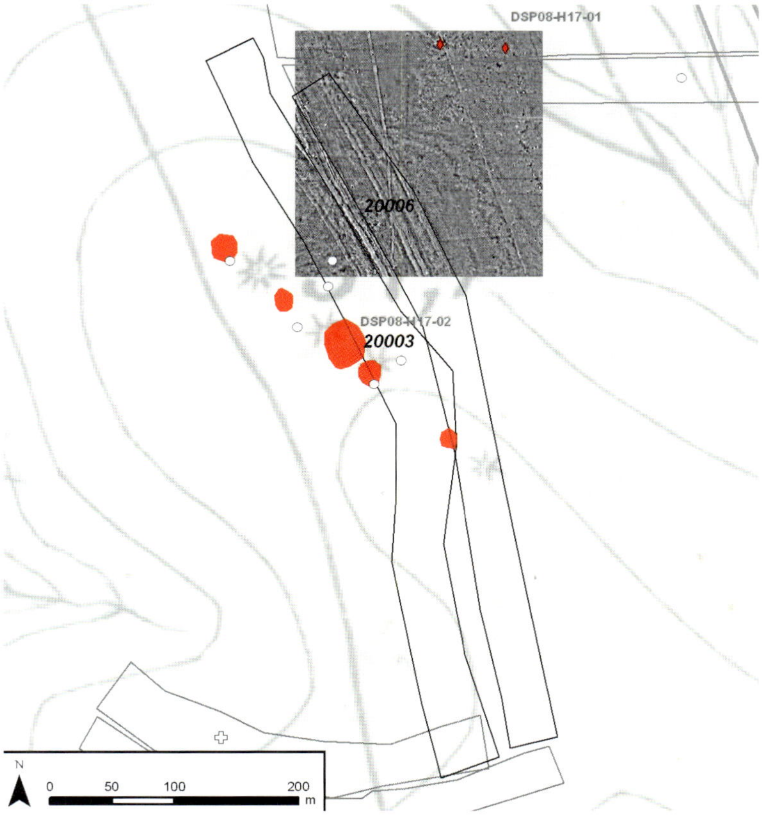

Fig. 4.239 • Distribution map of Units 20003 and 20006.

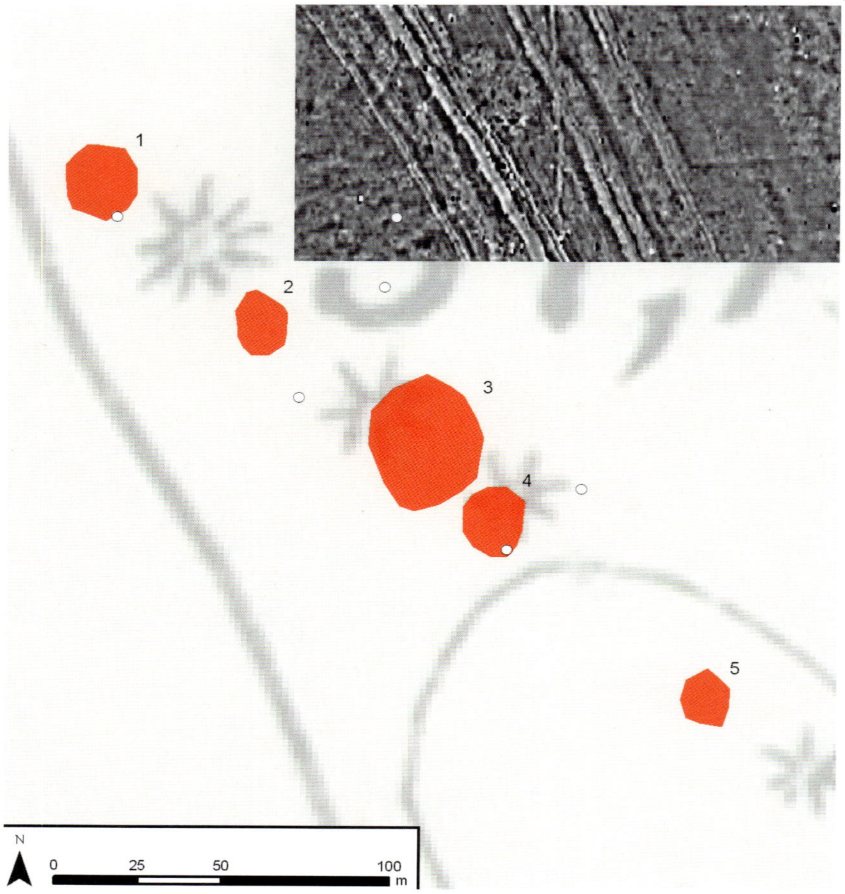

Fig. 4.240 • Distribution map of DSP08-H17-02.

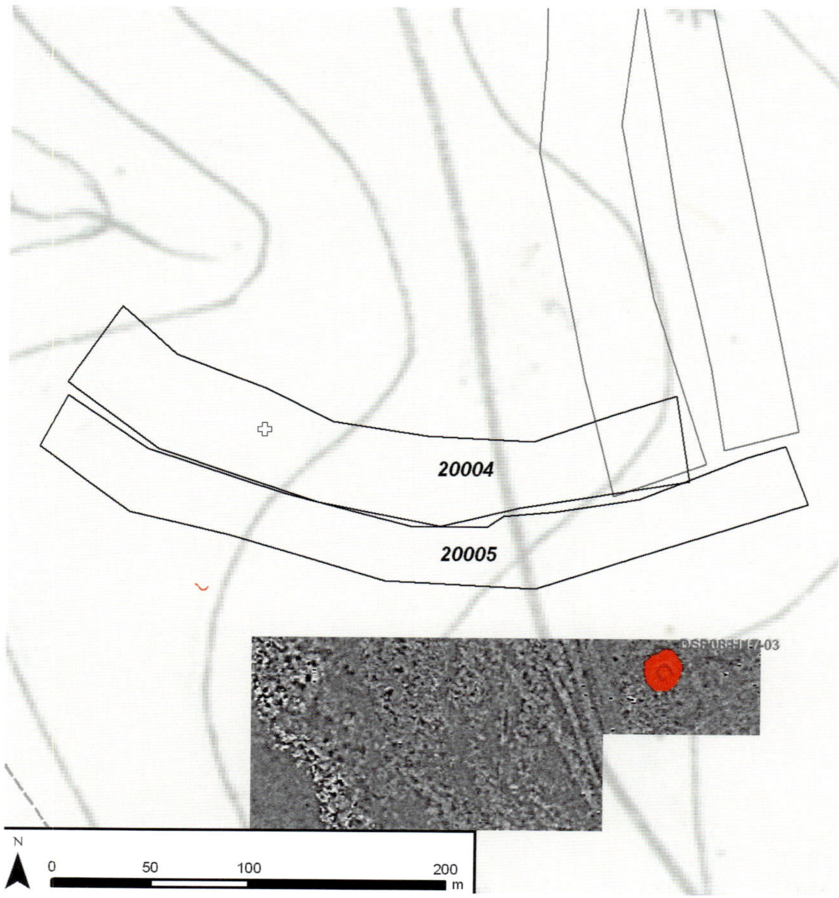

Fig. 4.241 • Distribution map of Units 20004-20005.

Fig. 4.242 • Stone circle on Hill 17.

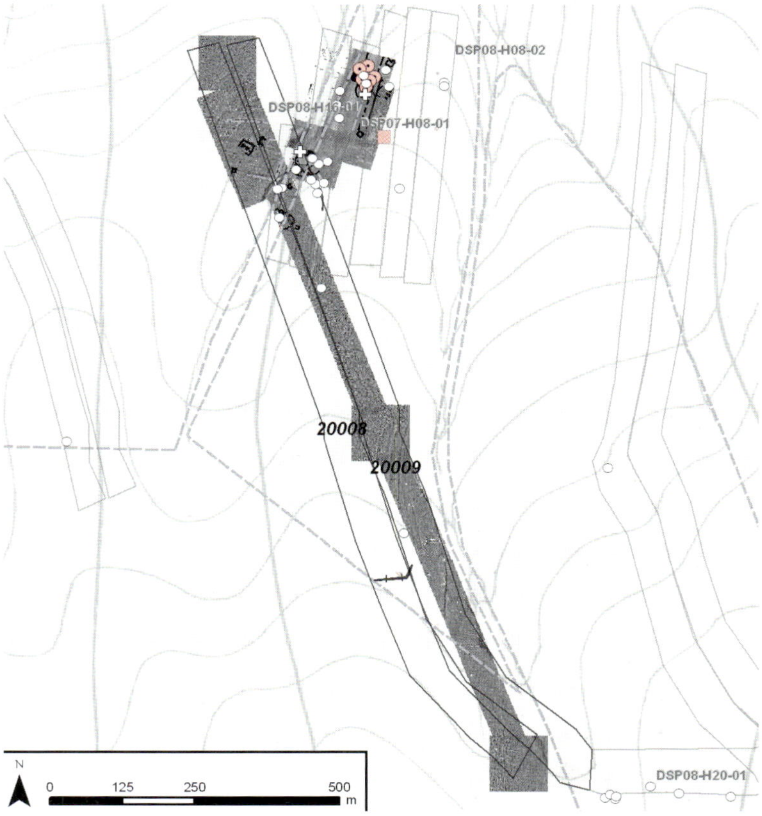

Fig. 4.243 • Distribution map of Units 20008-20009.

Fig. 4.244 • Evidence for recent cultivation (field clearance and vegetation).

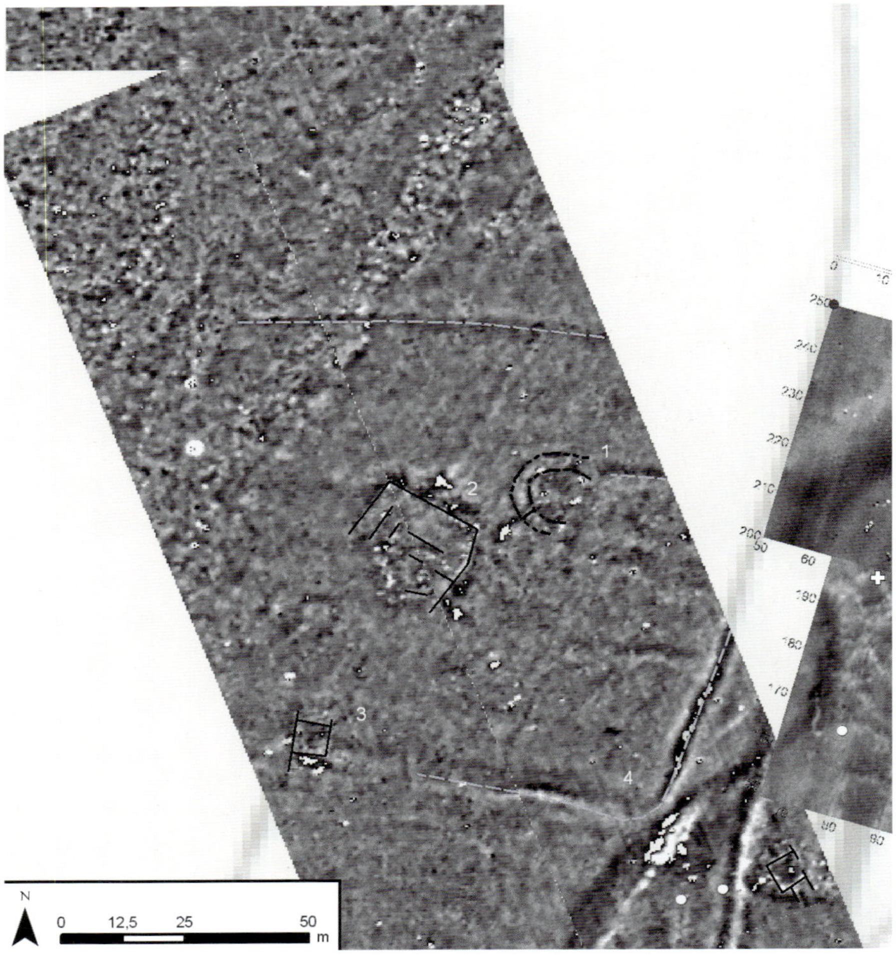

Fig. 4.245 • Distribution map of DSP08-H16-01.

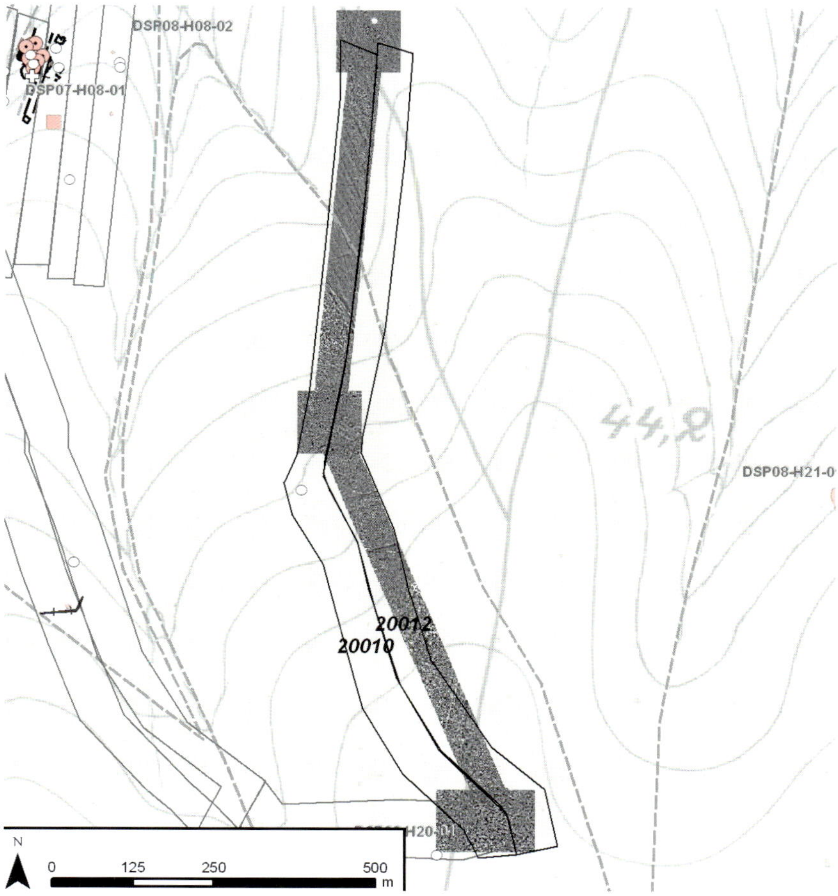

Fig. 4.246 • Distribution map of Units 20010 and 20012.

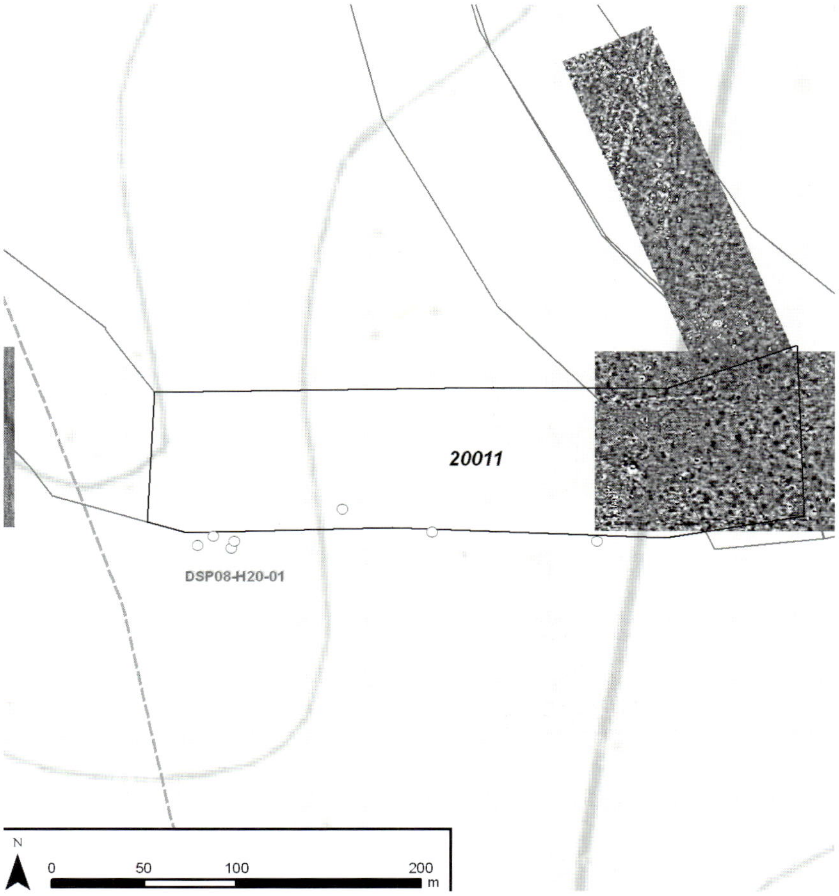

Fig. 4.247 • Distribution map of Unit 20011.

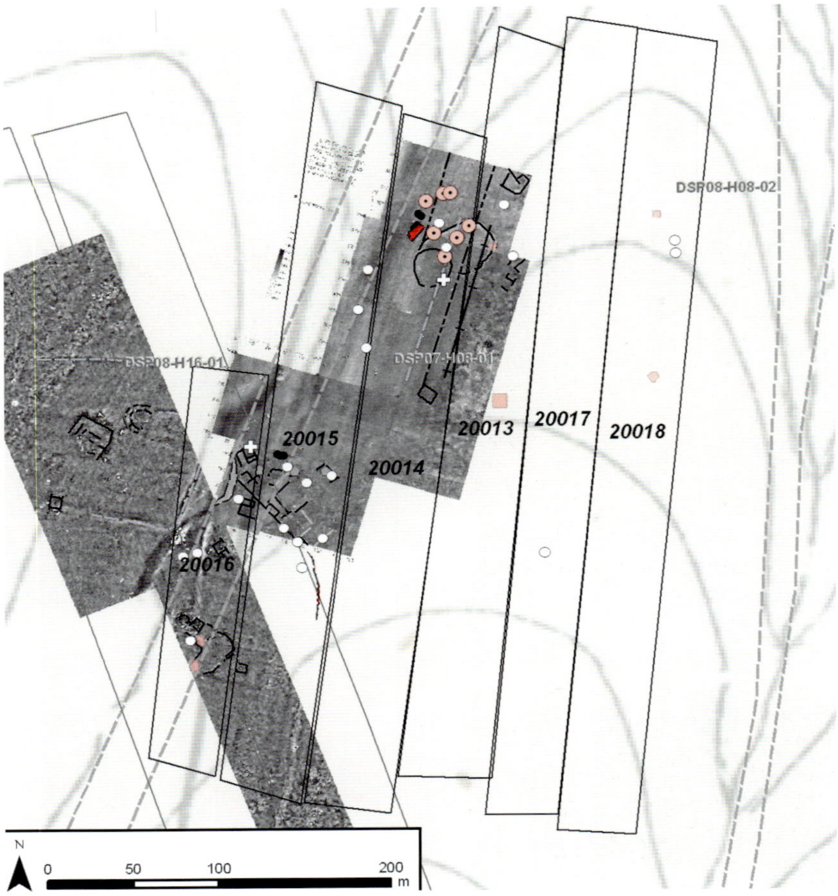

Fig. 4.248 • Distribution map of Units 20013-20018.

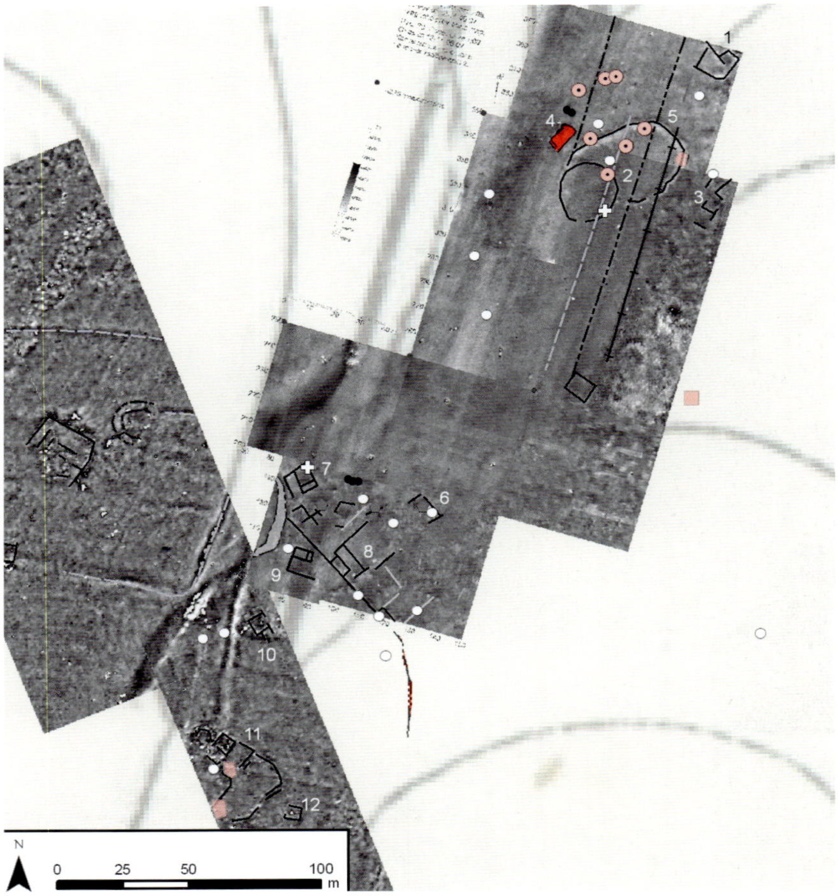

Fig. 4.249 • Distribution map of DSP07-H08-01.

Fig. 4.250 • View of northern part of DSP07-H08-01 from the west.

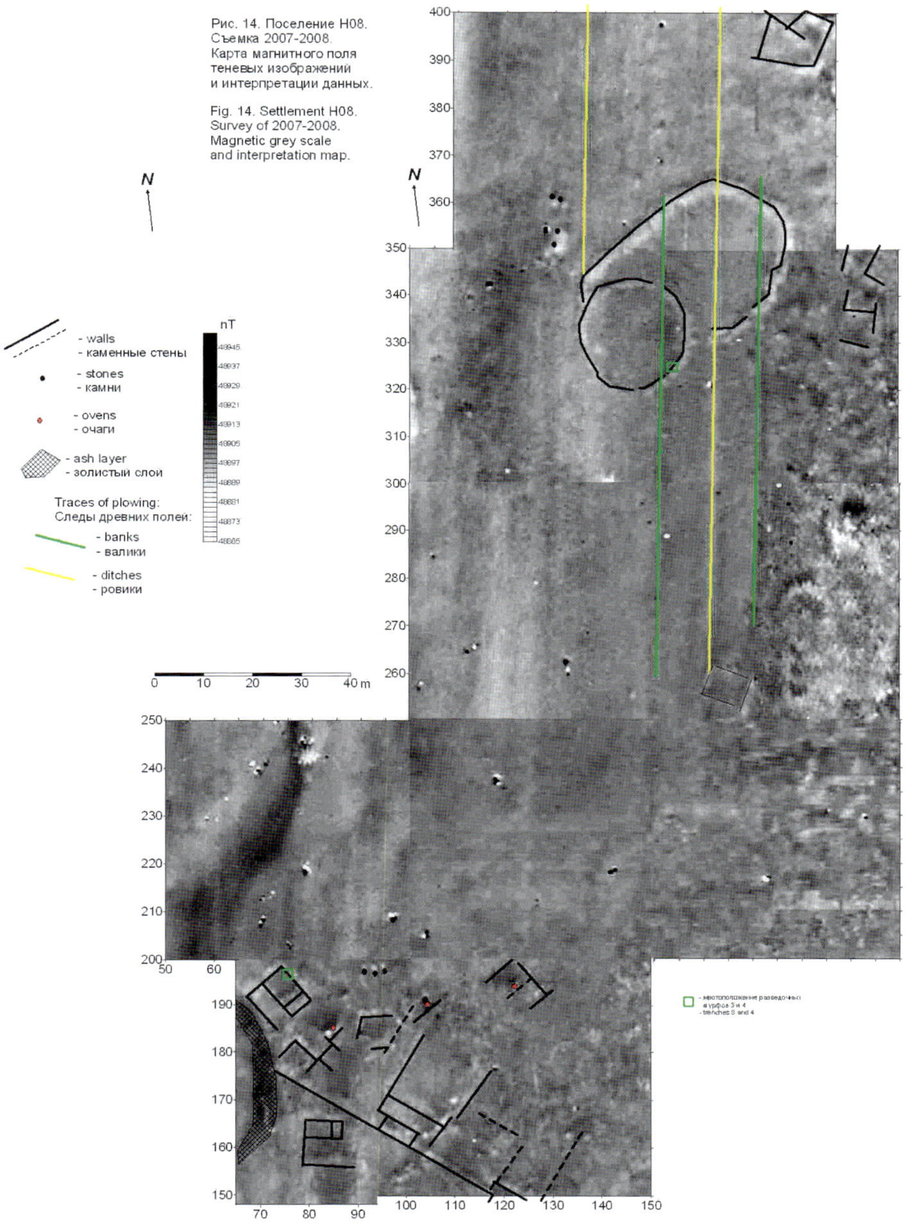

Fig. 4.251 • Geomagnetic map of DSP07-H08-01, interpretation.

Fig. 4.252 • Trial trench DSP07-H08-01 Feature 2.

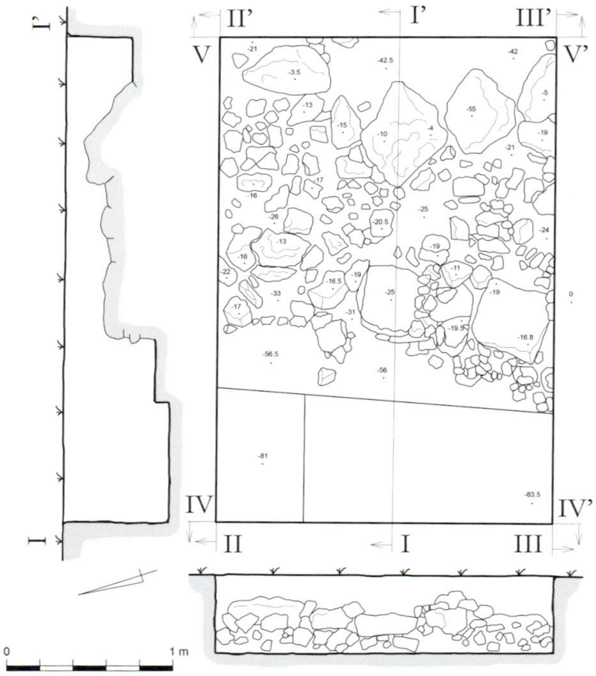

Fig. 4.253 • Drawing of trial trench DSP07-H08-01 Feature 2.

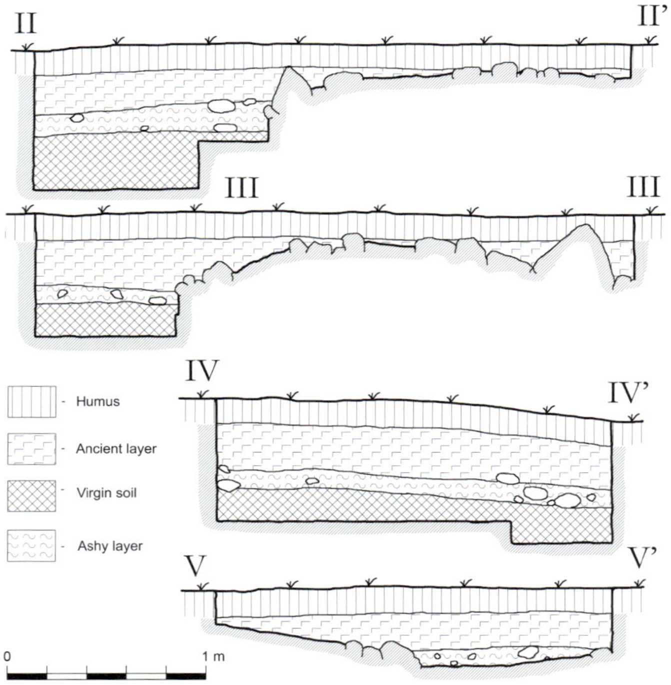

Fig. 4.254 • Profile drawings of trial trench DSP07-H08-01 Feature 2.

Fig. 4.255 • Trial trench DSP07-H08-01 Feature 7.

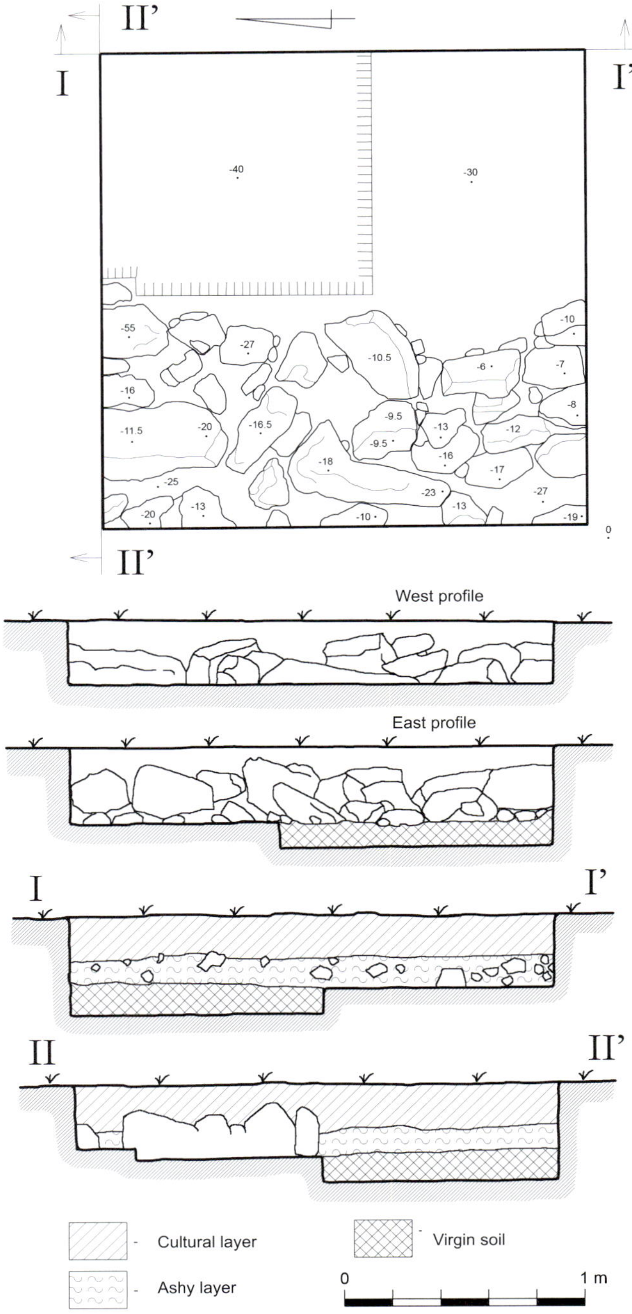

Fig. 4.256 • Drawings of trial trench DSP07-H08-01 Feature 7.

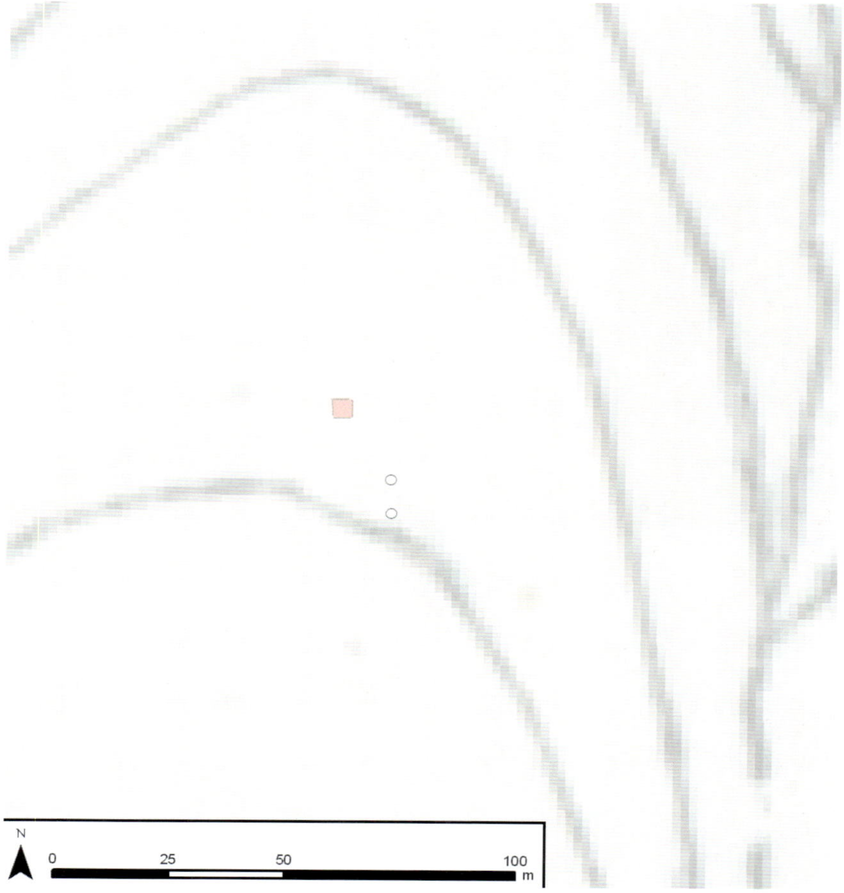

Fig. 4.257 • Distribution map of DSP08-H08-02.

Fig. 4.258 • Square stone foundation on the eastern edge of Hill 8 (DSP08-H08-02).

Fig. 4.259 • Distribution map of Units 20019-20020.

Fig. 4.260 • Core at Foundation 47.

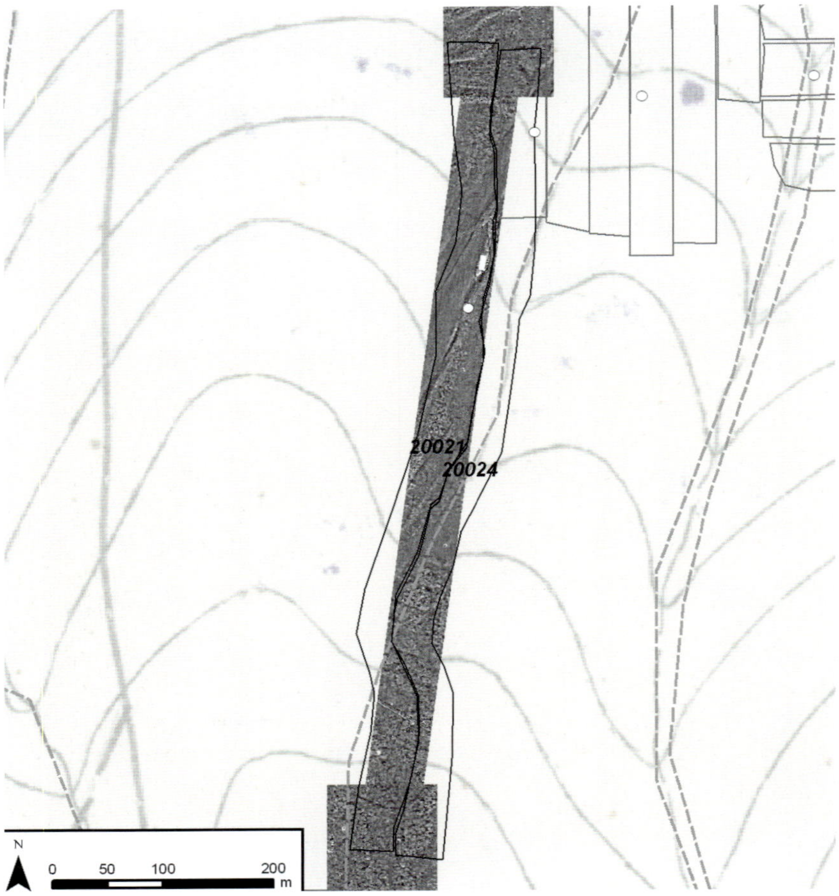

Fig. 4.261 • Distribution map of Units 20021 and 20024.

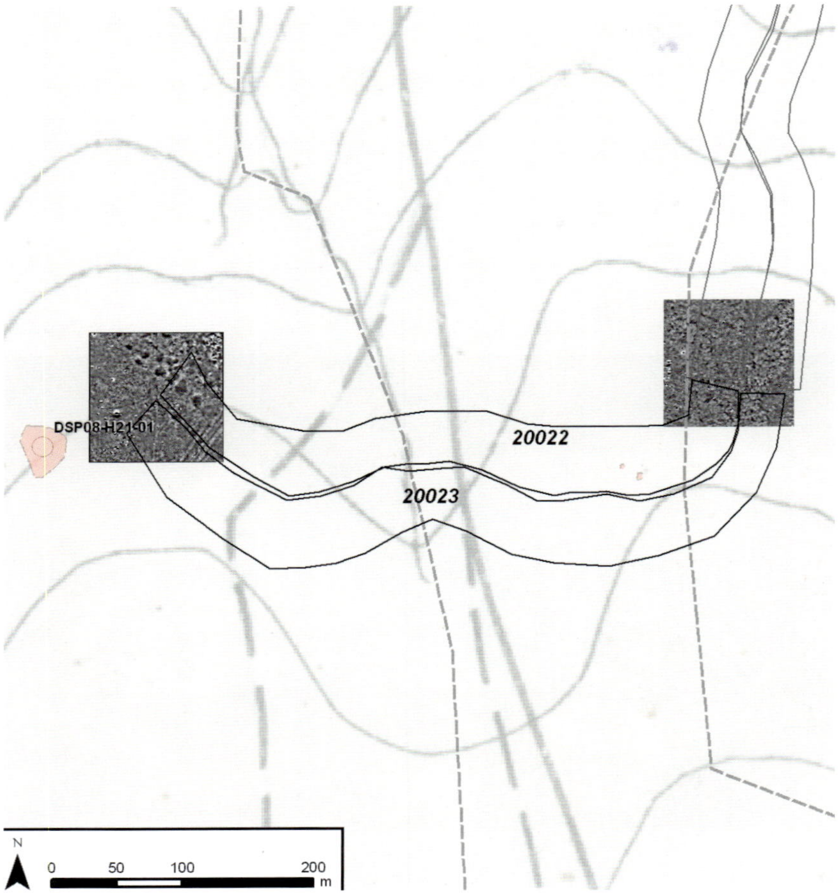

Fig. 4.262 • Distribution map of Units 20022-20023.

Fig. 4.263 • DSP08-H21-01 from the southwest.

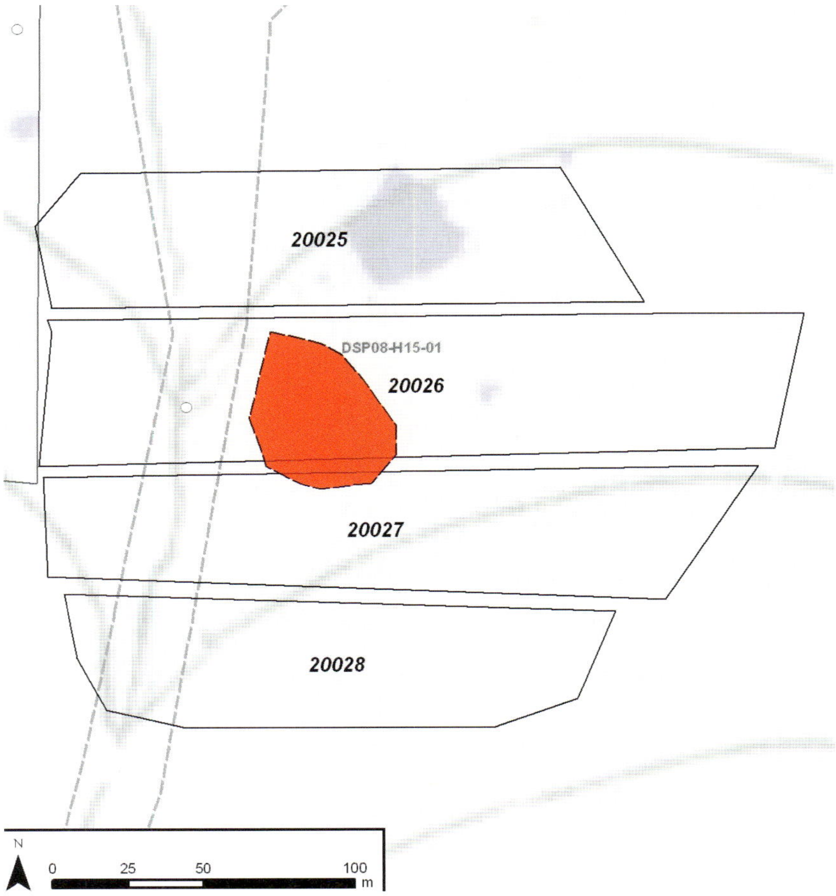

Fig. 4.264 • Distribution map of Units 20025-20028.

Fig. 4.265 • Northwestern corner of DSP08-H15-01 from the northeast.

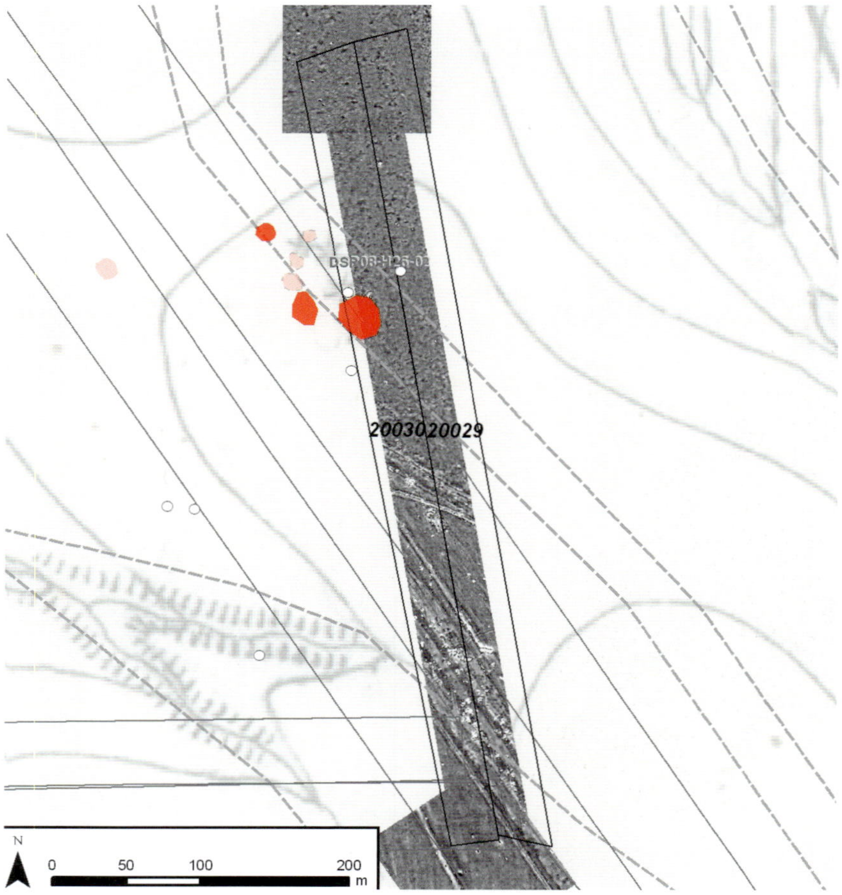

Fig. 4.266 • Distribution map of Units 20029-20030.

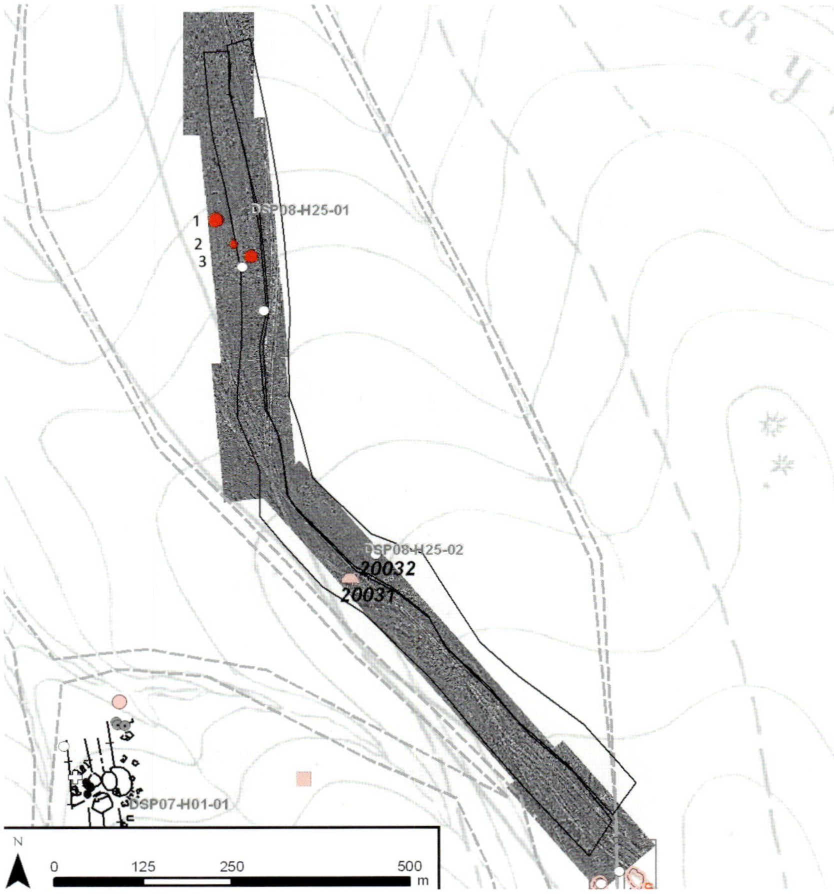

Fig. 4.267 • Distribution map of Units 20031-20032.

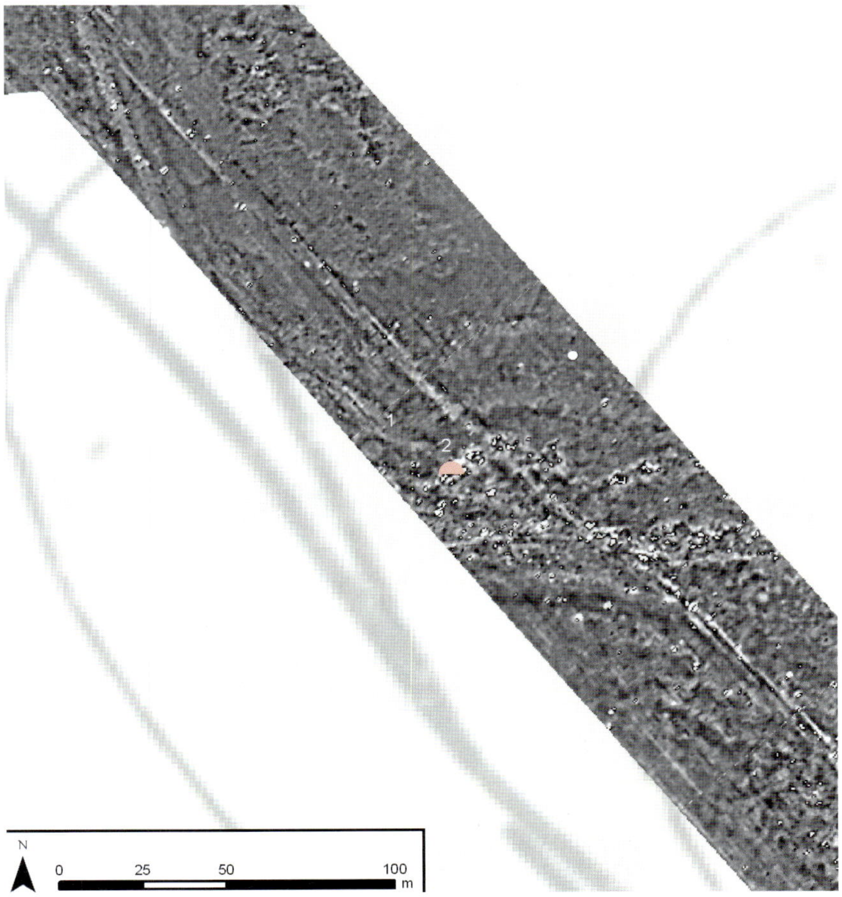

Fig. 4.268 • Distribution map of DSP08-H25-02.

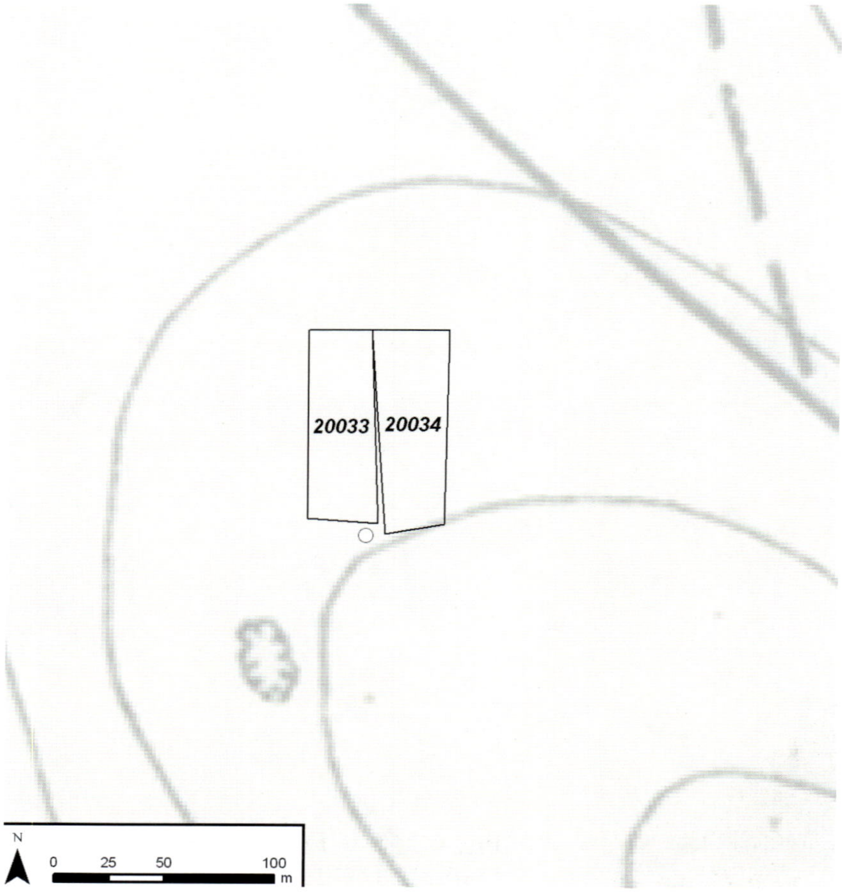

Fig. 4.269 • Distribution map of Units 20033-20034.

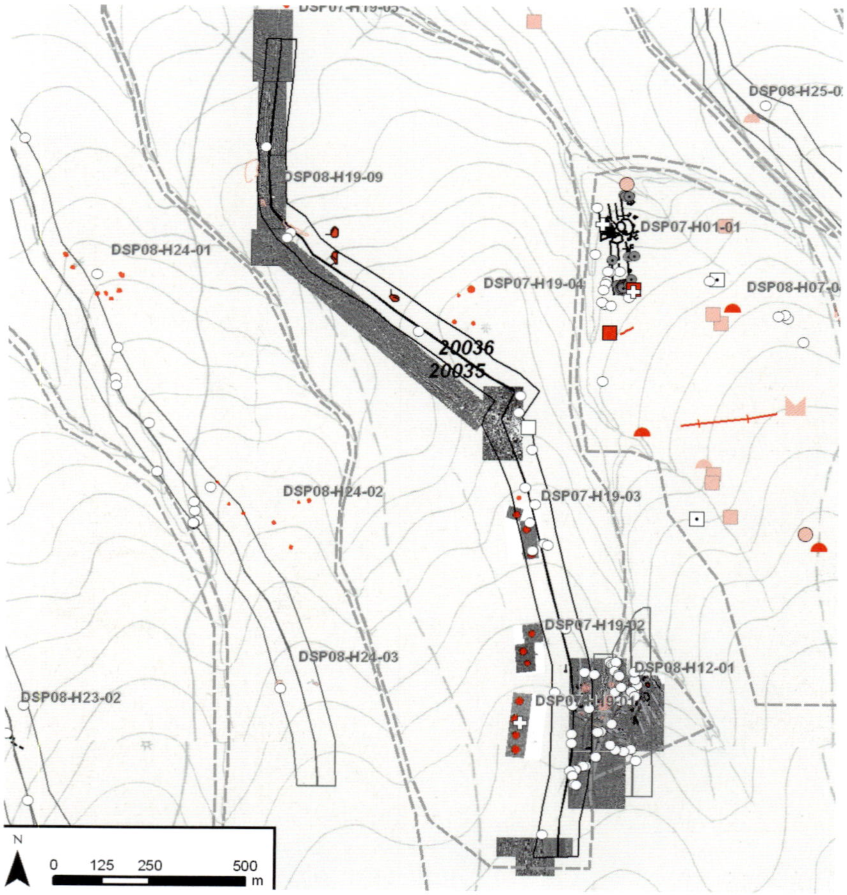

Fig. 4.270 • Distribution map of Units 20035-20036.

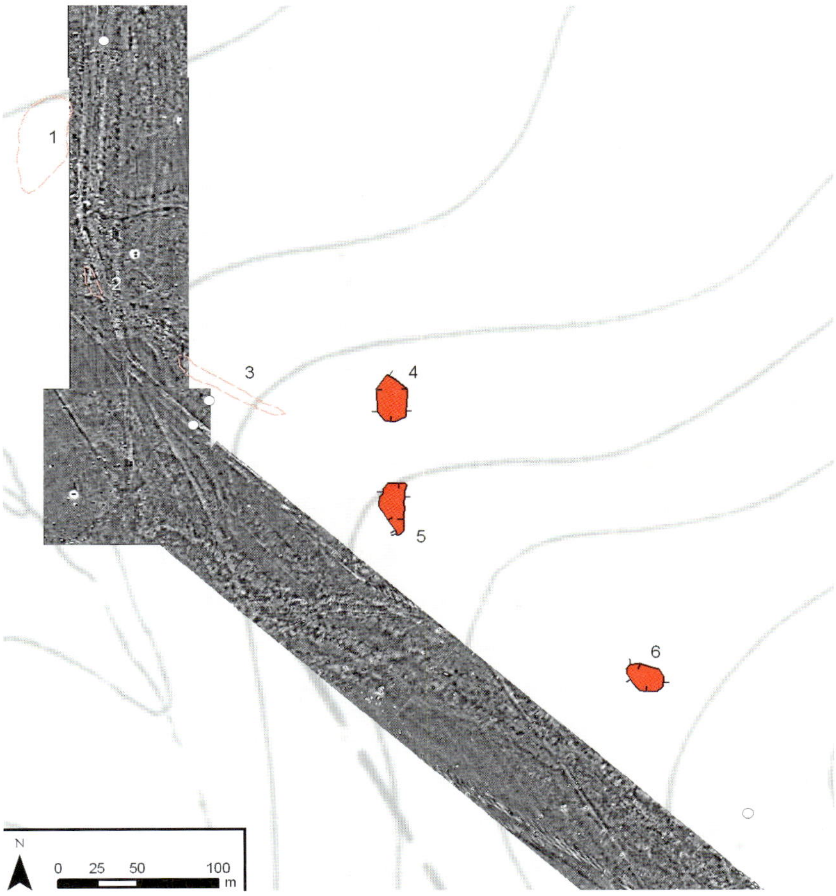

Fig. 4.271 • Distribution map of DSP08-H19-09.

Fig. 4.272 • Evidence for quarrying (DSP08-H19-09).

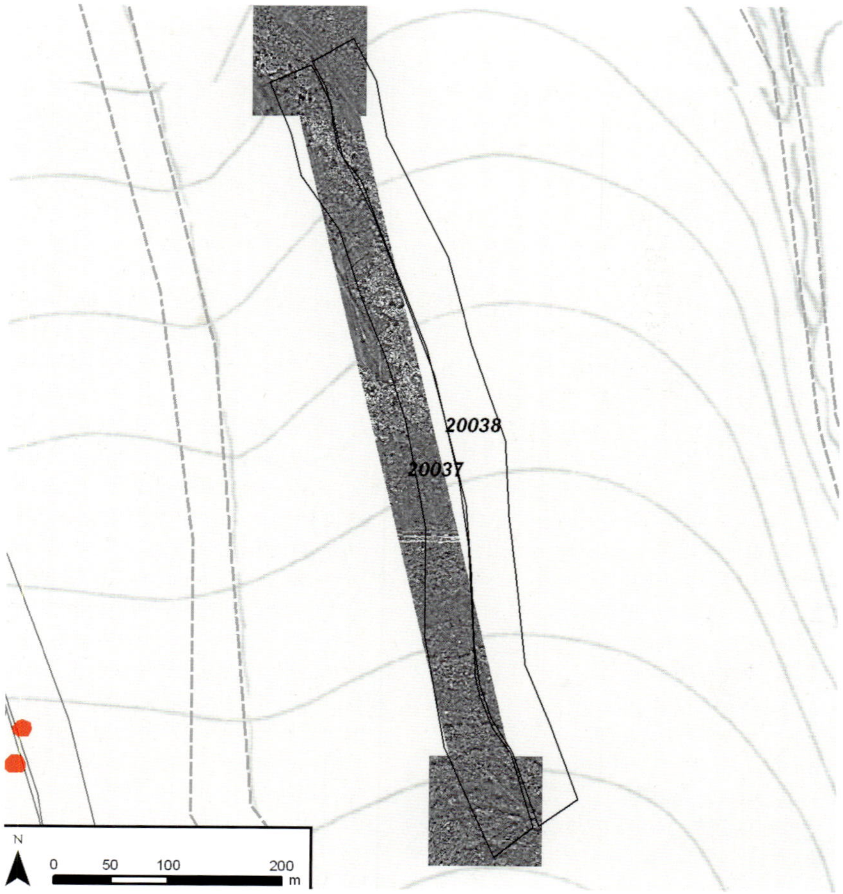

Fig. 4.273 • Distribution map of Units 20037-20038.

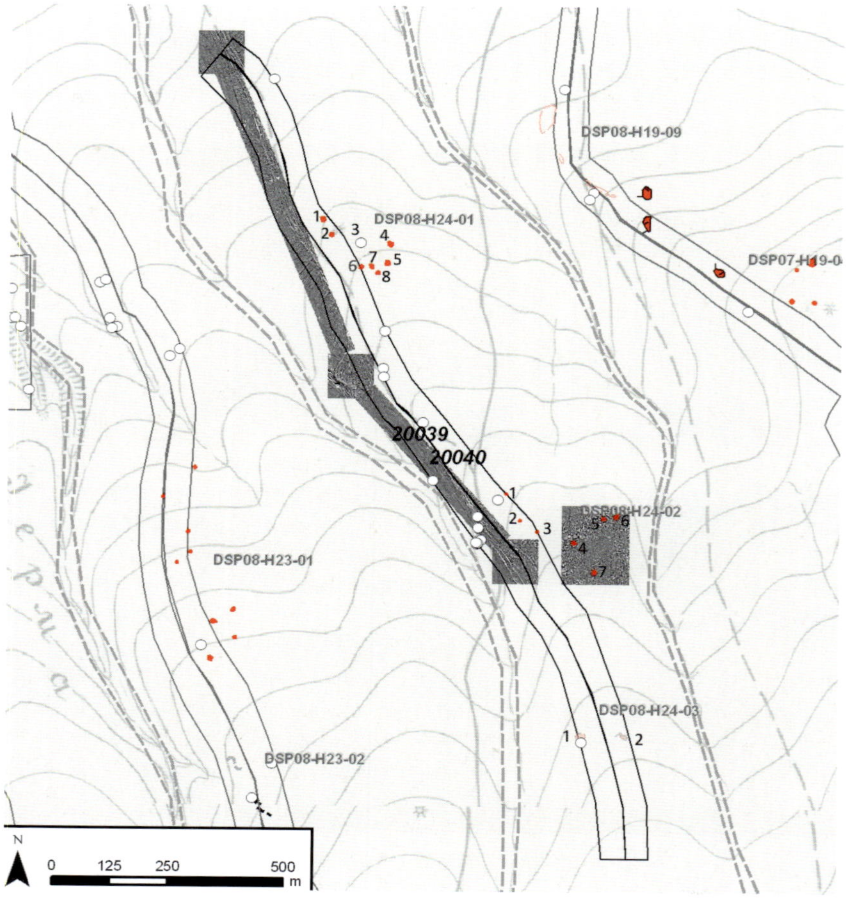

Fig. 4.274 • Distribution map of Units 20039-20040.

Fig. 4.275 • Distribution map of DSP08-H24-03.

Fig. 4.276 • Possible quarry near DSP08-H24-03 from the southeast, kurgans in the background.

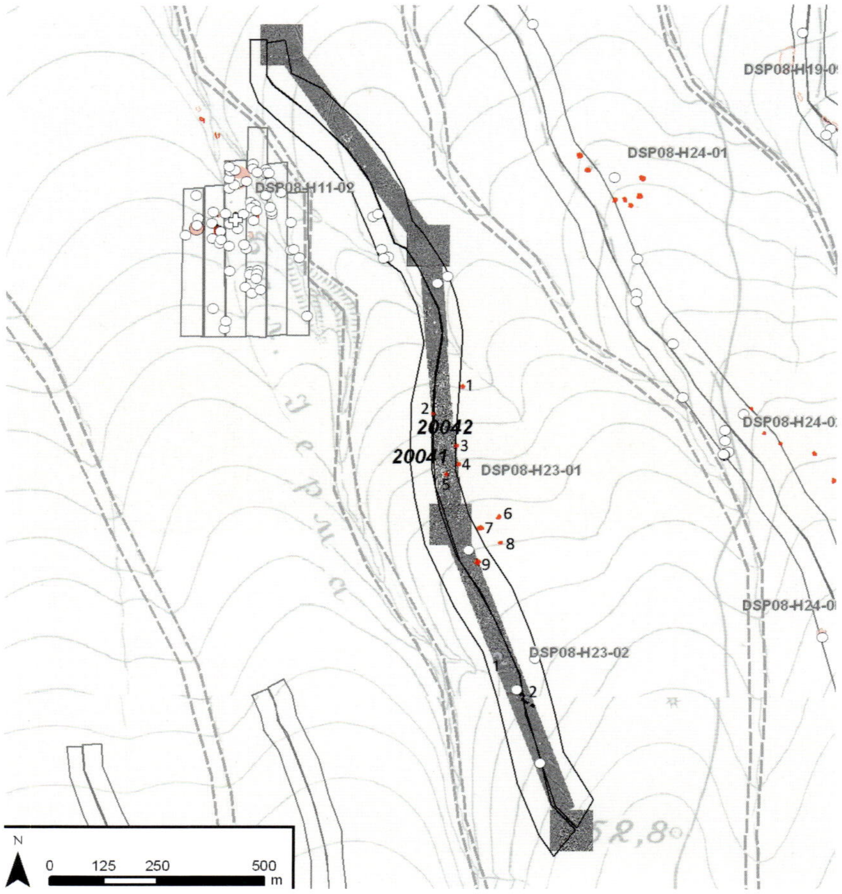

Fig. 4.277 • Distribution map of Units 20041-20042.

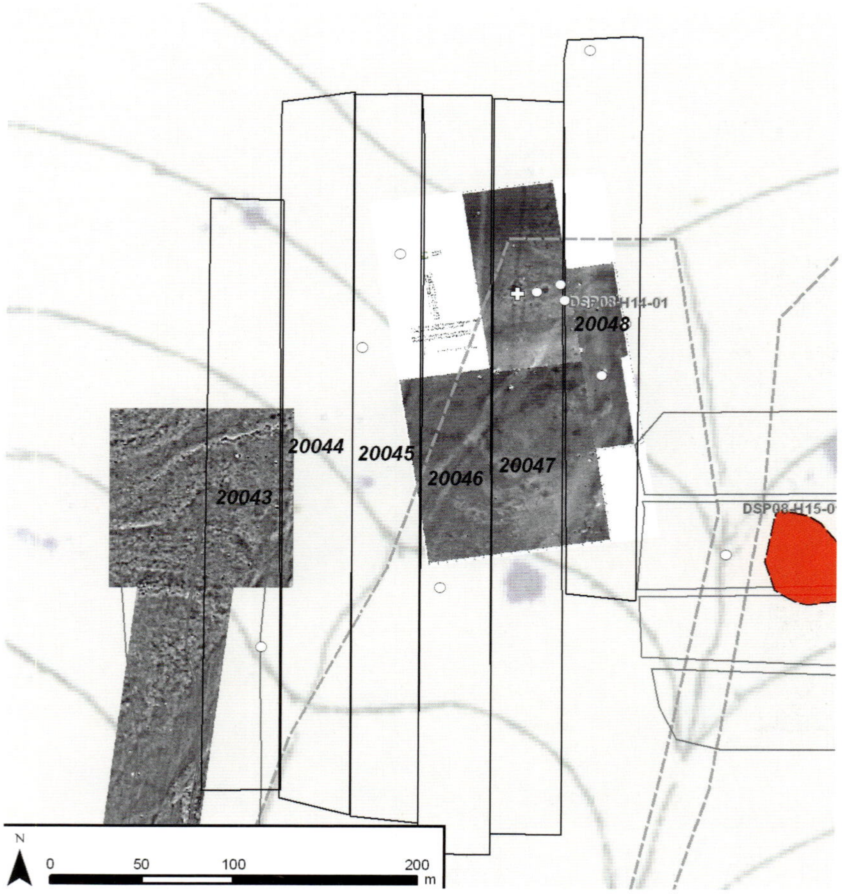

Fig. 4.278 • Distribution map of Units 20043-20048.

Fig. 4.279 • DSP08-H14-01 from the north.

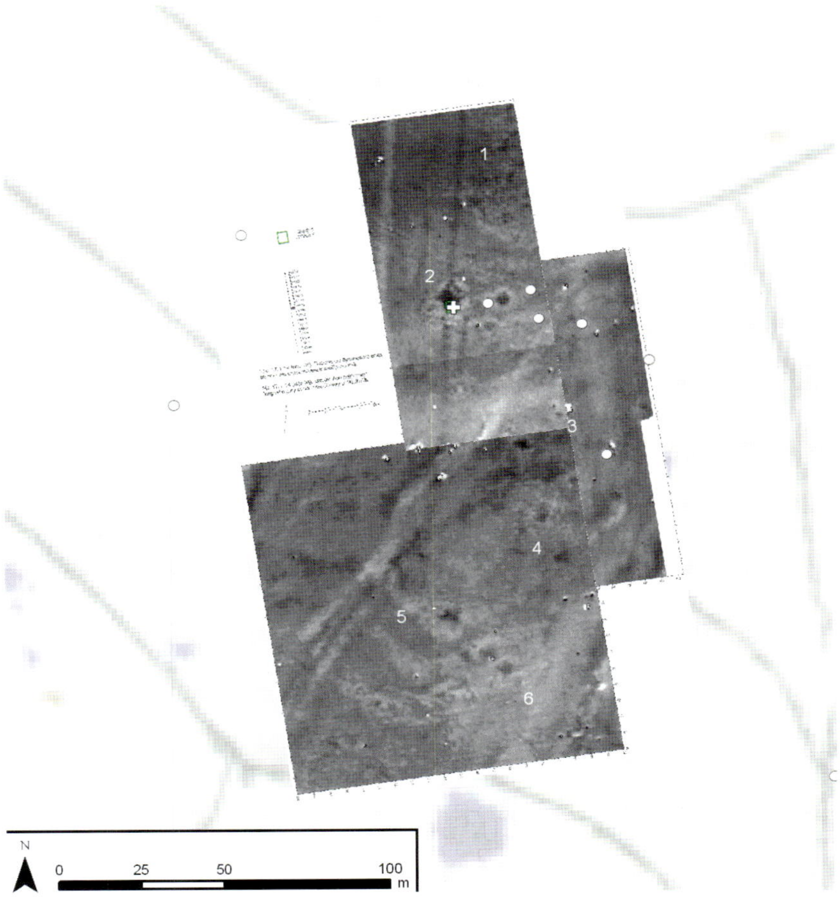

Fig. 4.280 • Distribution map of DSP08-H14-01.

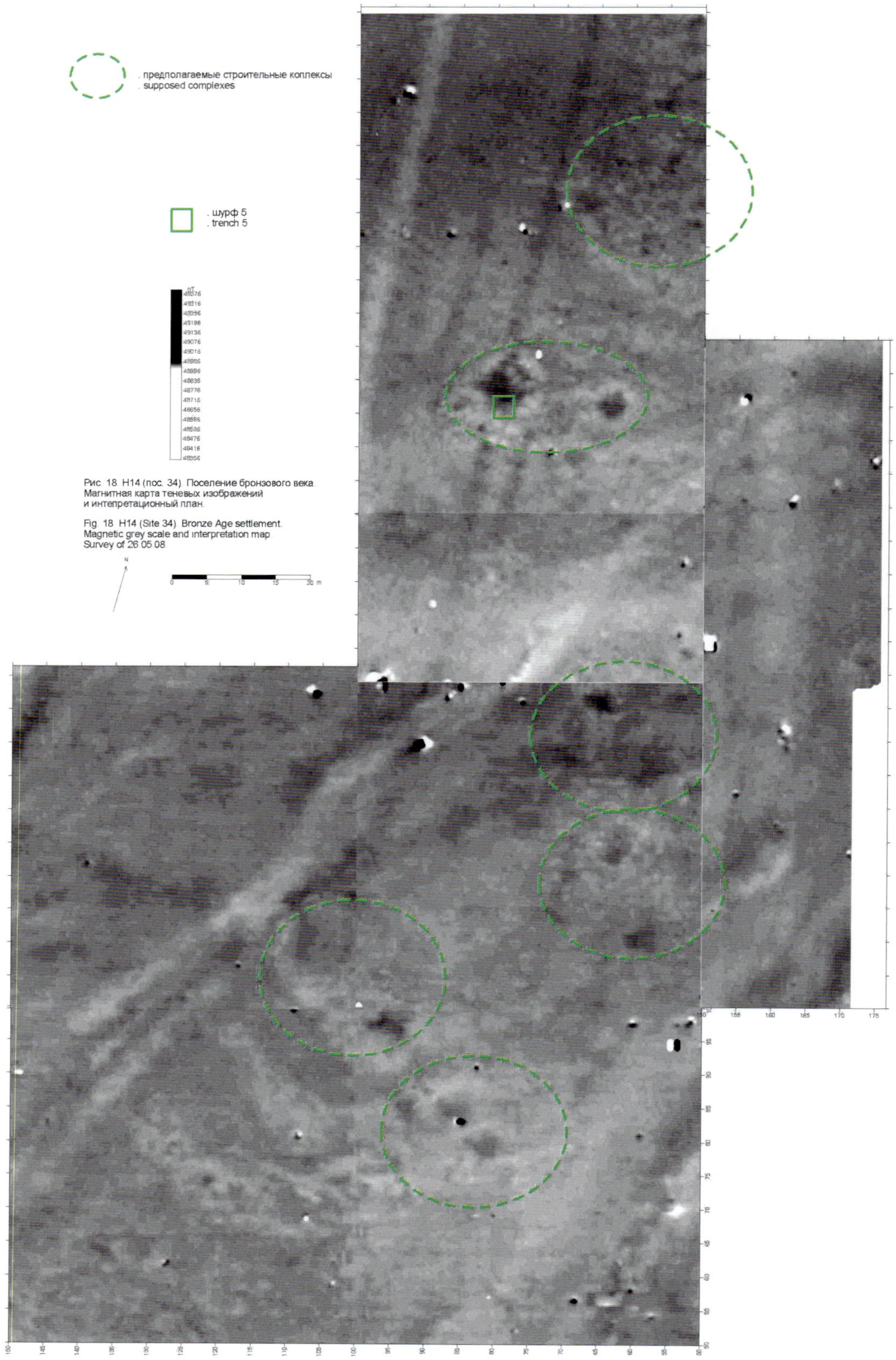

Fig. 4.281 • Geomagnetic map of DSP08-H14-01, interpretation.

Fig. 4.282 • Trial trench DSP08-H14-01 Feature 2.

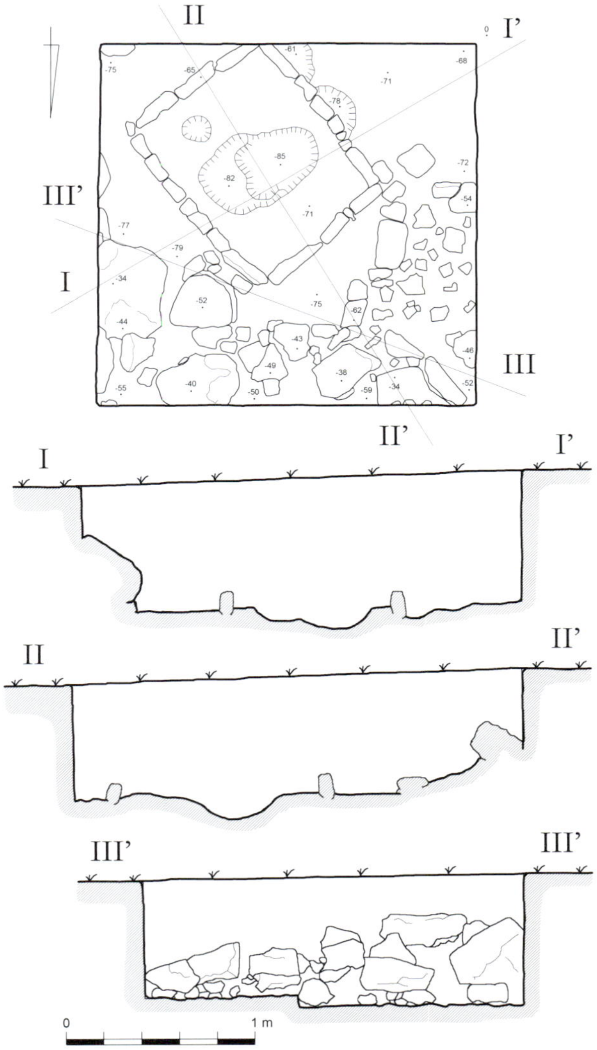

Fig. 4.283 • Drawings of trial trench DSP08-H14-01 Feature 2.

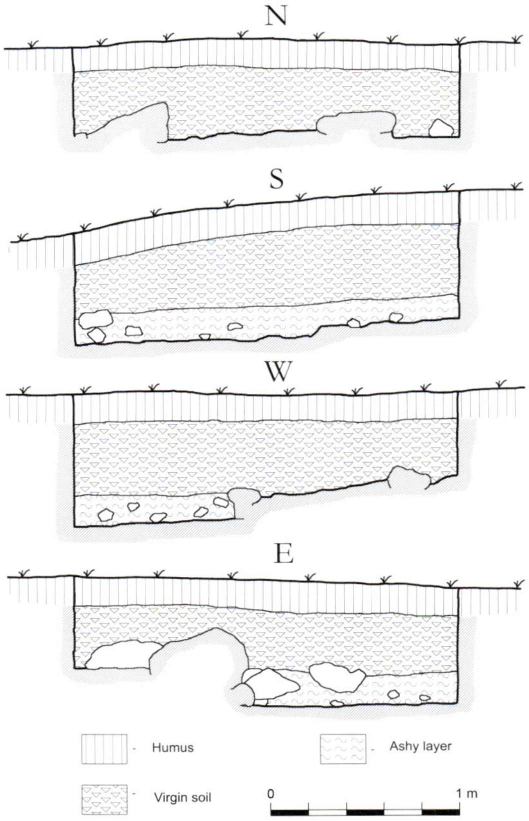

Fig. 4.284 • Profile drawings of trial trench DSP08-H14-01 Feature 2.

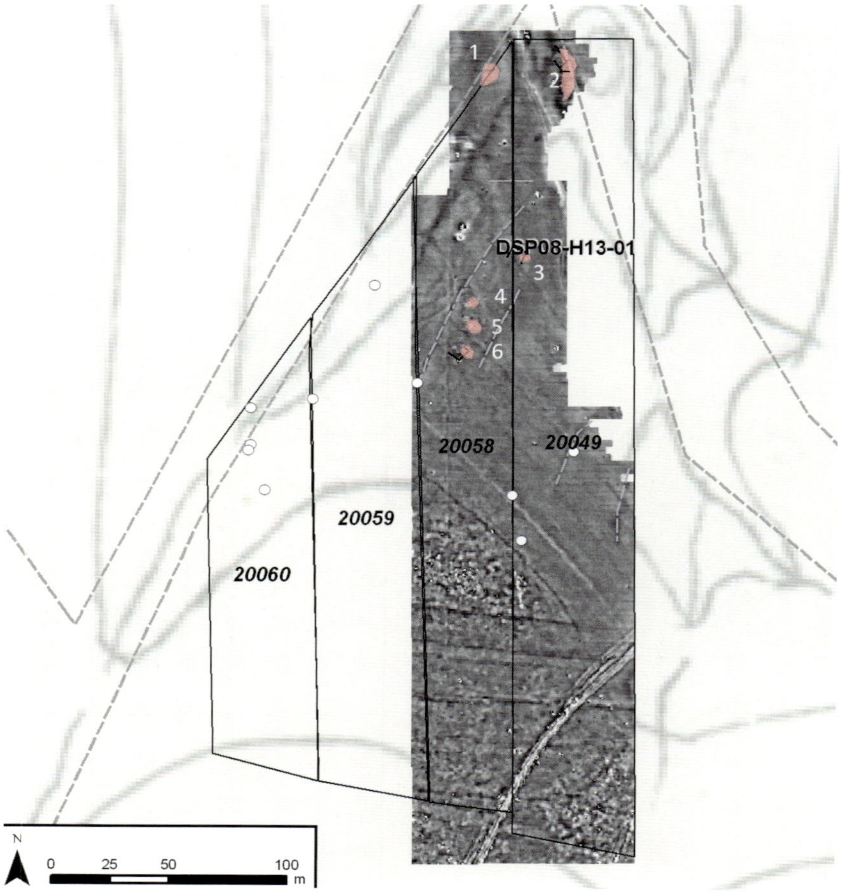

Fig. 4.285 • Distribution map of Units 20049 and 20058-20060.

Fig. 4.286 • Twin peaks of DSP08-H13-01 seen from the northwest.

Fig. 4.287 • View across DSP08-H13-01 towards the sea.

Fig. 4.288 • Possible structure on the peaks of DSP08-H13-01.

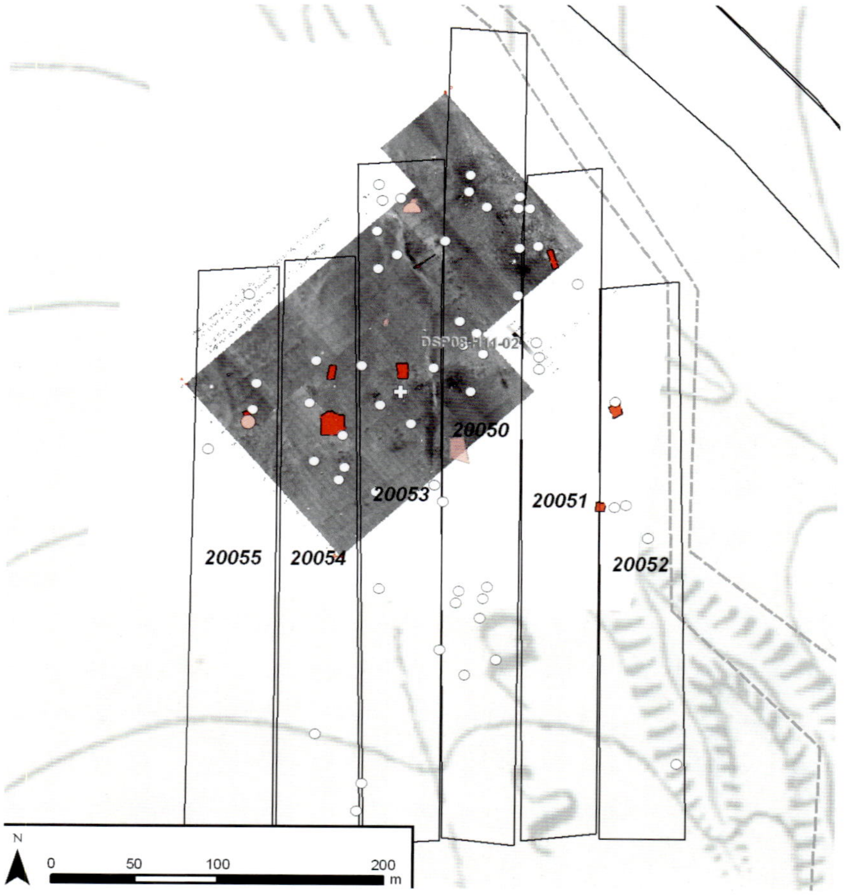

Fig. 4.289 • Distribution map of Units 20050-20055.

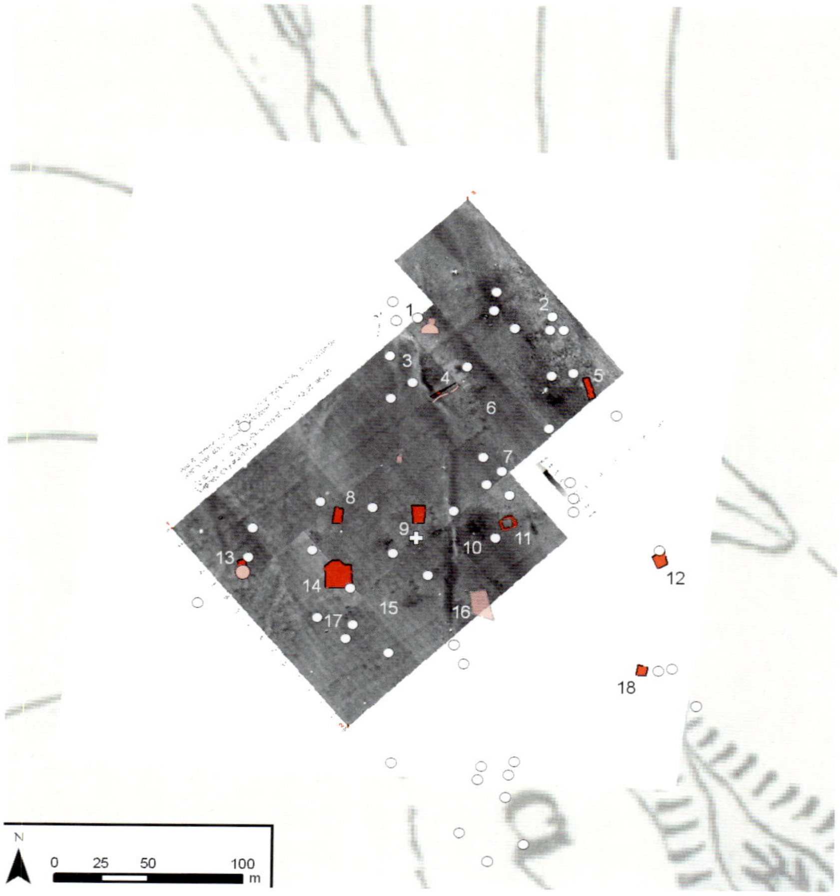

Fig. 4.290 • Distribution map of DSP08-H11-02.

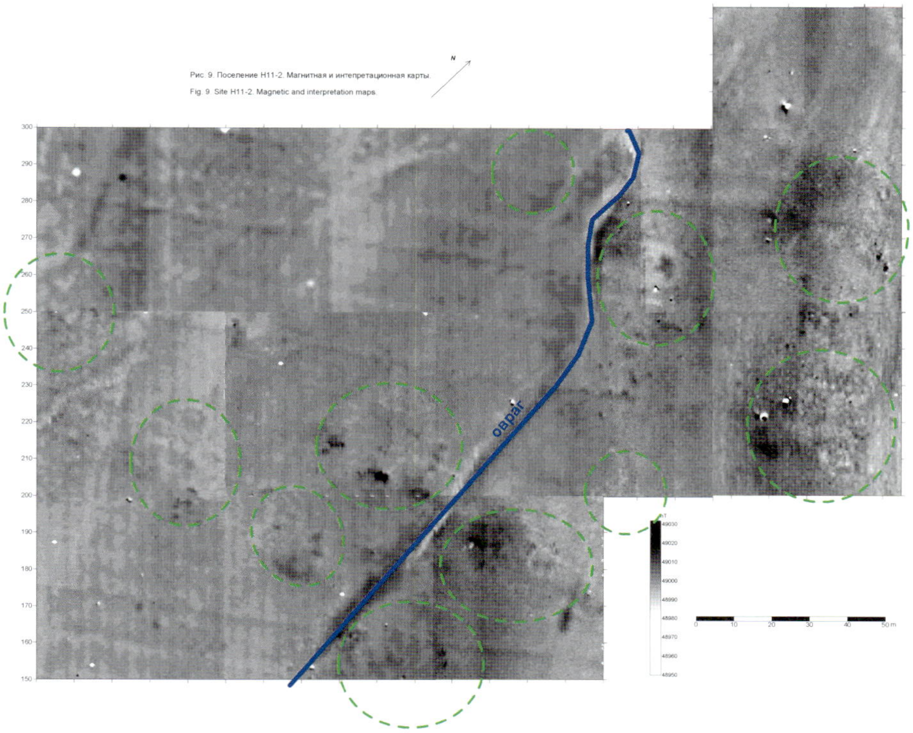

Fig. 4.291 • Geomagnetic map of DSP08-H11-02, interpretation.

Fig. 4.292 • Trial trench DSP08-H11-02 Feature 9.

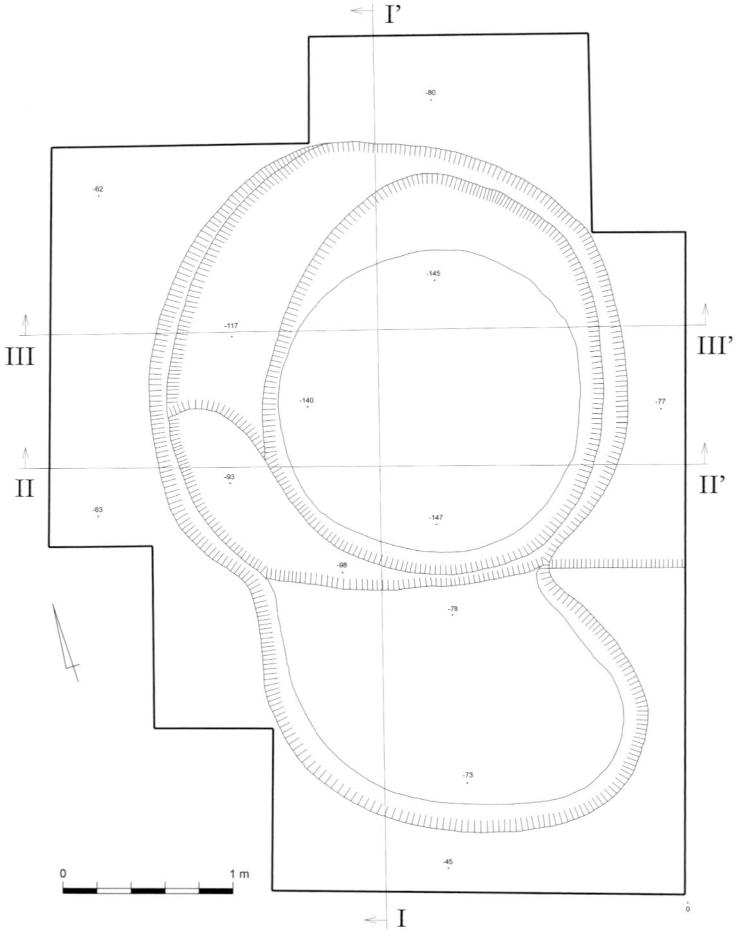

Fig. 4.293 • Drawing of trial trench DSP08-H11-02 Feature 9.

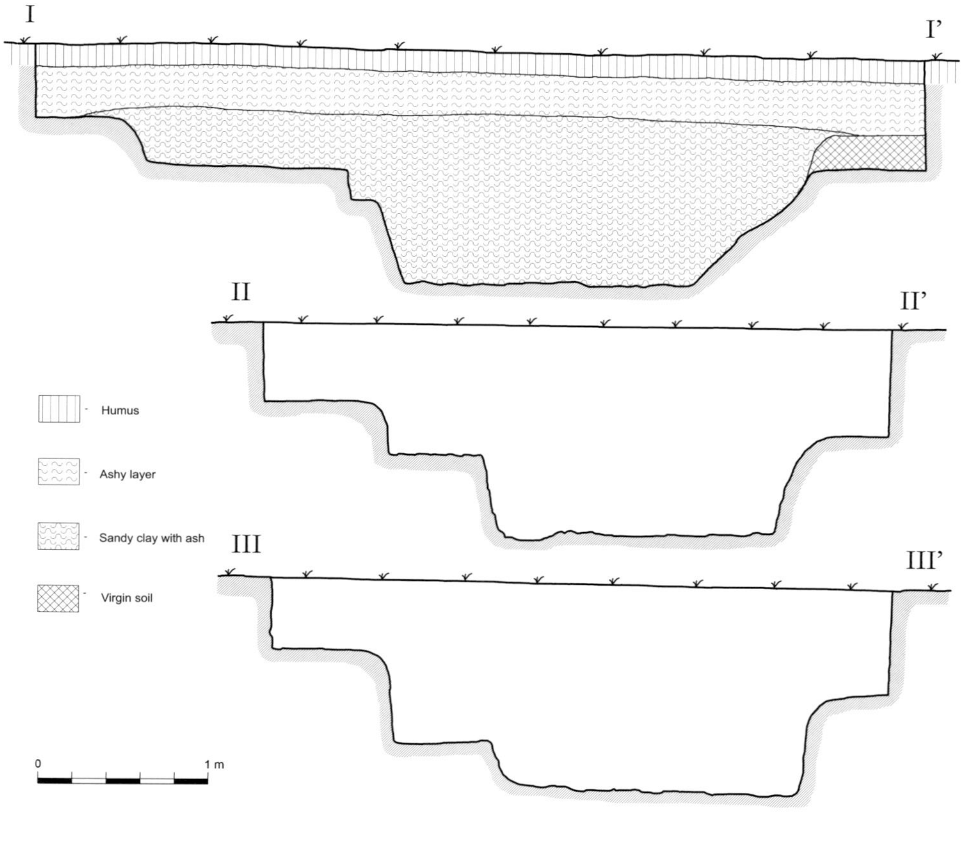

Fig. 4.294 • Profile drawing trial trench at DSP08-H11-02 Feature 9.

Fig. 4.295 • Distribution map of Units 20056-20057.

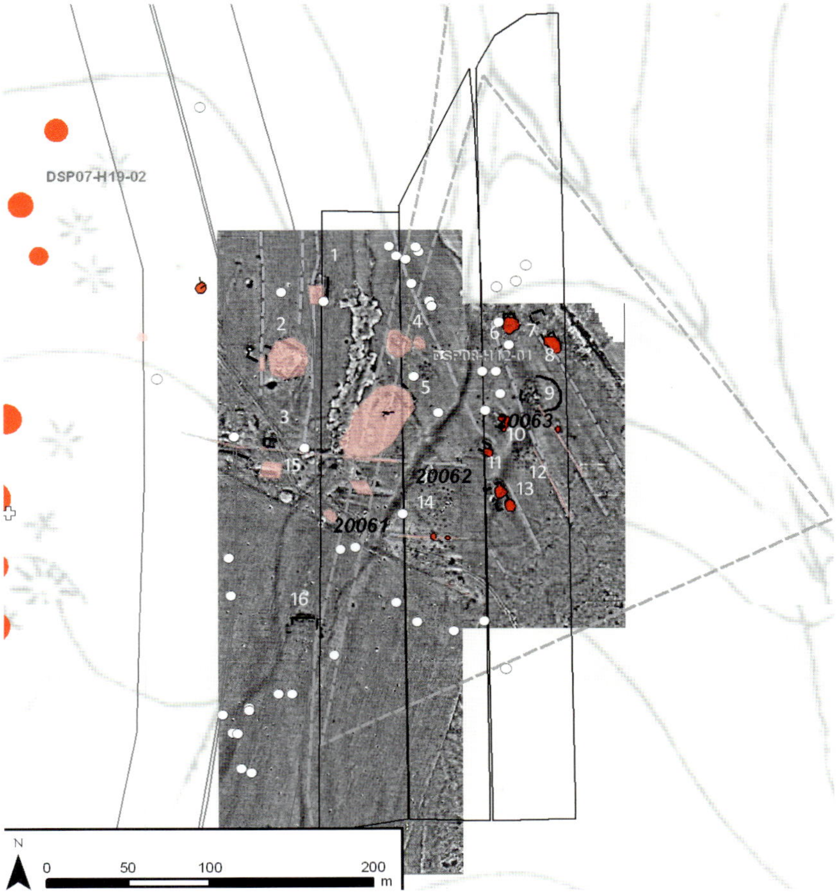

Fig. 4.296 • Distribution map of DSP08-H12-01.

Fig. 4.297 • View across DSP08-H12-01 from the south towards the sea. The dark green vegetation indicates the location of the structures.

Fig. 4.298 • View across DSP08-H12-01 from the northwest. Hill 12 in the centre with the loess built-up at the tip. The dark green vegetation indicates the location of the structures.

Fig. 4.299 • DSP08-H12-01 Feature 11 from the north.

Fig. 4.300 • DSP08-H12-01 Feature 10 from the northeast.

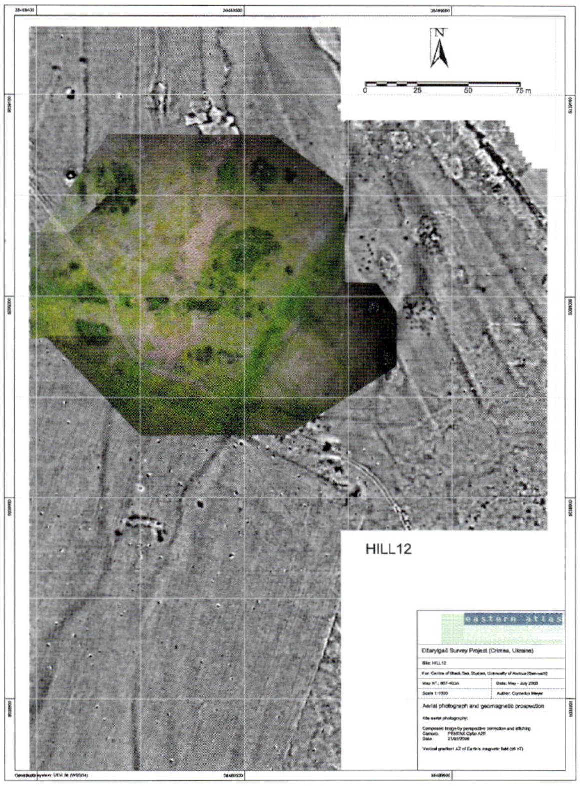

Fig. 4.301 • Aerial photo of DSP08-H12-01.

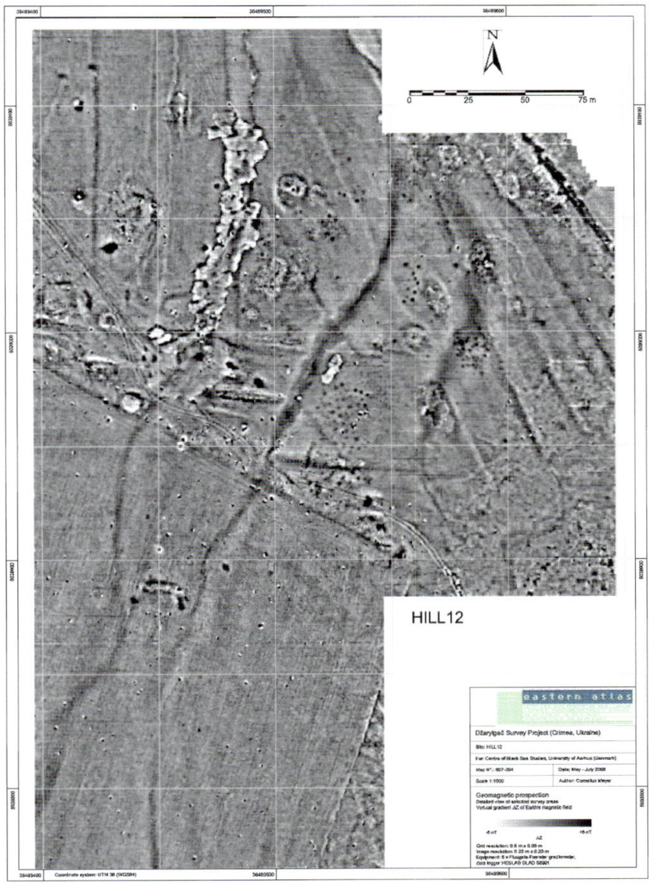

Fig. 4.302 • Geomagnetic map of DSP08-H12-01.

Fig. 4.303 • Geomagnetic map of DSP08-H12-01, interpretation.

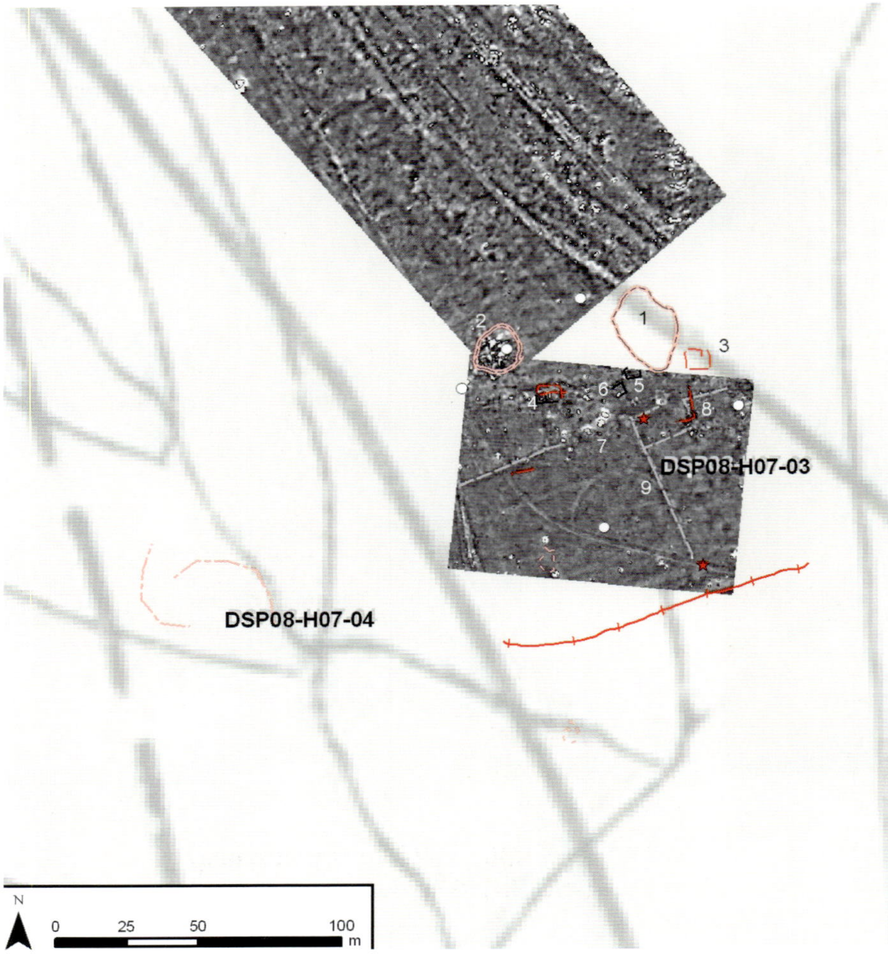

Fig. 4.304 • Distribution map of Units 20064-20065. The borders of the units are not marked on the plan, only the recorded observations.

Fig. 4.305 • DSP08-H07-03 from the southwest.

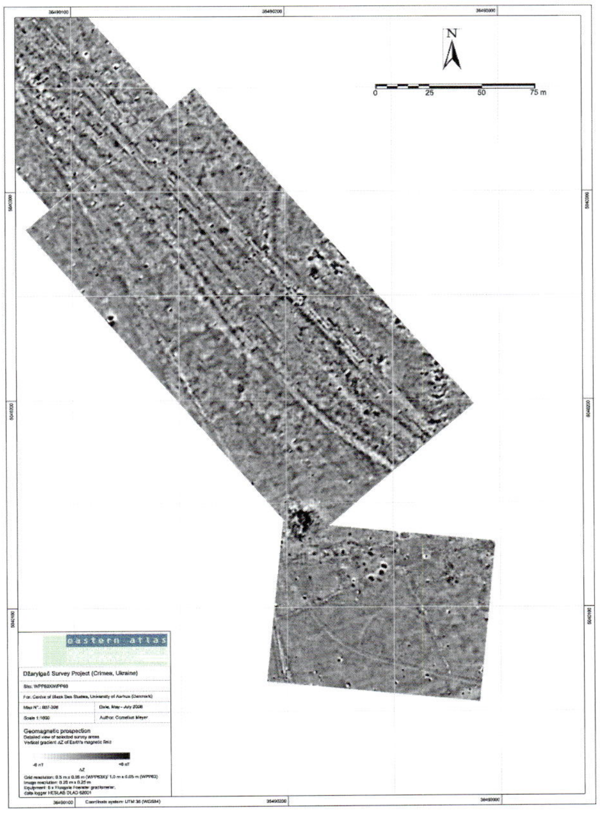

Fig. 4.306 • Geomagnetic map of DSP08-H07-03.

Fig. 4.307 • Geomagnetic map of DSP08-H07-03, interpretation.

Fig. 4.308 • DSP08-H07-04 from the southeast.

Fig. 5.1 • Communal well at Snežnoe.

Fig. 5.2 • A single platform core for the production of blades (1579/02). Two sides are shown, the front (left, with flake-scars) and the back of the core which consists of weathered cortex (right). Probably Late Upper Palaeolithic or Mesolithic.

Fig. 5.3 • Brandon gunflint (9050/01) which was used during the Crimean-war (1853-1856).

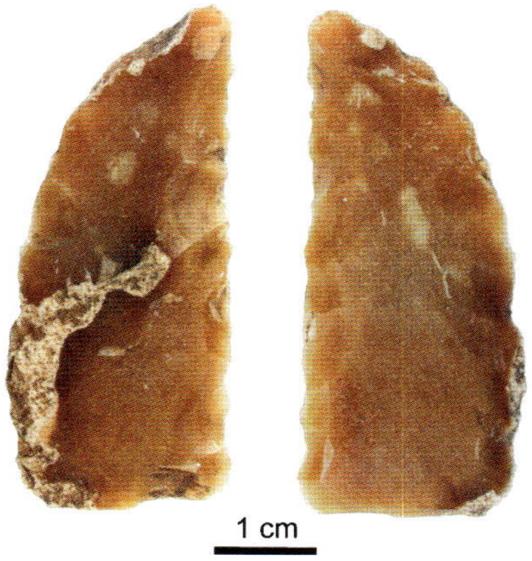

Fig. 5.4 • Asymmetrically shaped Bronze Age "knife" from trial trench 4 (DSP07-H08-01) seen from two sides. The bifacial working of the edges is clearly visible.

Fig. 5.5 • Fragment (cutting edge) of a battle-axe with a cylindrical perforation from DSP07-H08-01. Bronze Age.

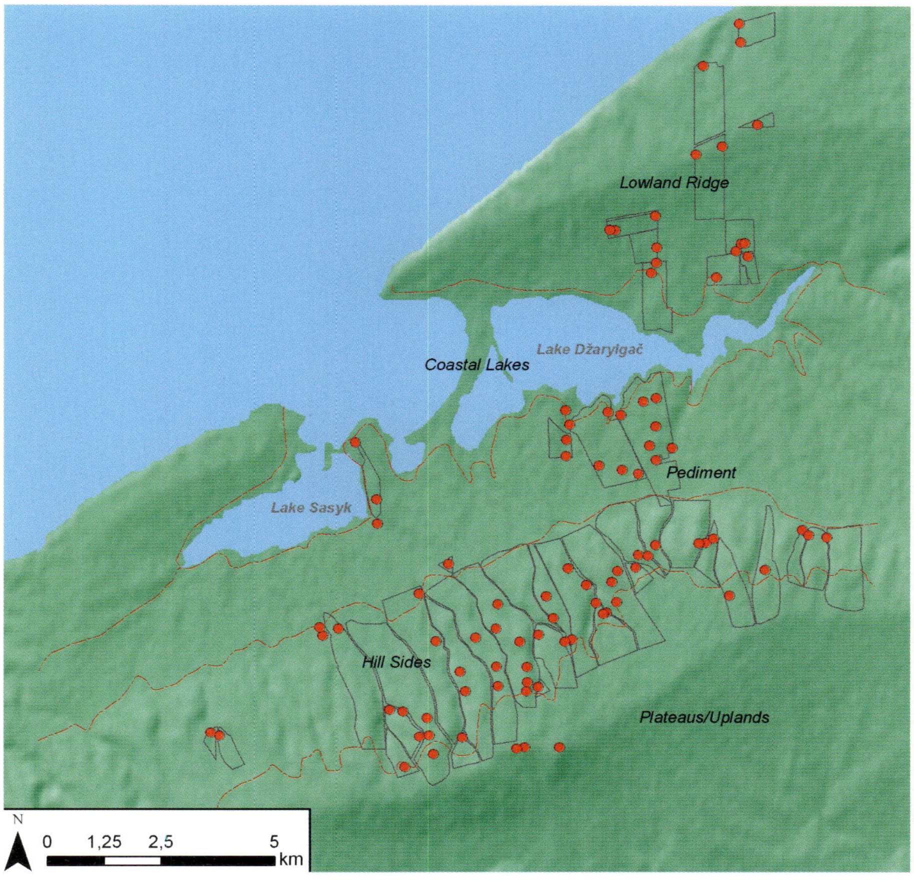

Fig. 6.1 • Landscape types, surveyed fields and hills and sites found.

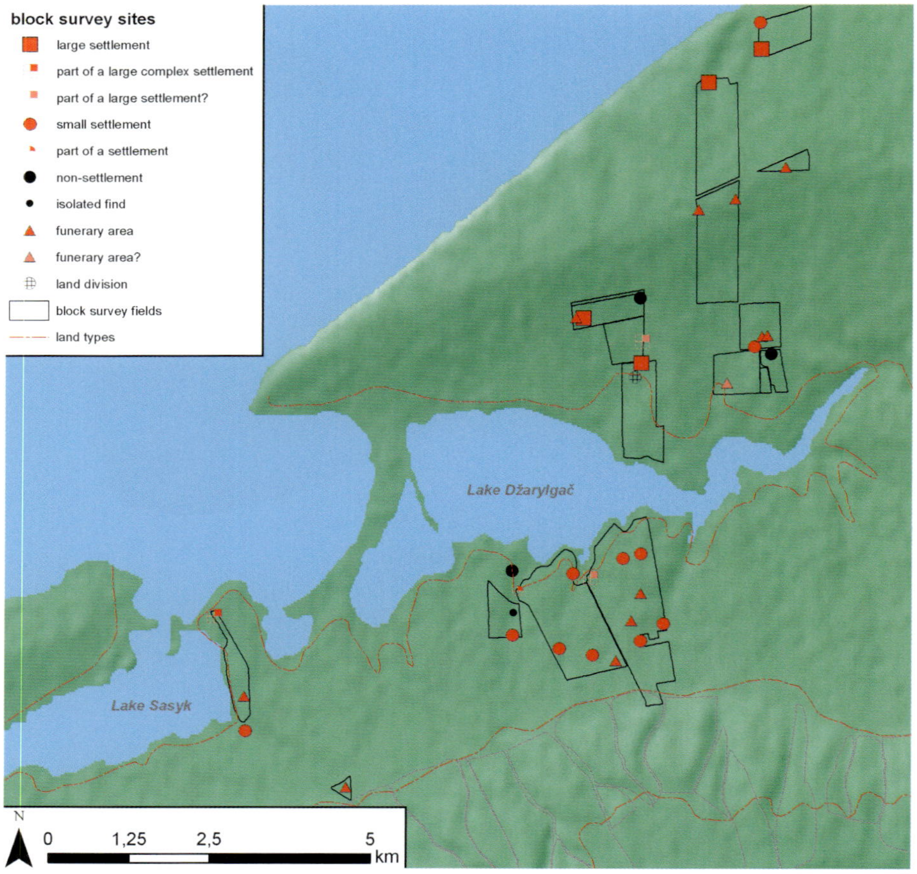

Fig. 6.2 • Map showing the distribution of the various classes over the low-lying areas.

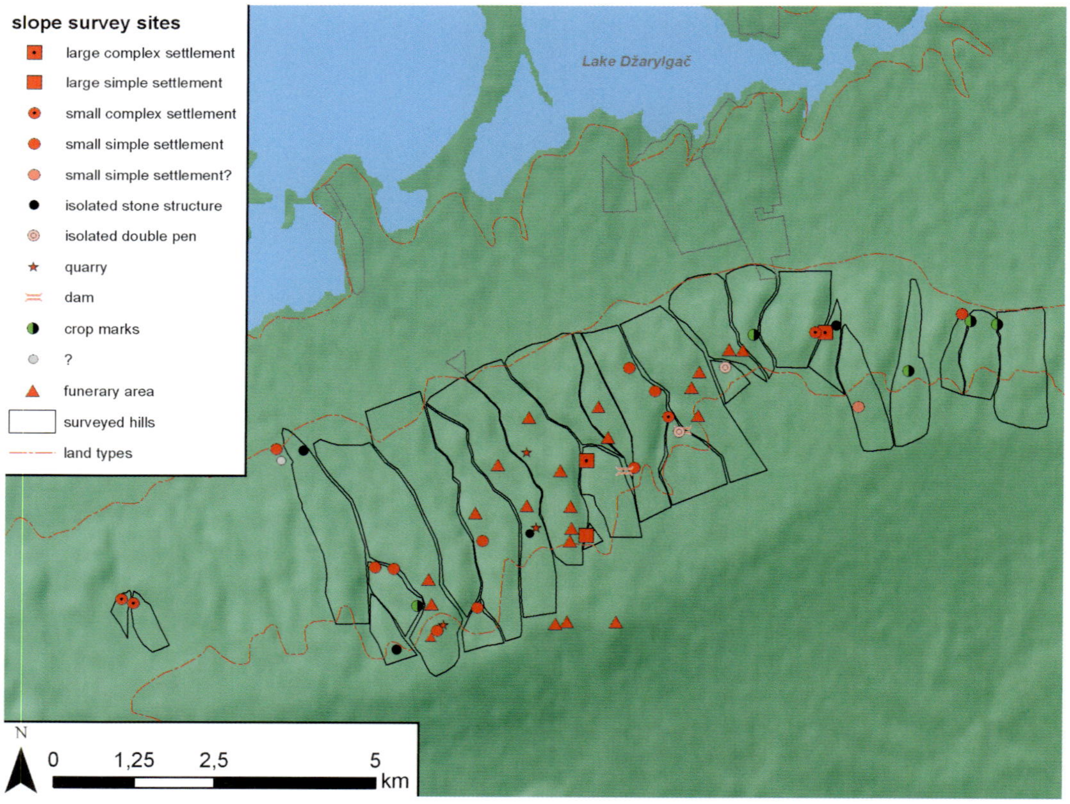

Fig. 6.3 • Map showing the distribution of the various classes over the higher parts.

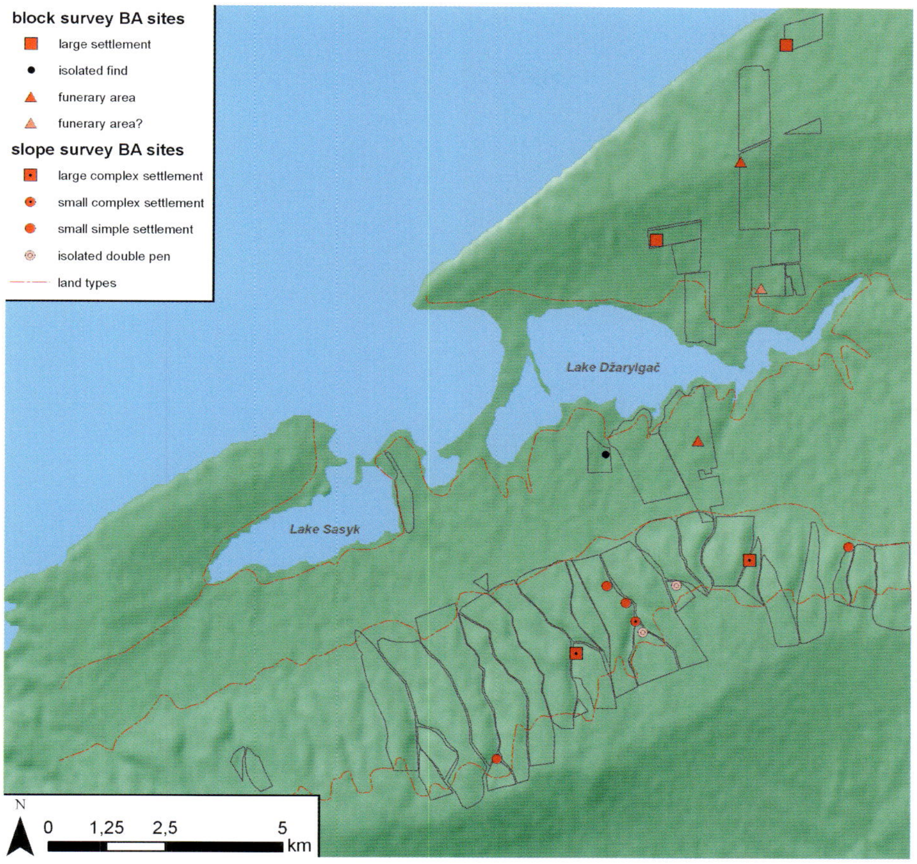

Fig. 6.4 • Bronze Age sites of the DSP transect.

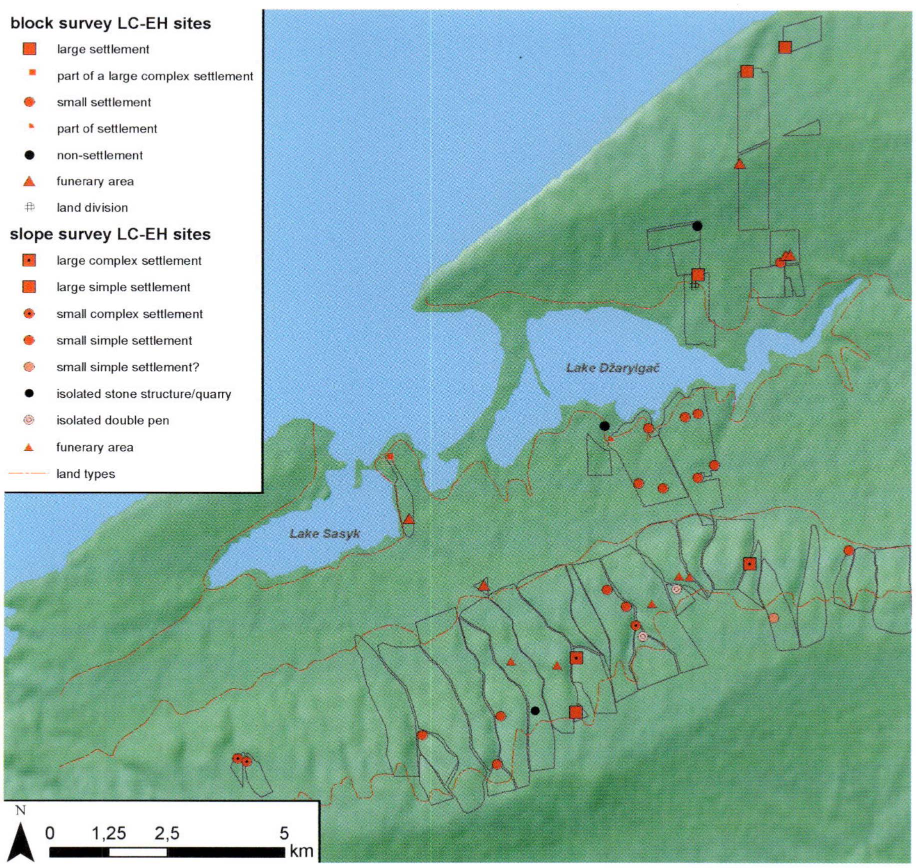

Fig. 6.5 • Late Classical/Early Hellenistic sites of the DSP transect.

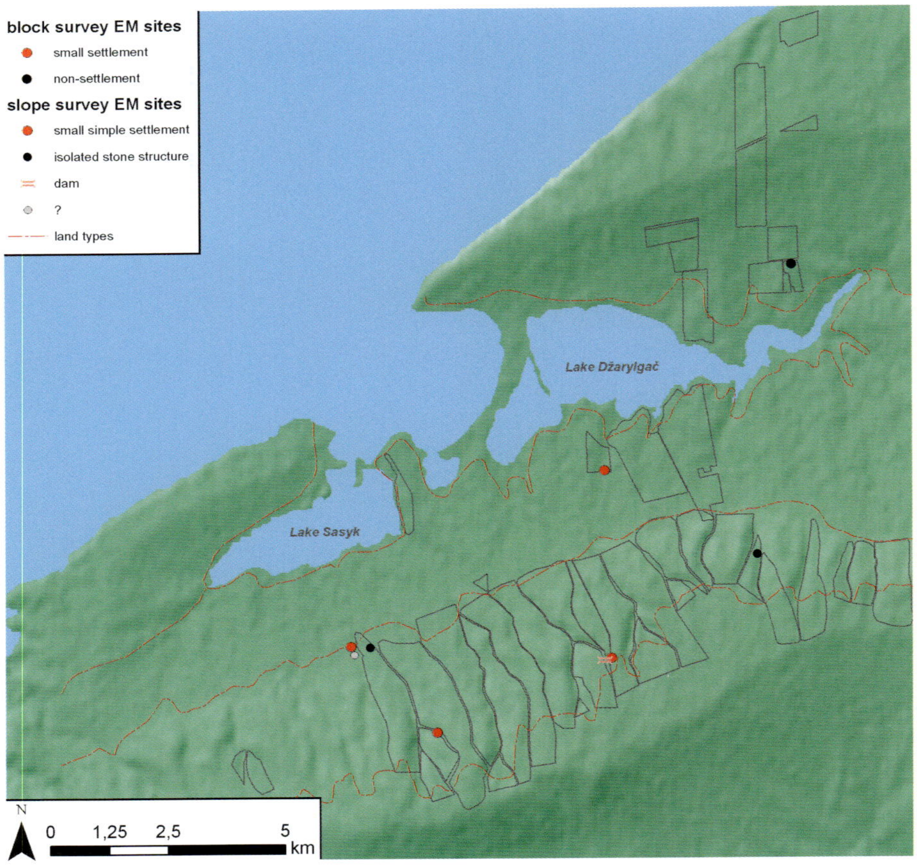

Fig. 6.6 • Medieval and Early Modern sites of the DSP transect.

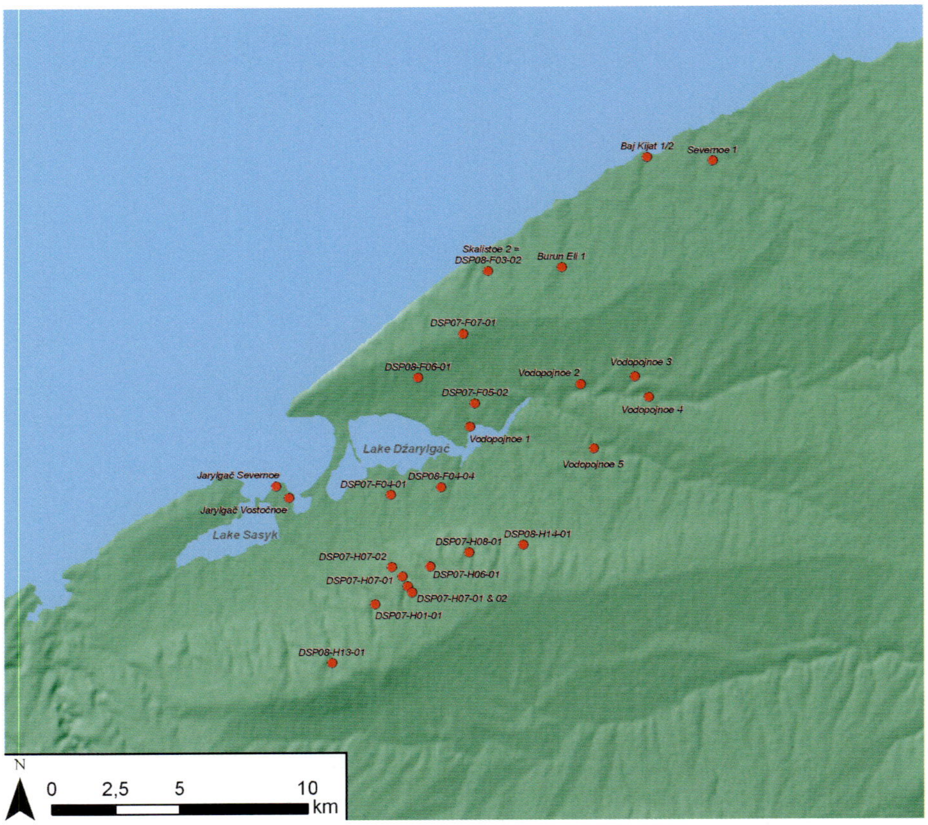

Fig. 6.7 • Prehistory: previously known sites and sites located by DSP between Černomorskoe and Masliny.

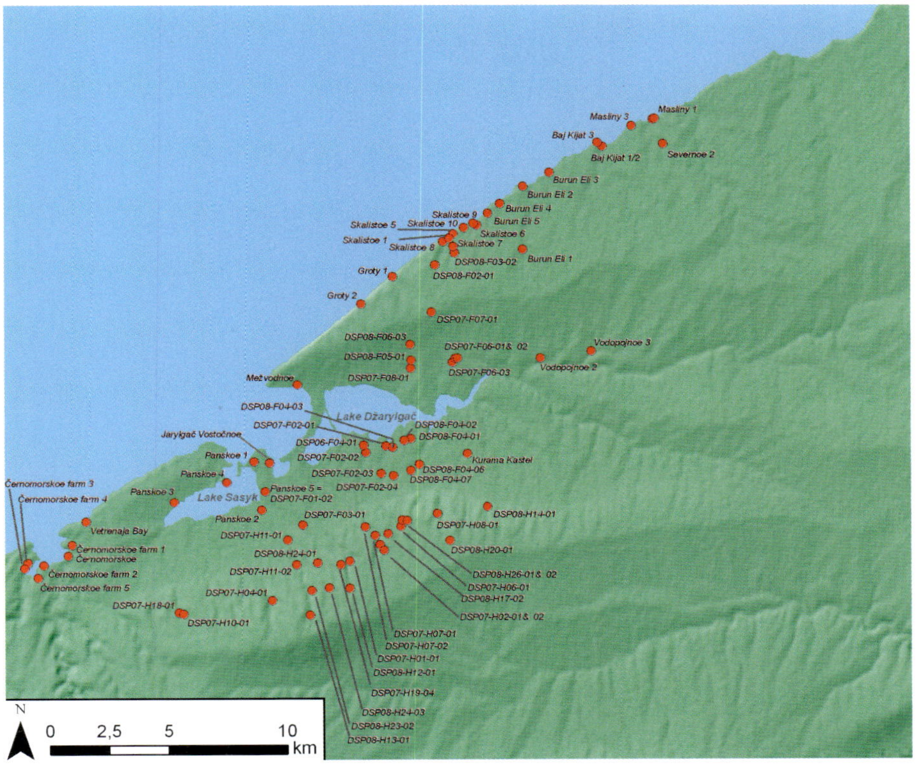

Fig. 6.8 • Greco-Roman period: previously known sites and sites located by DSP between Černomorskoe and Masliny.

Fig. 6.9 • Kurgan at Snežnoe functioning as present-day graveyard.

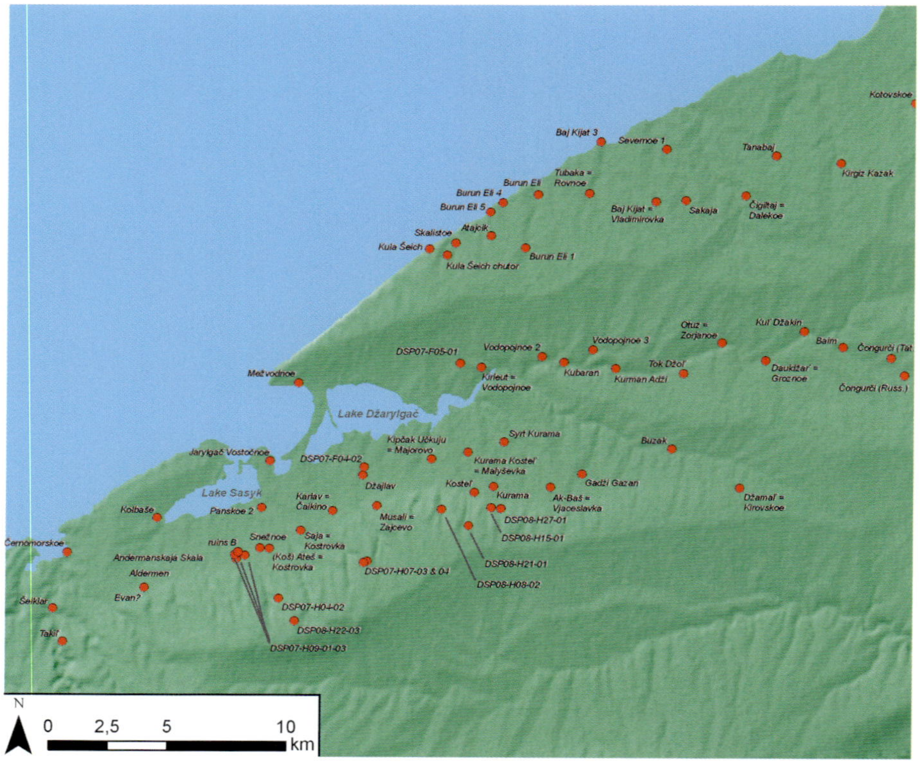

Fig. 6.10 • Post-Antique period: previously known sites and sites located by DSP between Černomorskoe and Masliny.

Fig. 6.11 • The Early Modern settlements of Kipčak Učkuju and Kurama Kostel' can easily be distinguished from satellite images due to the presence of ashy soils and building remains (Google Earth).

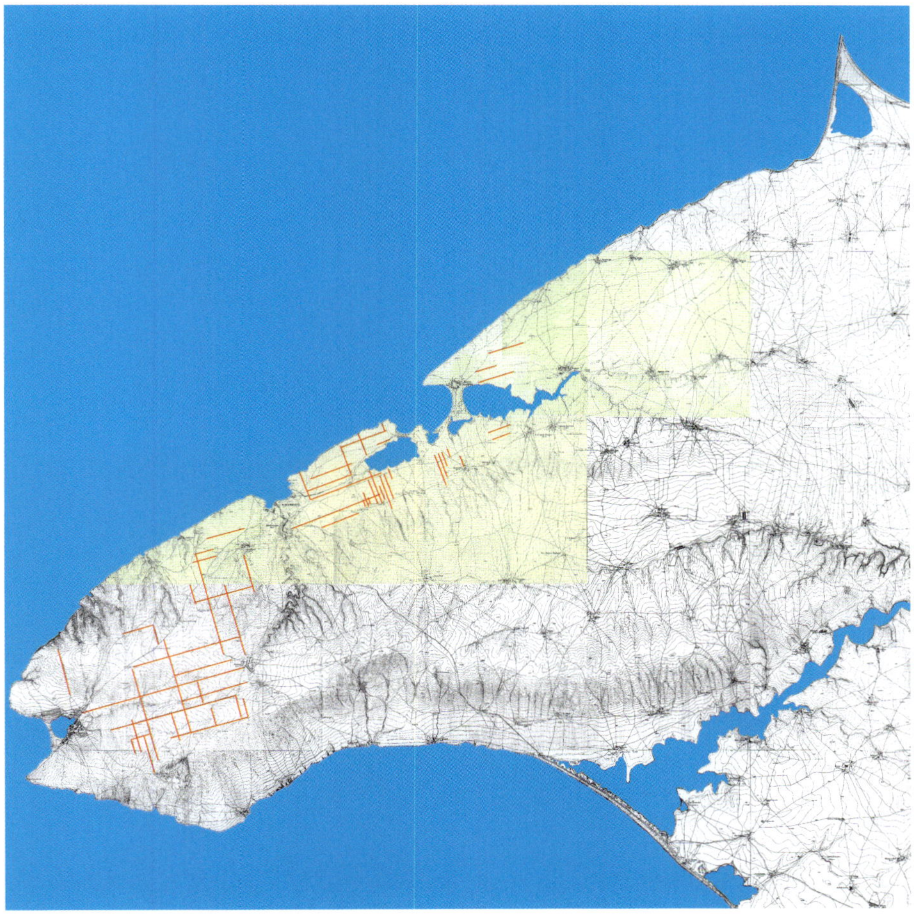

Fig. 6.12 • Tarchankut, cadastral system reconstructed on the basis of tracks on the half-verst map unrelated to contemporary settlement structure.

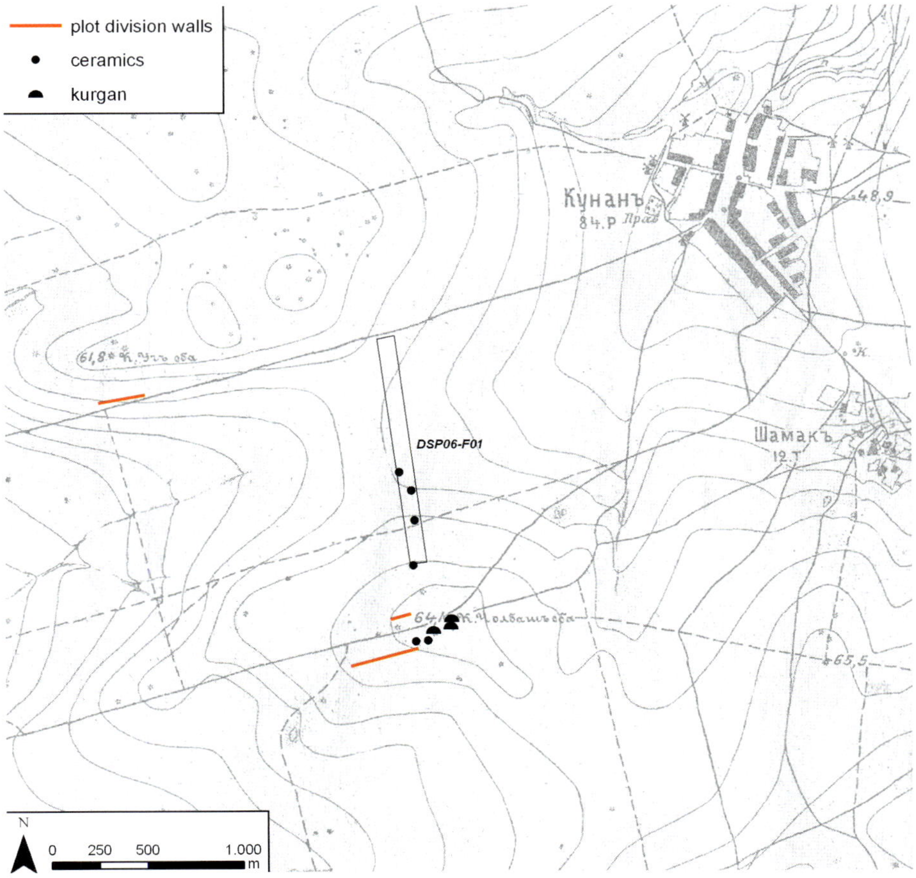

Fig. 6.13 • Plot divisions, western Tarchankut.

Fig. 6.14 • Southernmost plot division wall in form of straight soil bank from west towards east.

Fig. 6.15 • Dry stone plot division wall east of the Karadža Settlement.

Fig. 6.16 • Area between the Černomorskoe Settlement and the Vetrenaja farmhouse with indication of plot lines (Google Earth).

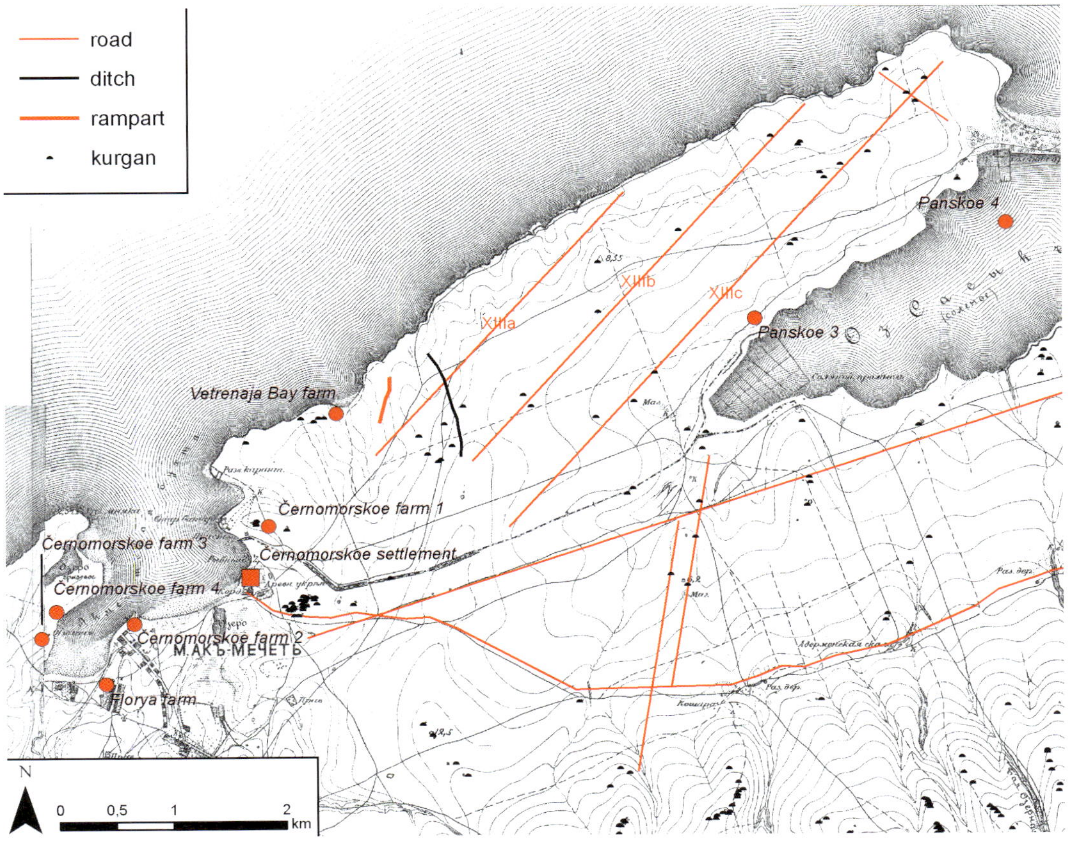

Fig. 6.17 • Sasyk promontory.

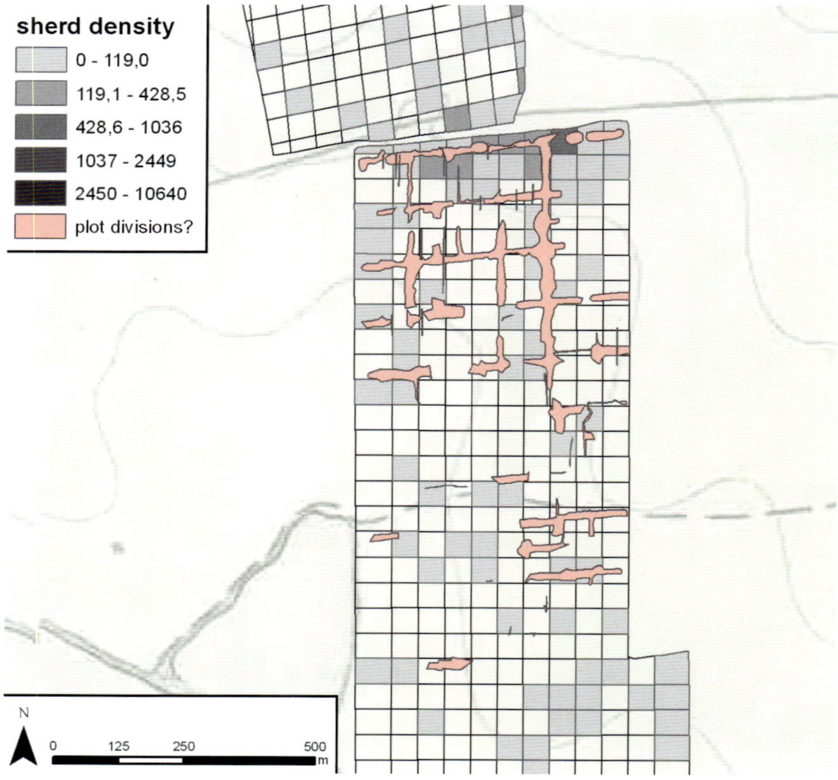

Fig. 6.18 • Field DSP-07-08, plot divisions?

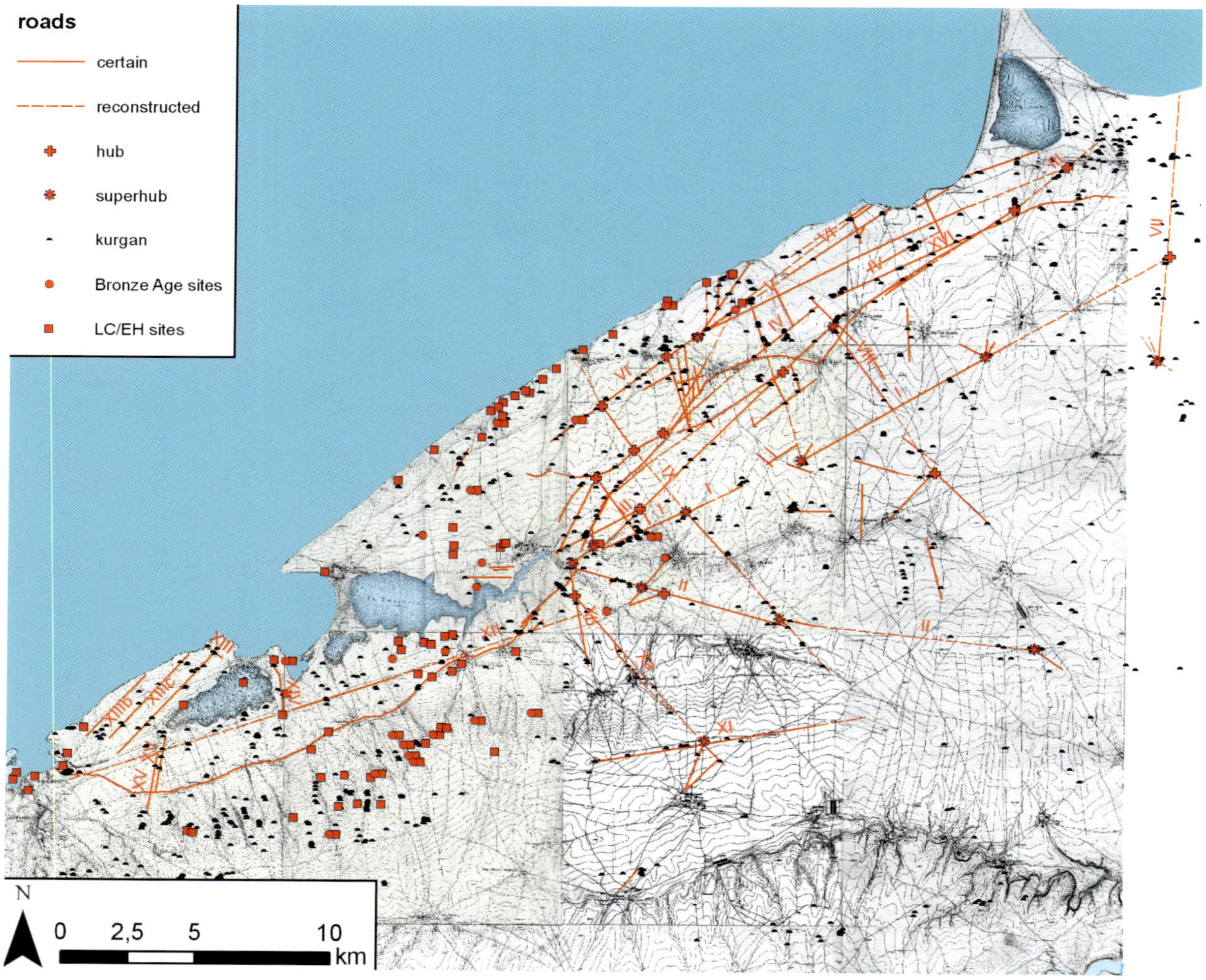

Fig. 6.19 • Overall map with key to roads.

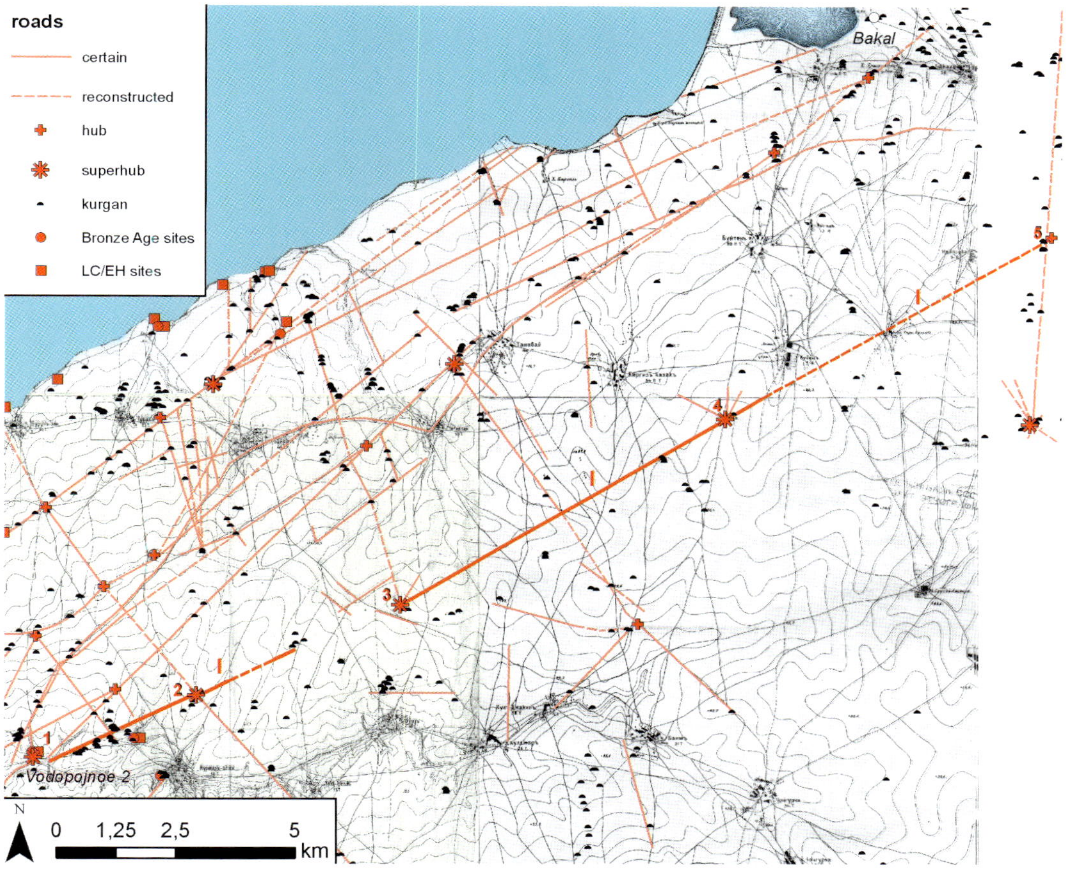

Fig. 6.20 • Road I.

Fig. 6.21 • Superhub I.1/III.1 (Google Earth).

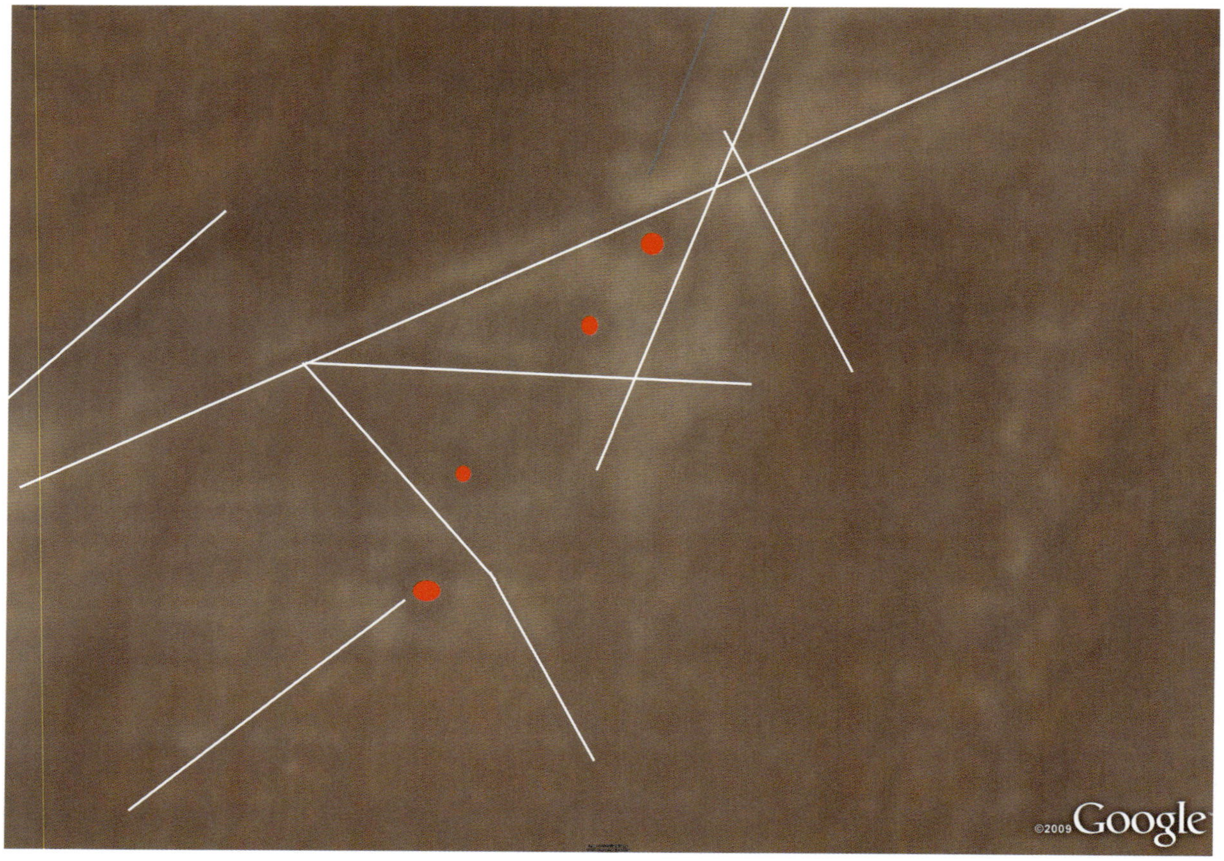

Fig. 6.22 • Superhub I.2/IX.3 (Google Earth).

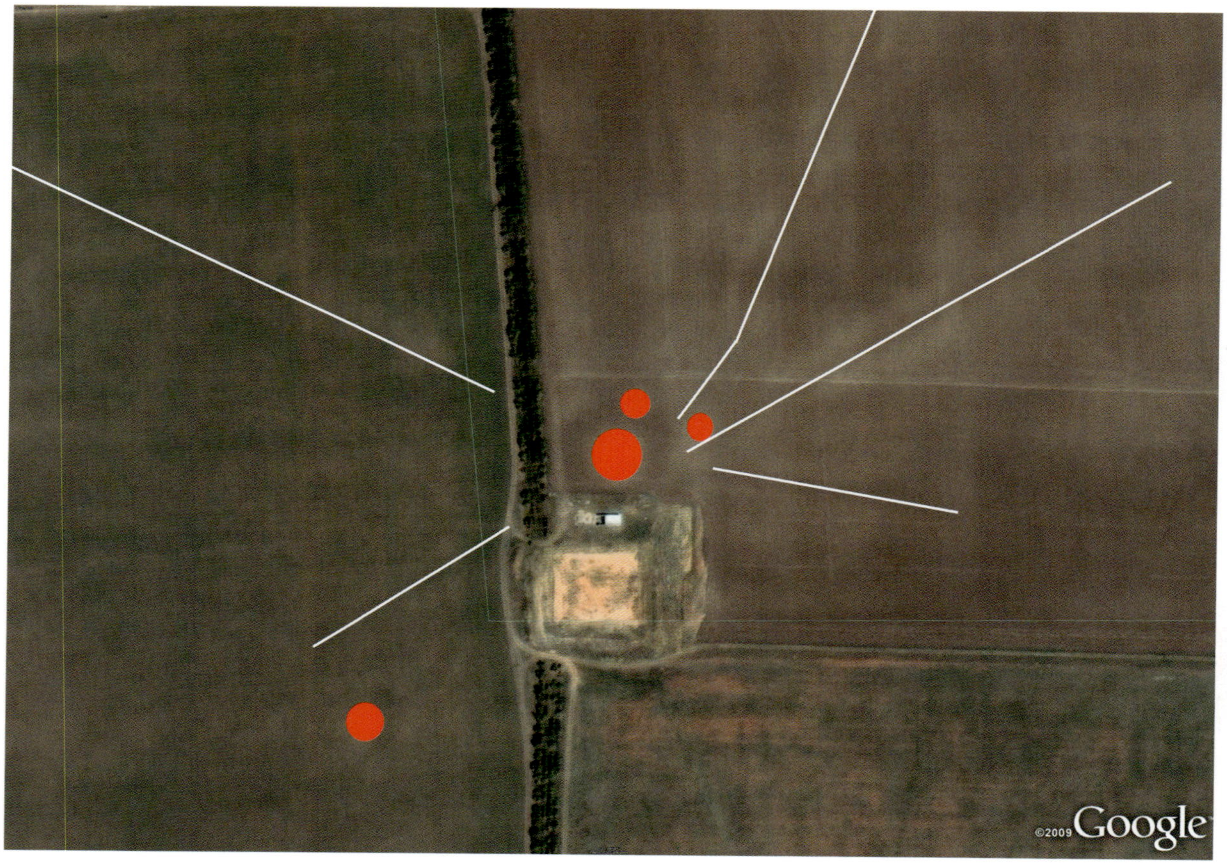

Fig. 6.23 • Superhub I.4 (Google Earth).

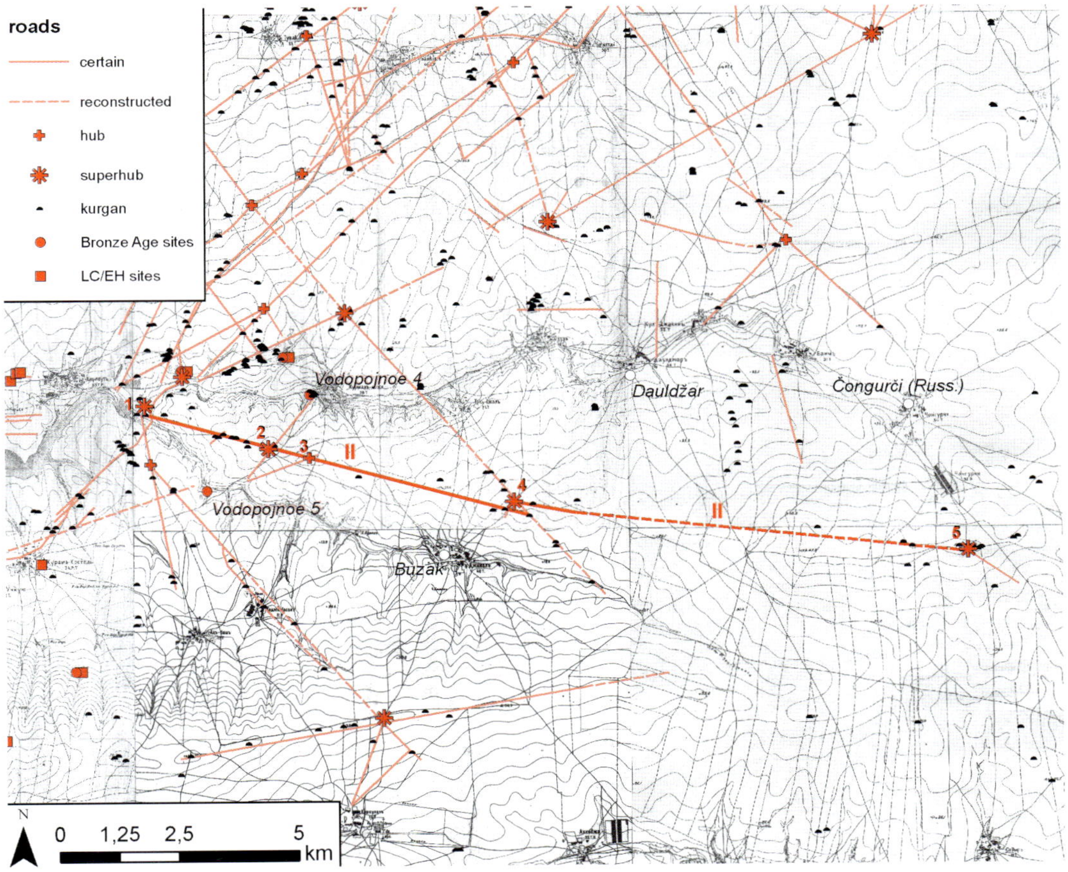

Fig. 6.24 • Road II.

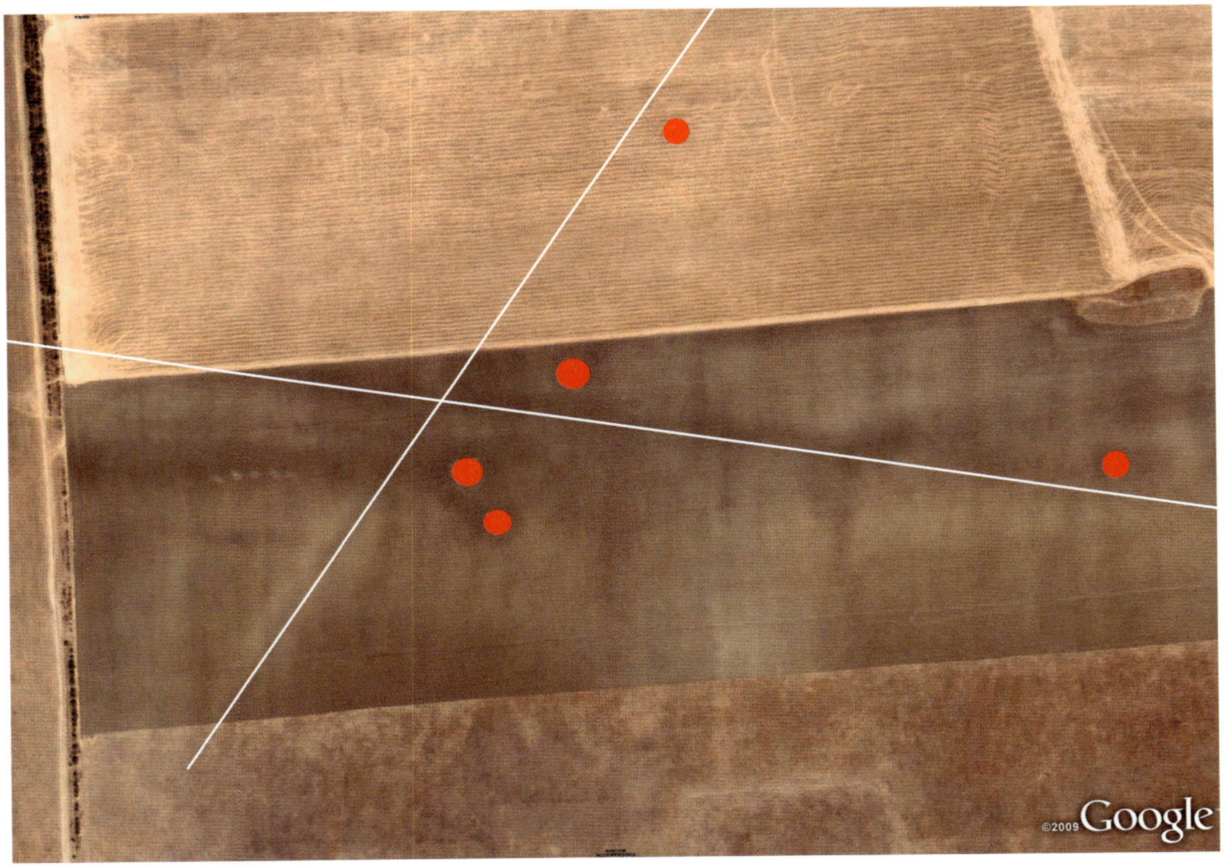

Fig. 6.25 • Superhub II.4/IX.4 (Google Earth).

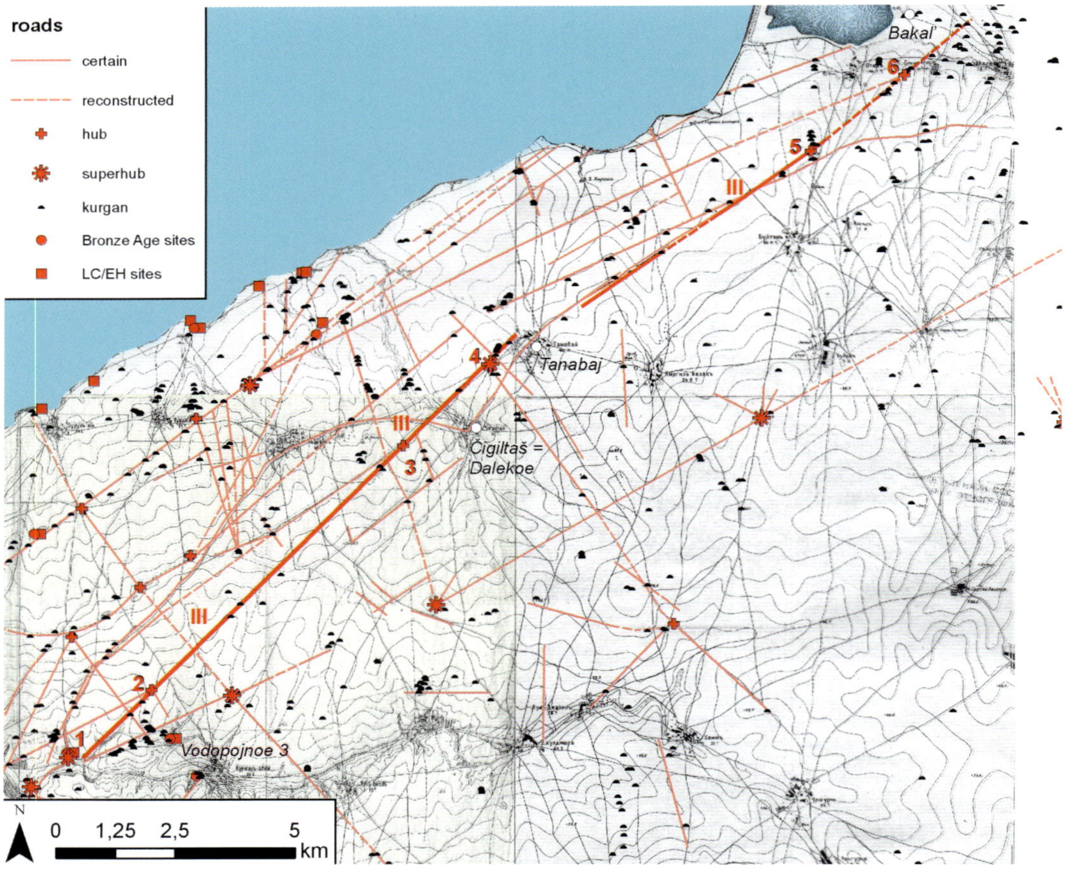

Fig. 6.26 • Road III.

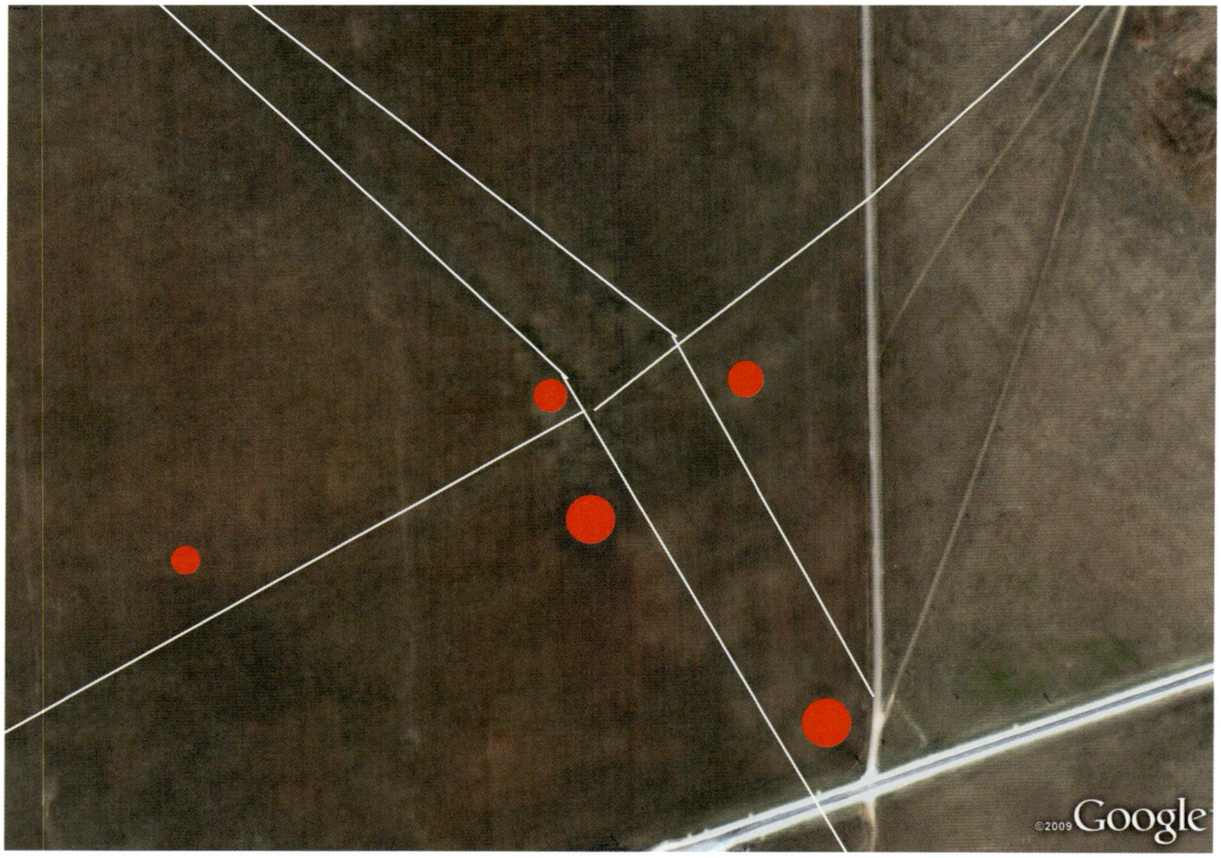

Fig. 6.27 • Superhub III.4 (Google Earth).

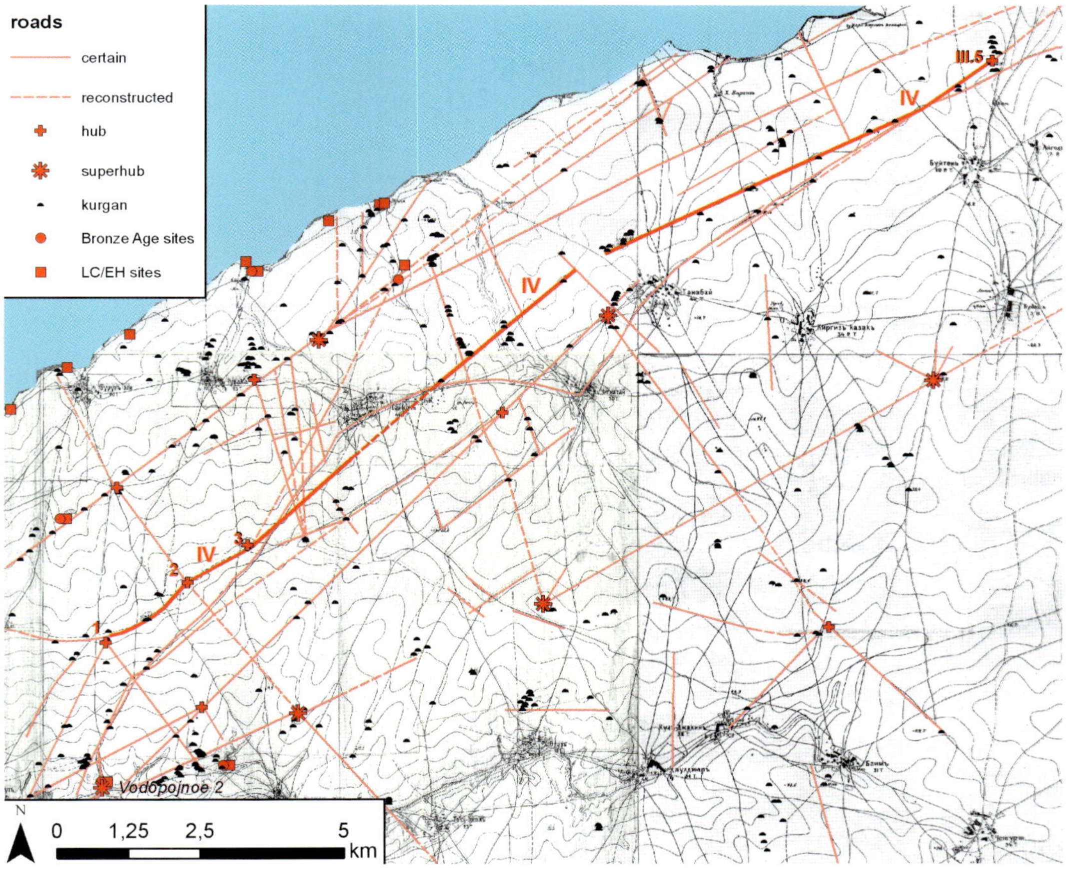

Fig. 6.28 • Road IV.

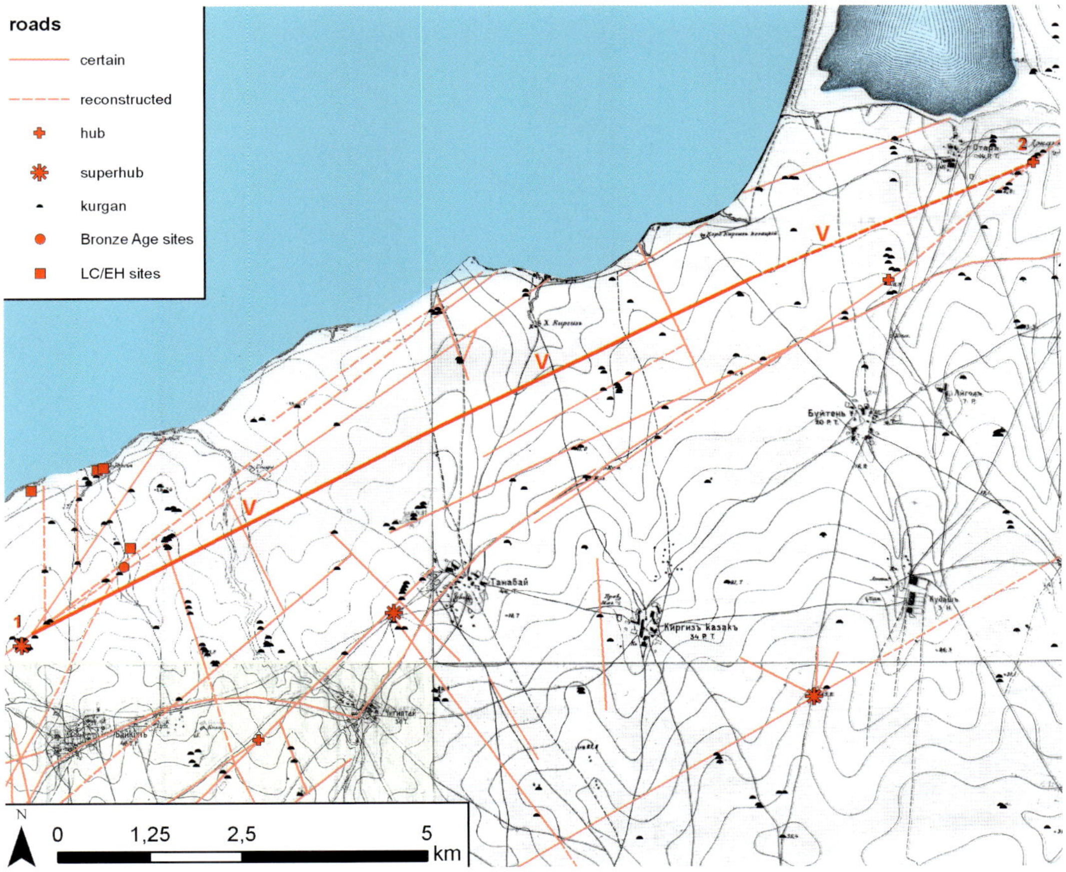

Fig. 6.29 • Road V.

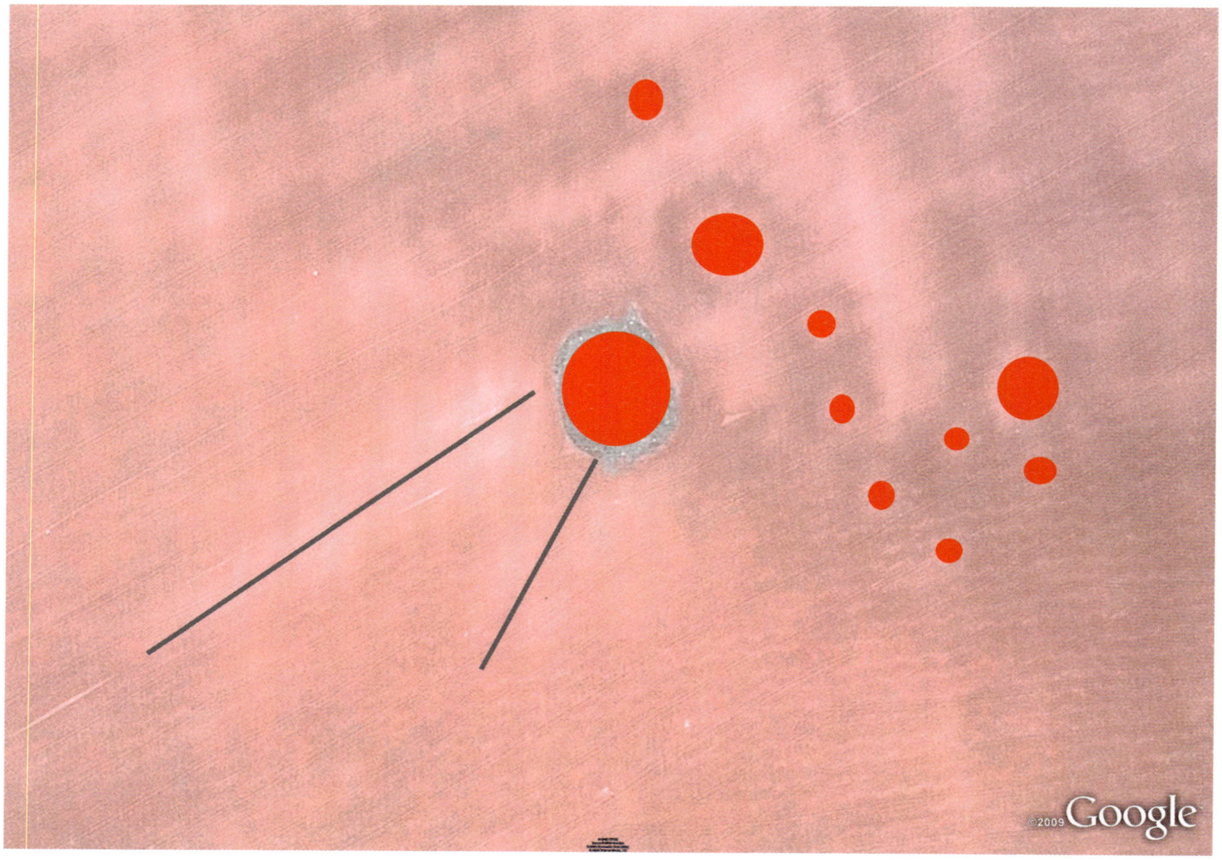

Fig. 6.30 • Hub V.1/VI.3 (Google Earth).

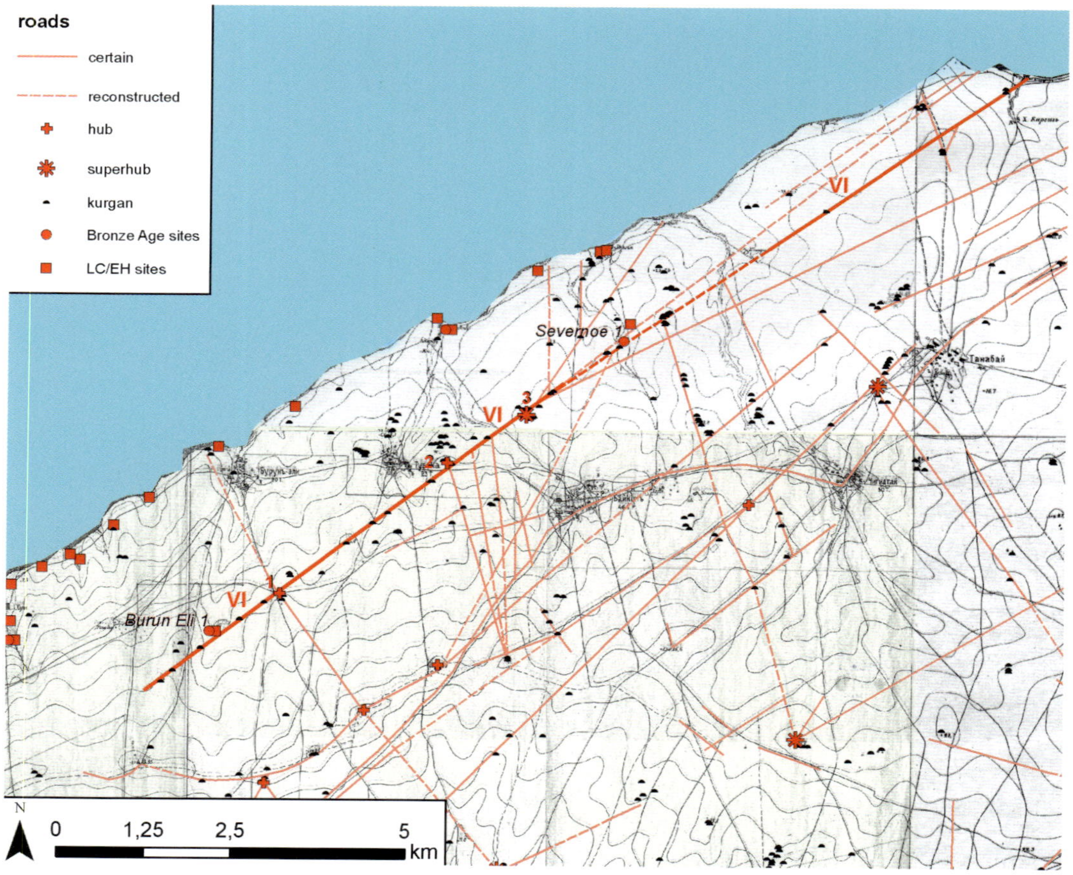

Fig. 6.31 • Road VI.

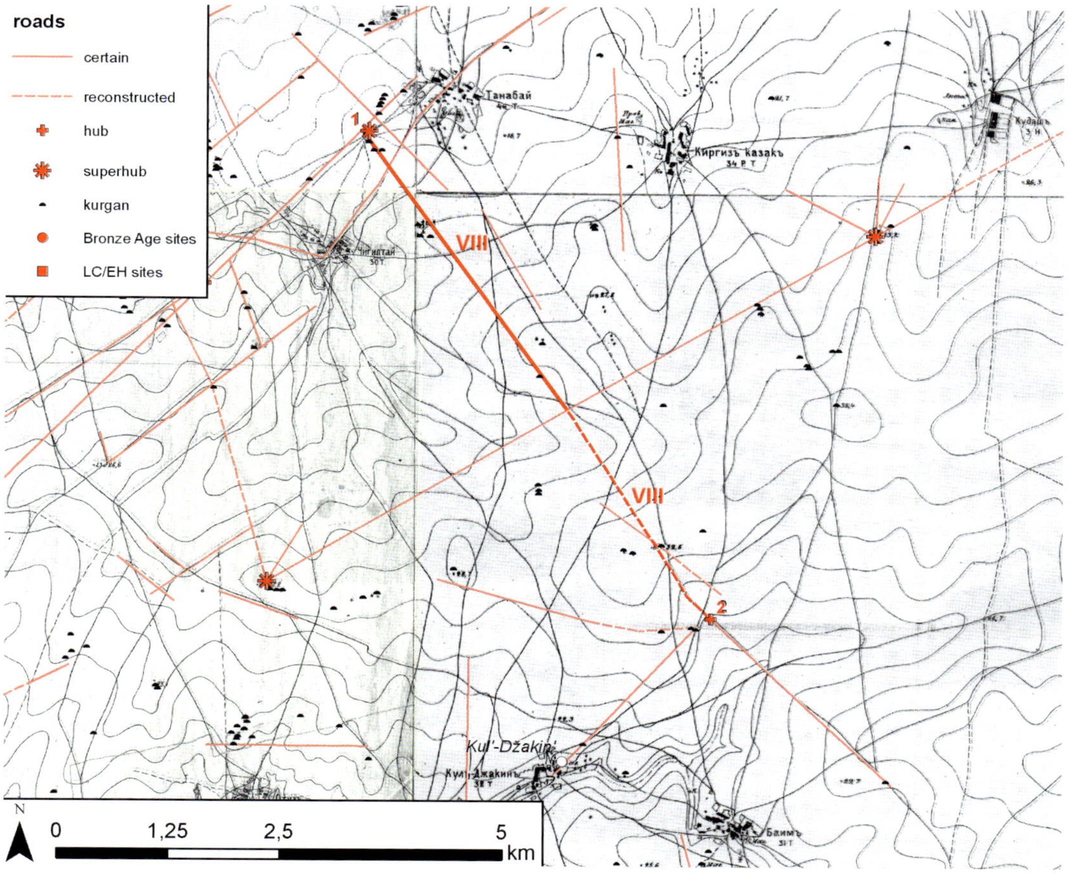

Fig. 6.32 • Road VIII.

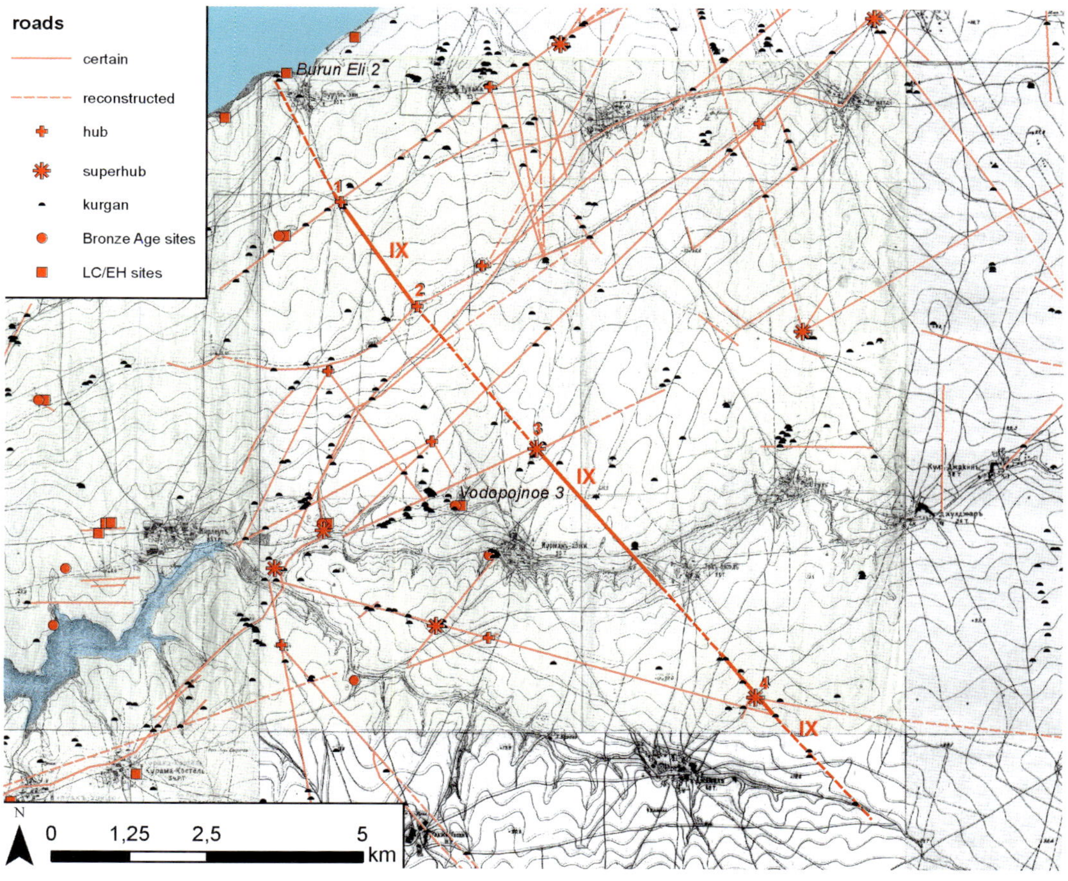

Fig. 6.33 • Road IX.

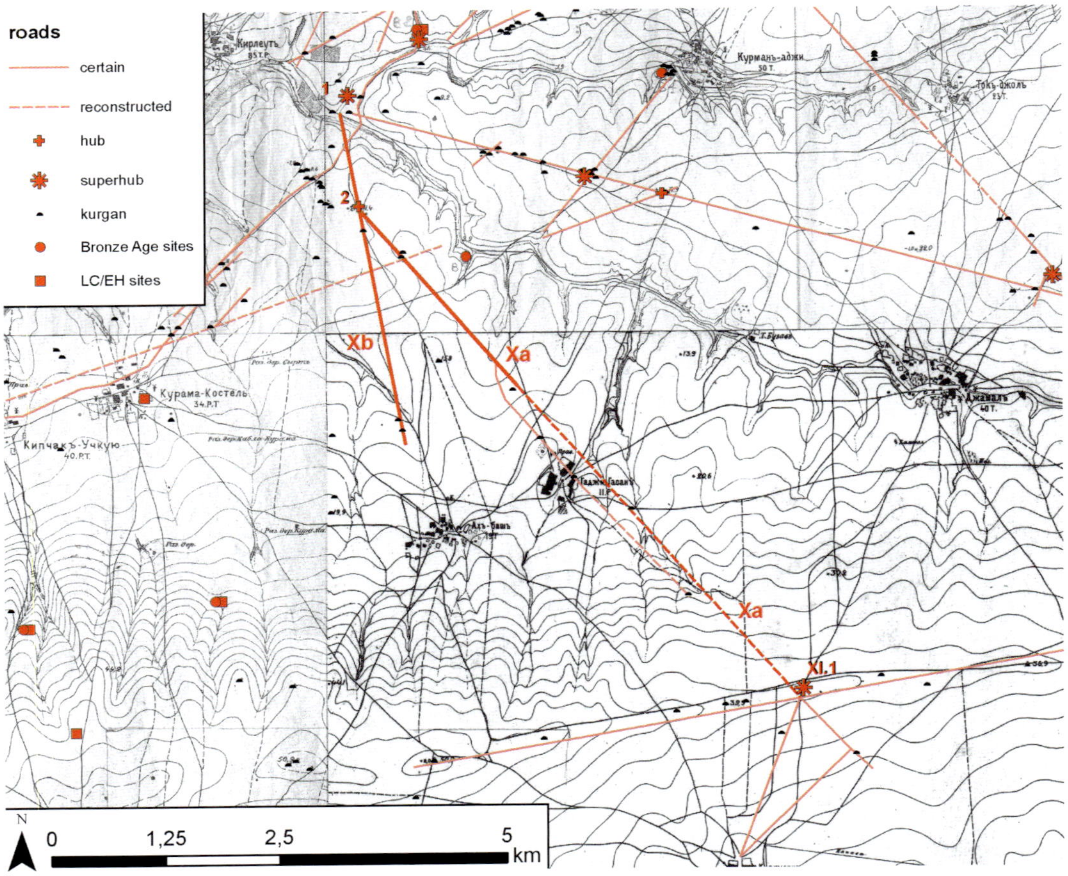

Fig. 6.34 • Road X.

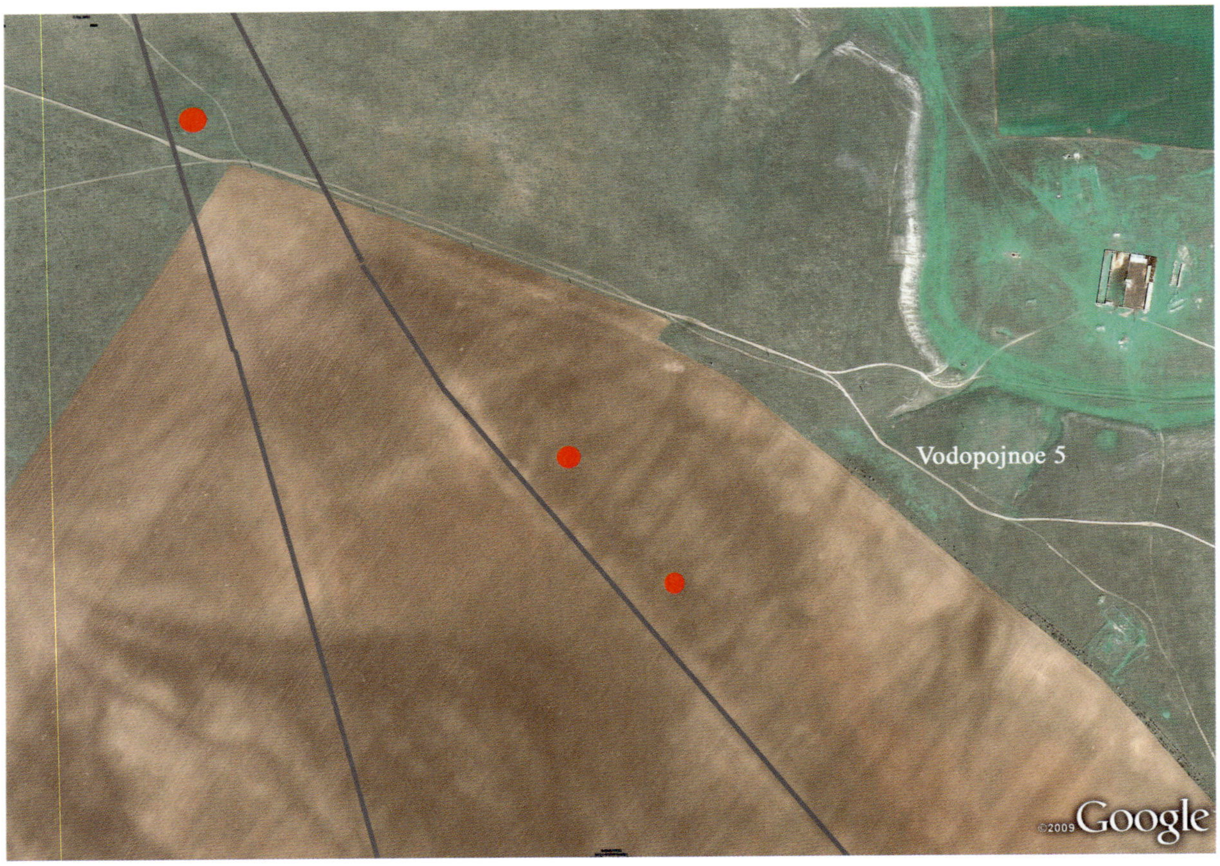

Fig. 6.35 • Road Xa and b (Google Earth).

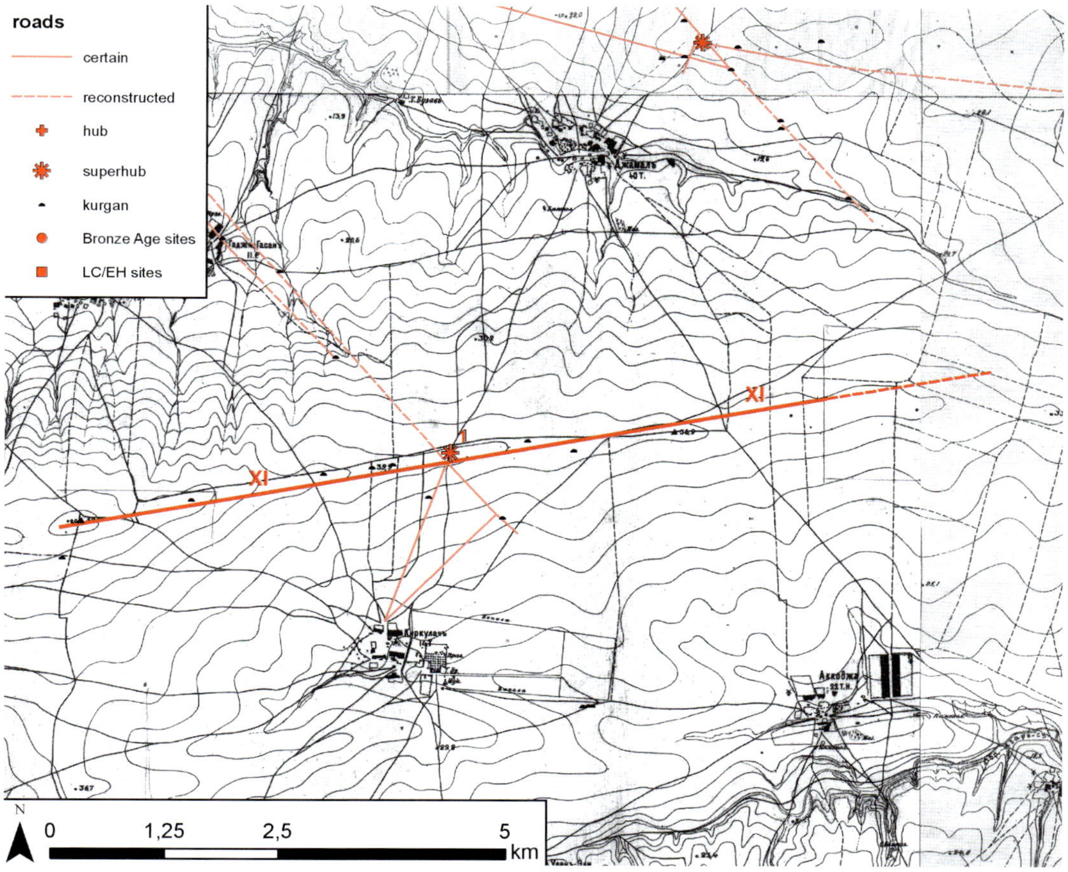

Fig. 6.36 • Road XI.

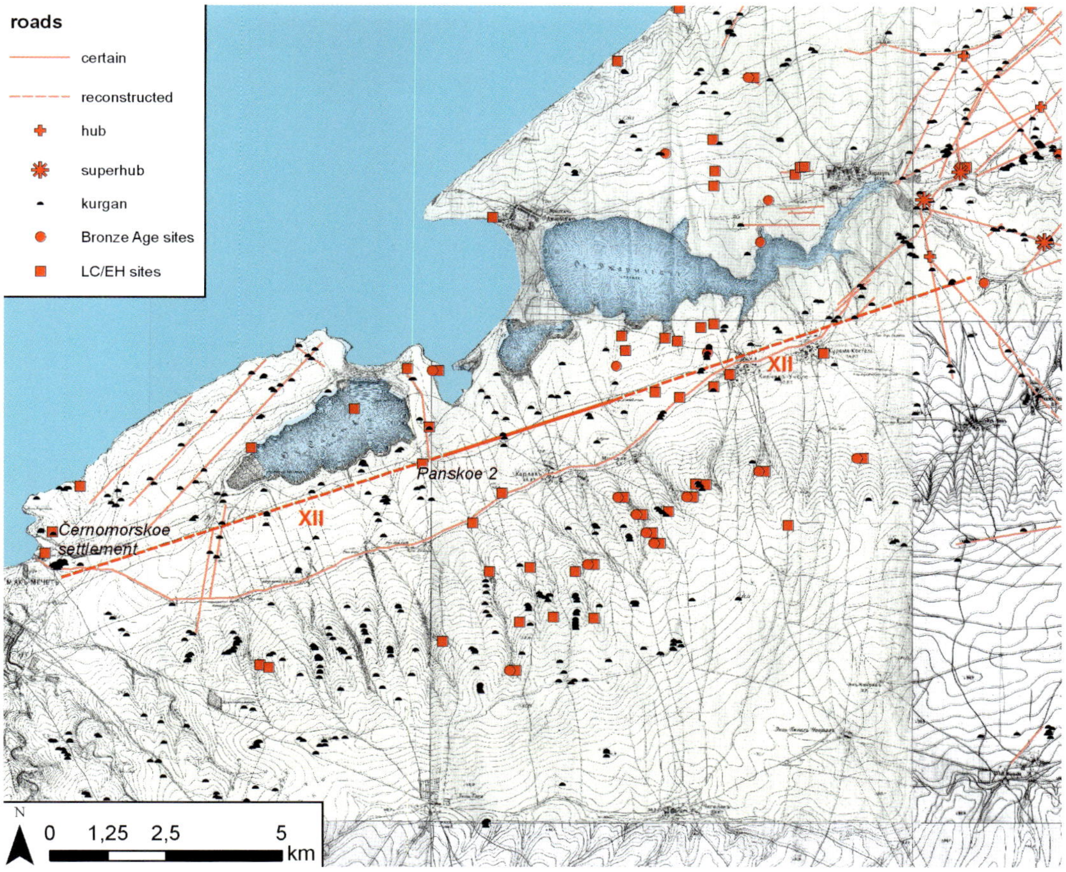

Fig. 6.37 • Road XII.

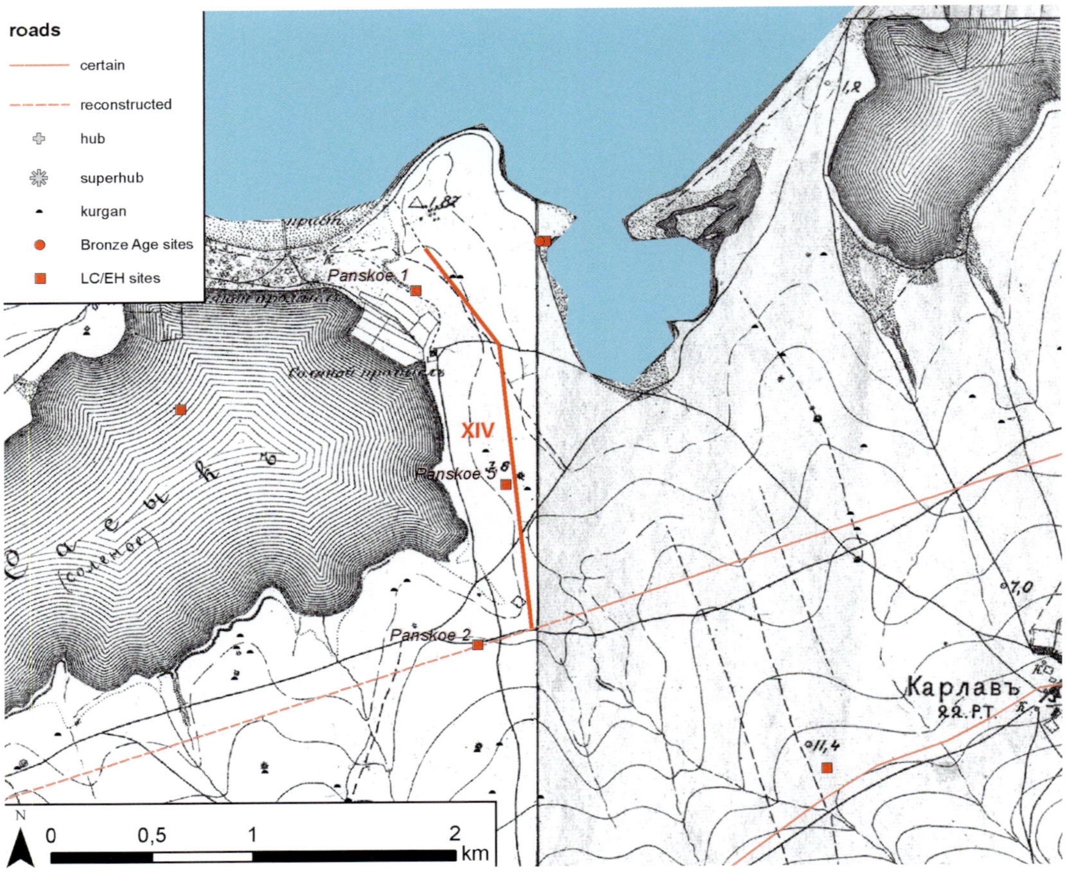

Fig. 6.38 • Road XIV.

Fig. 6.39 • Road XII and XIV (Google Earth).

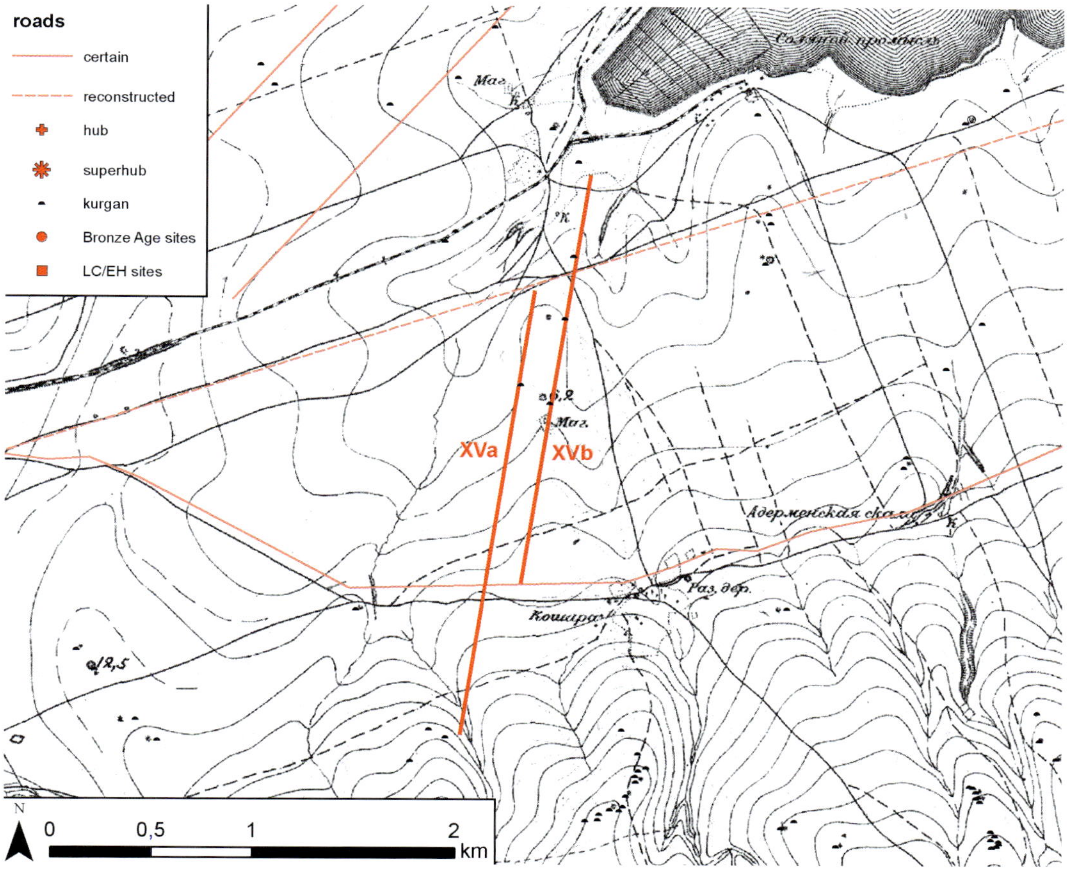

Fig. 6.40 • Road XV.

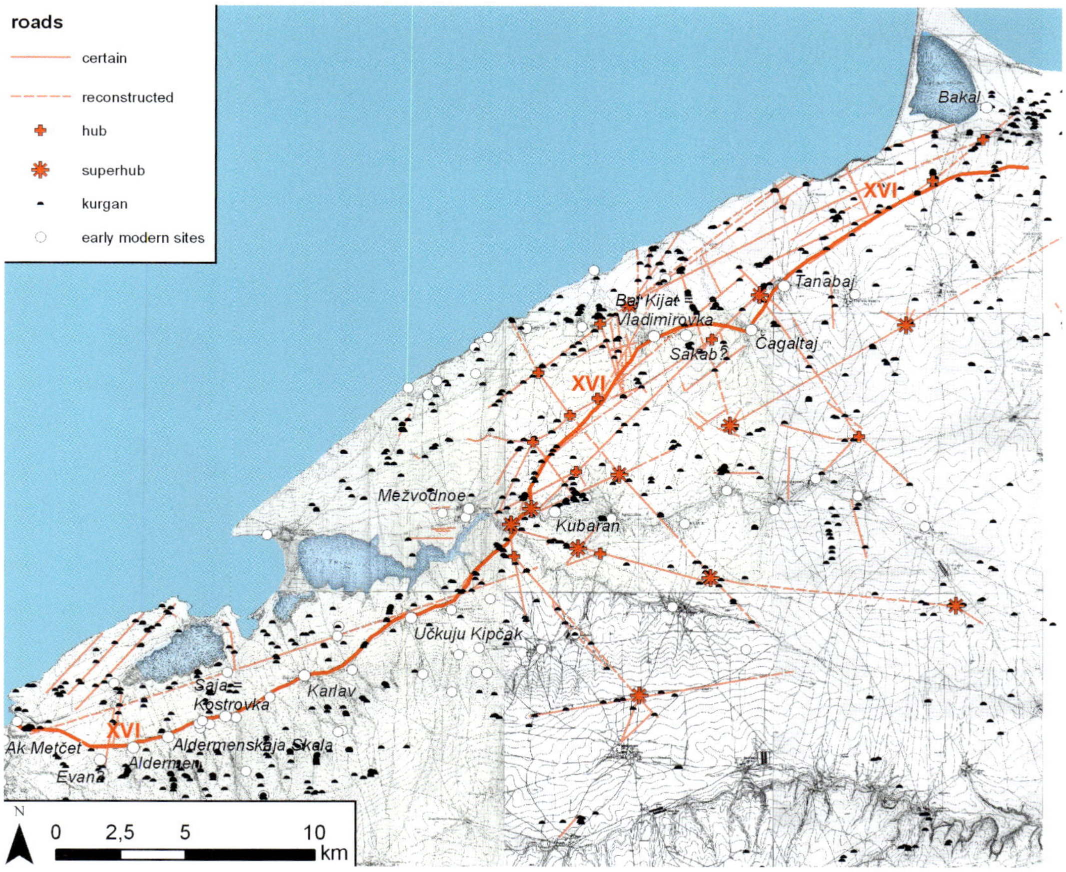

Fig. 6.41 • Road XVI.

Cat. 3

Cat. 5

Cat. 6

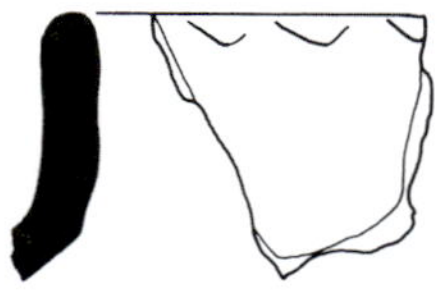

Cat. 7

Cat. 8

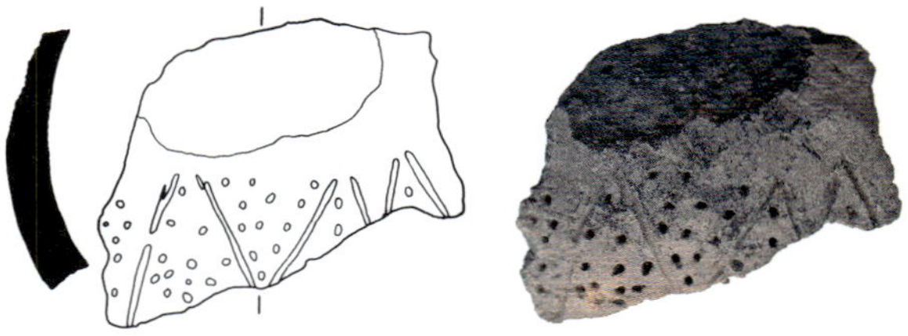

Cat. 9

Cat. 10

Cat. 11

Cat. 13

Cat. 15

Cat. 17

Cat. 21

Cat. 23

Cat. 24

Cat. 31

Cat. 32

SCALE 1:2

Cat. 34

Cat. 35

Cat. 37

Cat. 40

Cat. 41

Cat. 42

Cat. 44

Cat. 46

Cat. 47

Cat. 48

Cat. 50

Cat. 52

Cat. 53

Cat. 55

Cat. 56

Cat. 57

Cat. 58

Cat. 59

Cat. 60

Cat. 61

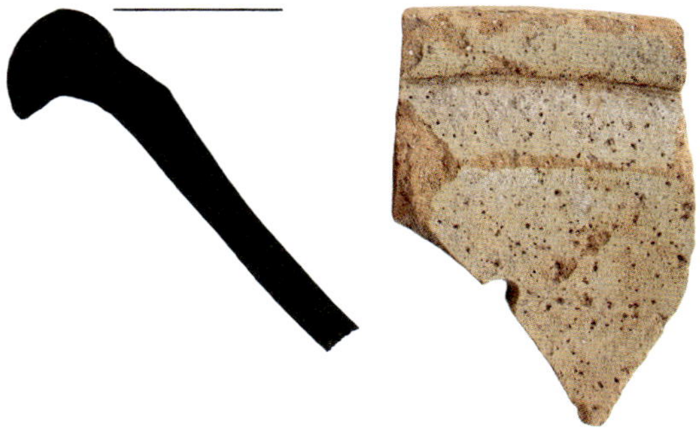

Cat. 64

SCALE 1:2

SITE F01-01

Cat. 65

Cat. 66

Cat. 68

Cat. 69

Cat. 72

Cat. 75

Cat. 79

DSP07-F02-01

Cat. 81

Cat. 82

Cat. 84

Cat. 85

Cat. 86

SCALE 1:2

DSP07-F02-02

Cat. 87

Cat. 88

DSP07-F02-03

Cat. 90

DSP07-F02-04

Cat. 92

Cat. 93

Cat. 94

Cat. 95

Cat. 96

DSP07-F03-01

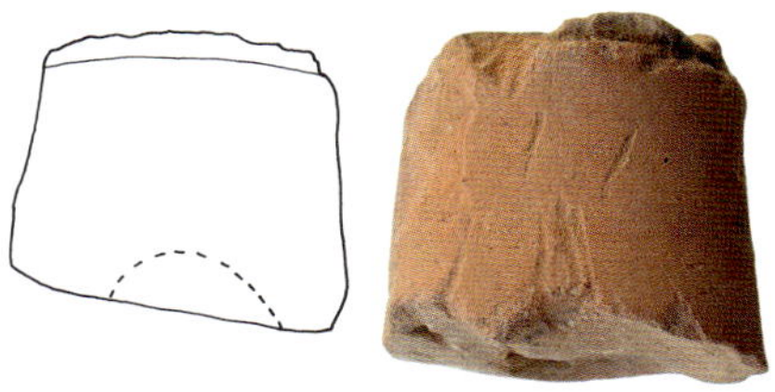

Cat. 97

Cat. 98

DSP07-F04-02

Cat. 101

Cat. 102

Cat. 104

Cat. 105

SCALE 1:2

DSP07-F02-04

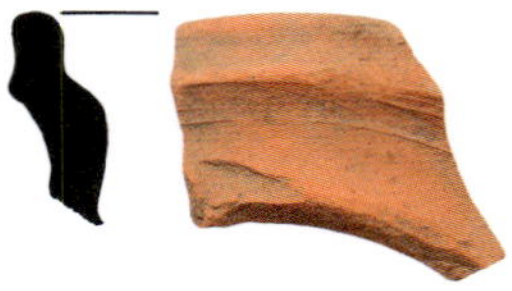

Cat. 106

Cat. 108

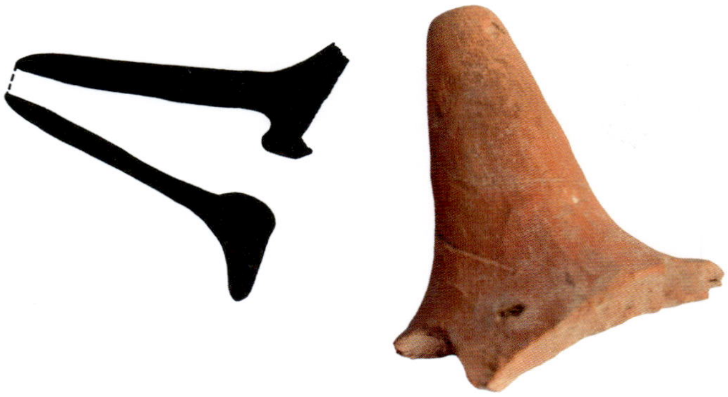

Cat. 109

DSP07-F05-01

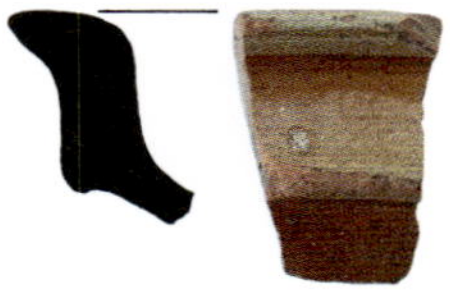

Cat. 110

Cat. 111

Cat. 113

Cat. 114

Cat. 115

DSP07-F06-01

Cat. 116

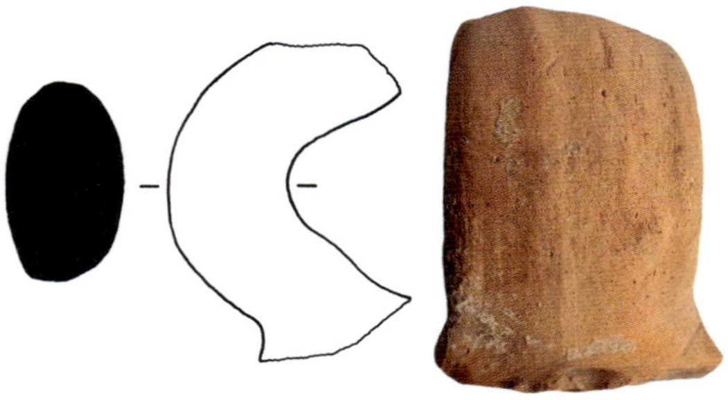

Cat. 117

SCALE 1:2

DSP07-F06-02

Cat. 119

Cat. 120

DSP07-F06-03

Cat. 122

Cat. 124

Cat. 125

DSP07-F08-01

Cat. 127

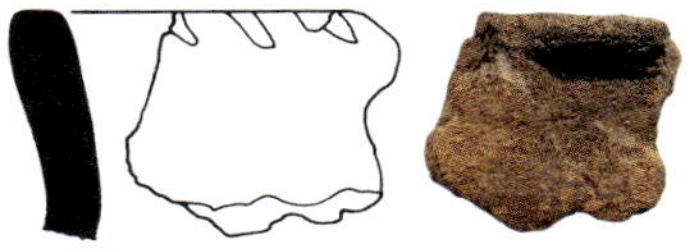

Cat. 131

Cat. 133

Cat. 135

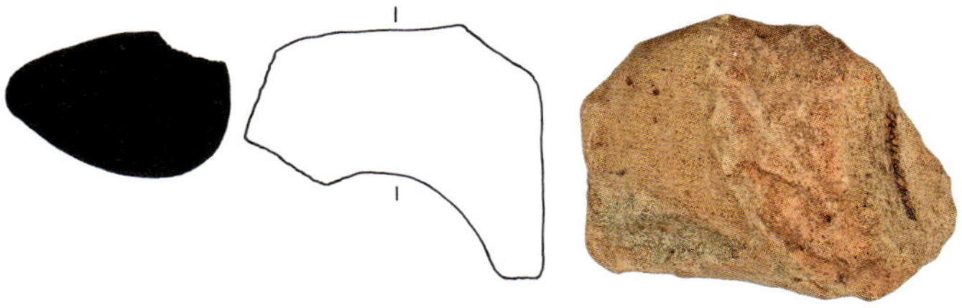

Cat. 136

SCALE 1:2

DSP08-F01-01

Cat. 138

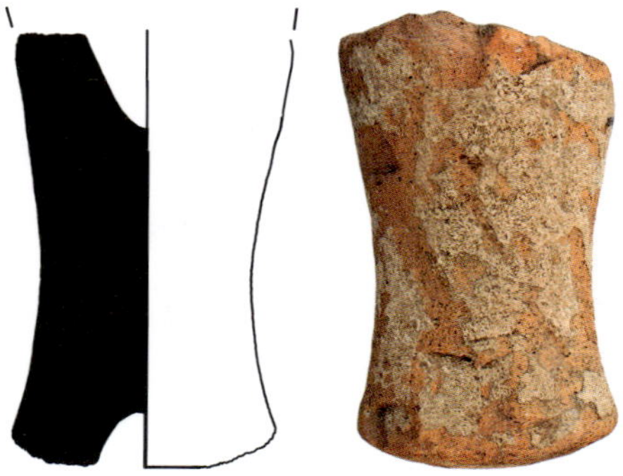

Cat. 140

Cat. 141

DSP08-F02-01

Cat. 142

Cat. 143

Cat. 144

Cat. 149

Cat. 152

Cat. 153

SCALE 1:2

DSP08-F02-01

Cat. 157

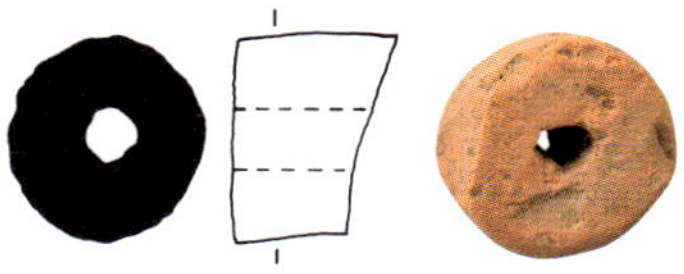

Cat. 158

Cat. 159

Cat. 161

Cat. 162

Cat. 164

DSP08-F03-01

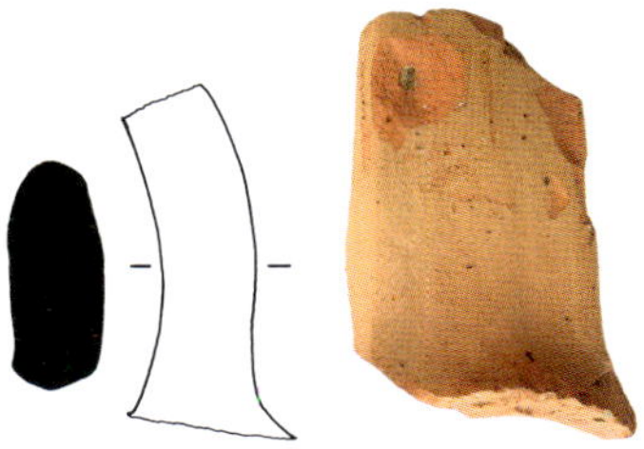

Cat. 166

Cat. 167

Cat. 170

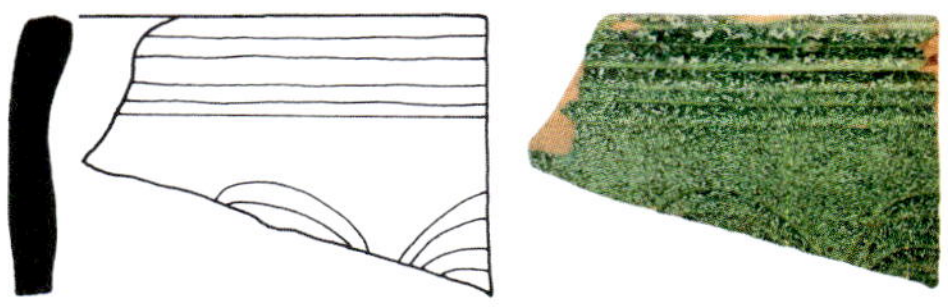

Cat. 171

Cat. 176

Cat. 177

SCALE 1:2

DSP08-F03-01

Cat. 179

Cat. 180

Cat. 181

Cat. 183

Cat. 186

Cat. 188

Cat. 190

DSP08-F04-01

Cat. 192

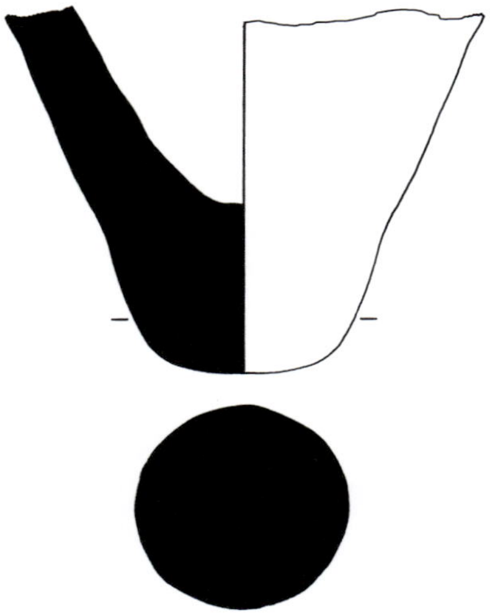

Cat. 194

Cat. 196

DSP08-F04-02

Cat. 198

Cat. 200

Cat. 204

Cat. 205

Cat. 207

SCALE 1:2

DSP08-F04-03

Cat. 212

Cat. 215

Cat. 216

Cat. 218

Cat. 221

Cat. 222

Cat. 223

Cat. 224

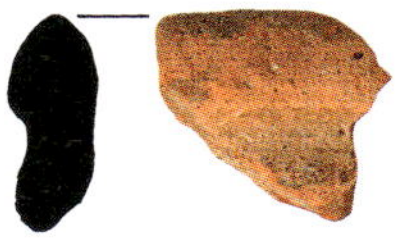

Cat. 225

Cat. 226

Cat. 231

Cat. 236

SCALE 1:2

DSP08-F04-03

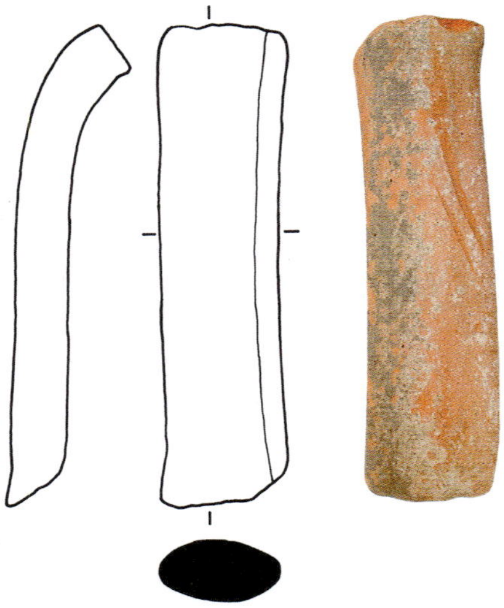

Cat. 237

Cat. 240

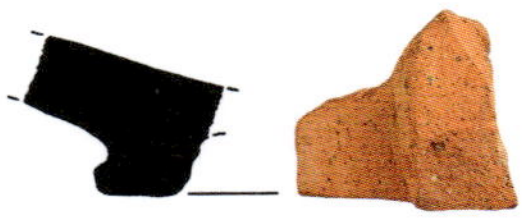

Cat. 241

Cat. 244

Cat. 246

Cat. 247

Cat. 248

Cat. 249

DSP08-F04-04

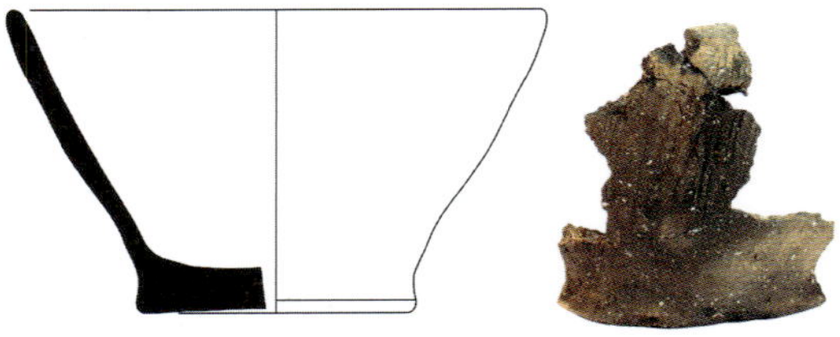

Cat. 250

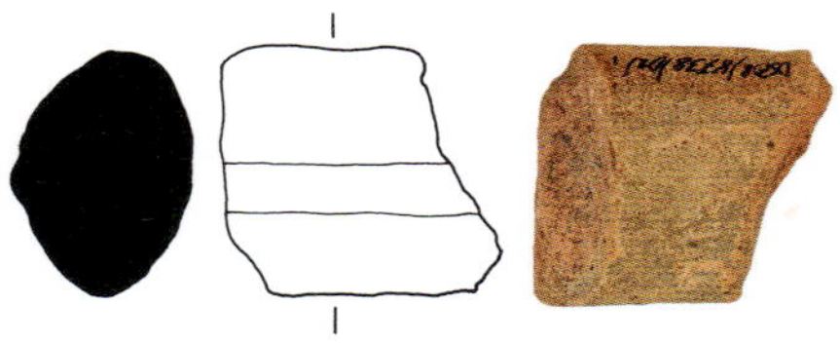

Cat. 251

Cat. 252

Cat. 253

SCALE 1:2

DSP08-F04-06

Cat. 257

Cat. 259

Cat. 261

Cat. 264

Cat. 265

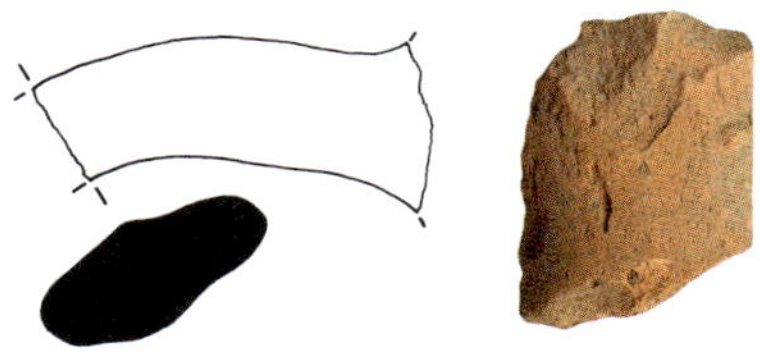

Cat. 266

DSP08-F04-07

Cat. 267

Cat. 268

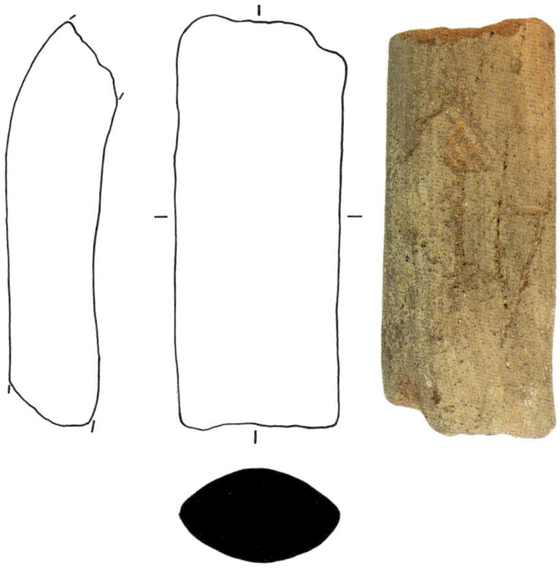

Cat. 270

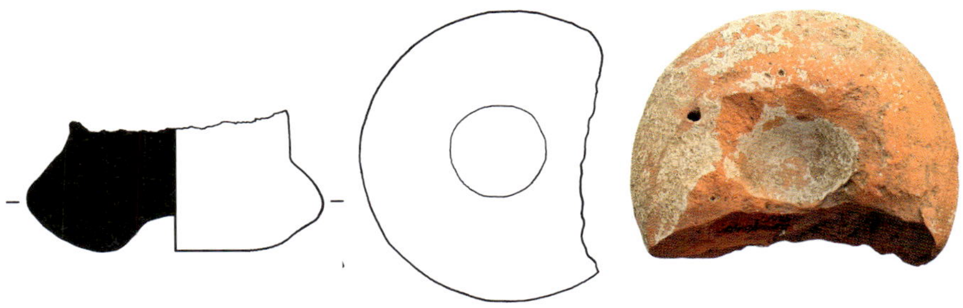

Cat. 271

Cat. 272

SCALE 1:2

DSP08-F04-07

Cat. 274

Cat. 275

Cat. 277

Cat. 280

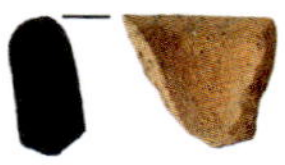

Cat. 282

DSP08-F05-01

Cat. 283

Cat. 284

Cat. 285

SCALE 1:2

DSP08-F06-01

Cat. 287

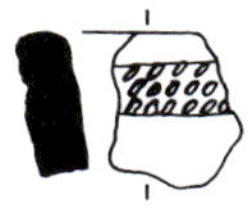

Cat. 289

Cat. 290

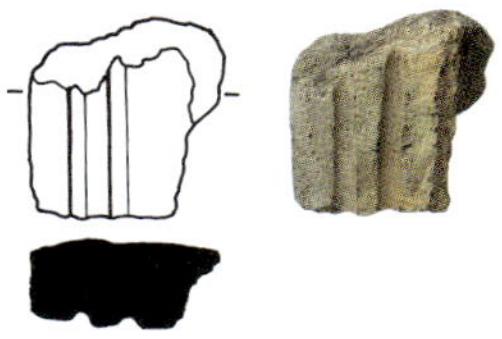

Cat. 292

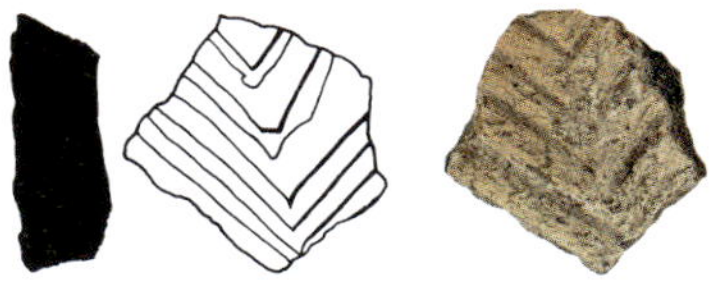

Cat. 295

Cat. 296

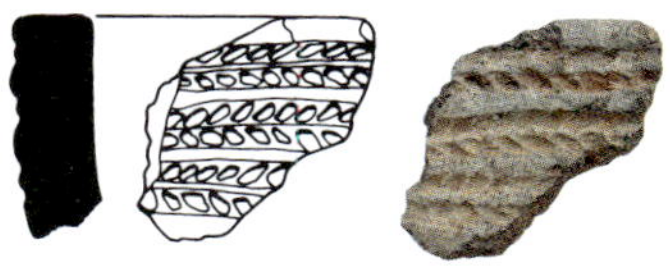

Cat. 297

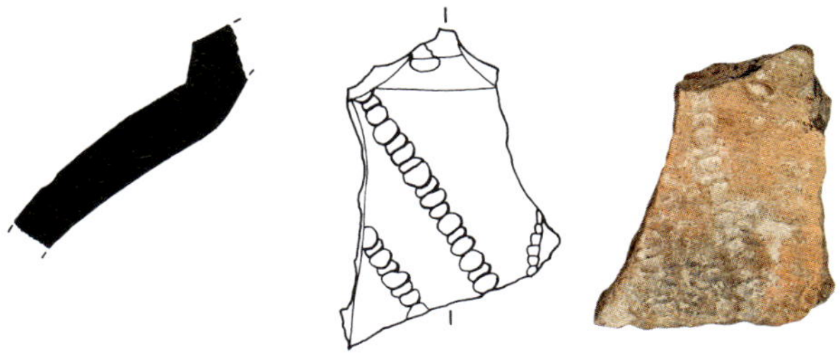

Cat. 298

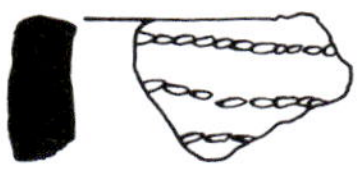

Cat. 299

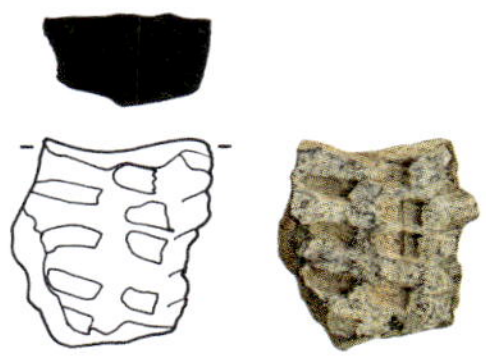

Cat. 301

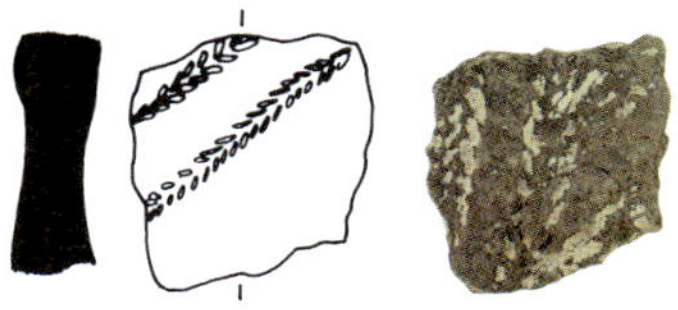

Cat. 304

Cat. 305

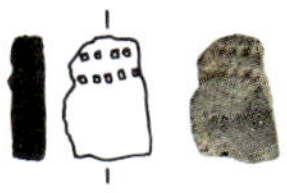

Cat. 309

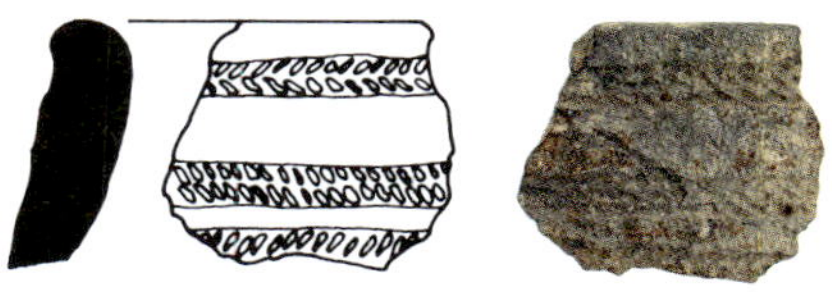

Cat. 310

Cat. 311

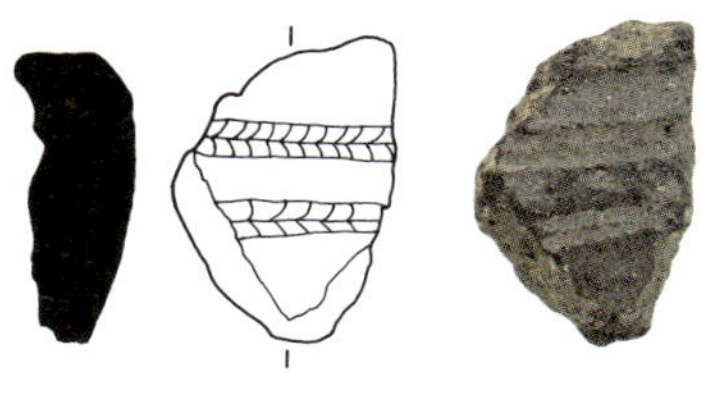

Cat. 313

DSP08-F06-01

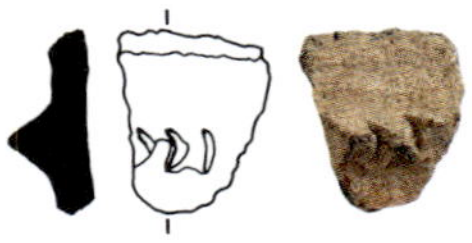
Cat. 319

Cat. 324

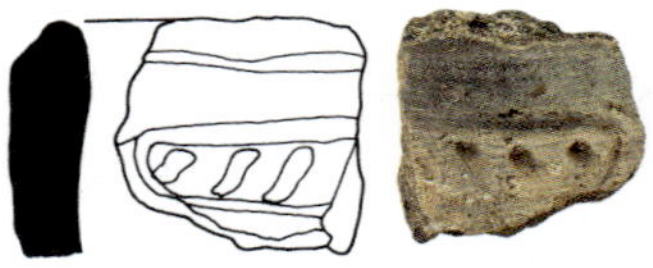
Cat. 325

Cat. 326

Cat. 328

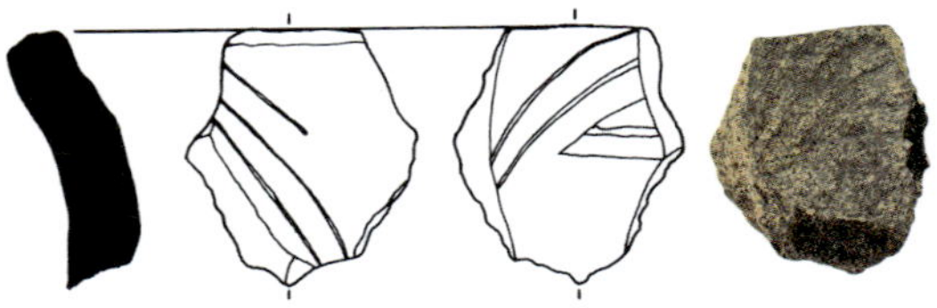
Cat. 330

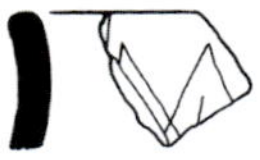
Cat. 333

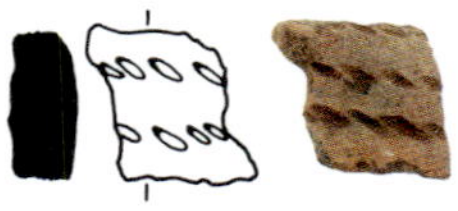
Cat. 336

Cat. 339

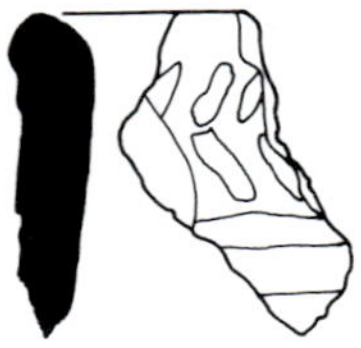
Cat. 346

Cat. 351

Cat. 356

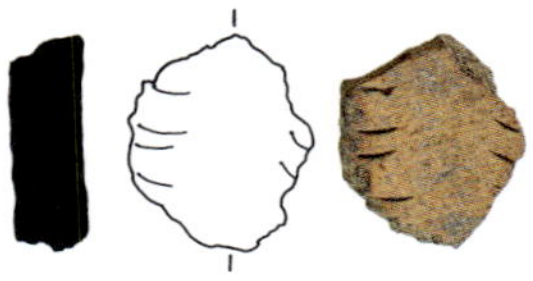
Cat. 359

Cat. 361

Cat. 371

Cat. 373

Cat. 374

DSP08-F06-03

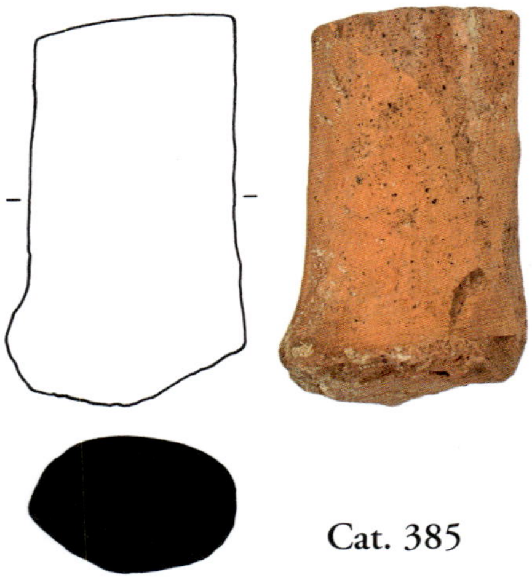
Cat. 385

FEATURE 18

Cat. 387

SCALE 1:2

FEATURE 18

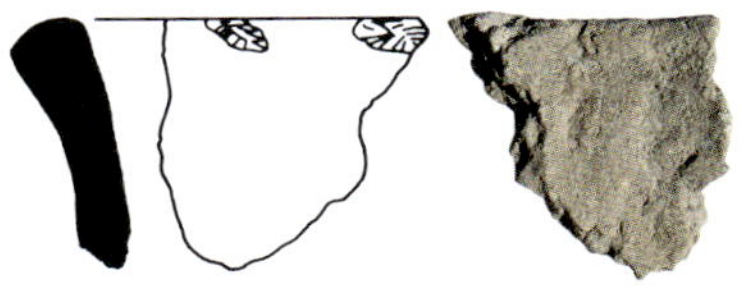

Cat. 388

Cat. 390

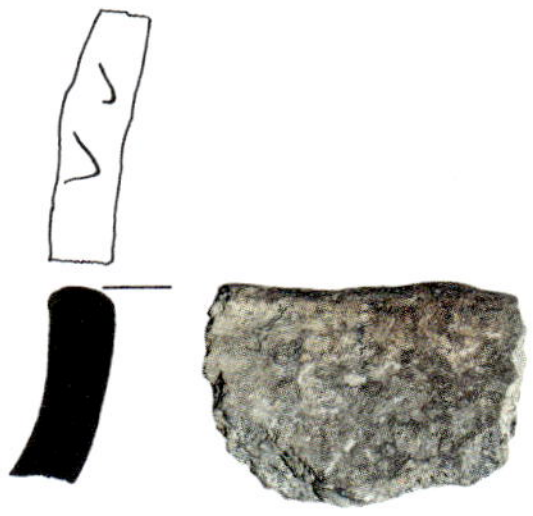

Cat. 391

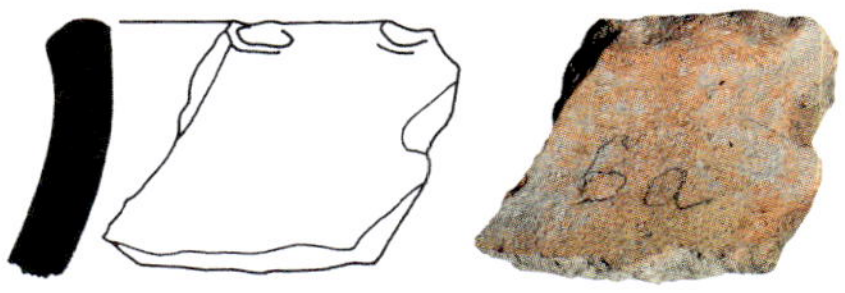

Cat. 393

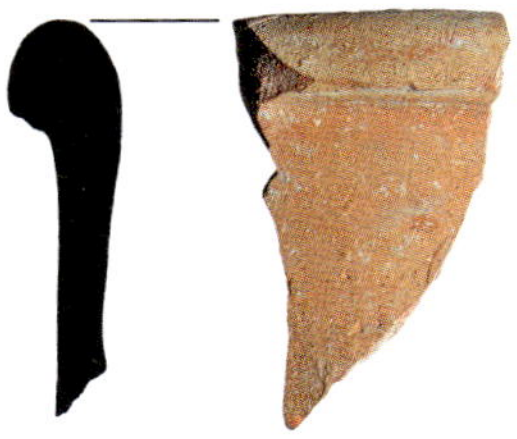

Cat. 395

Cat. 396

DSP07-H01-01

Cat. 397

Cat. 398

Cat. 399

DSP07-H04-02

Cat. 401

DSP08-H07-03

Cat. 402

DSP07-H08-02

Cat. 403

FEATURE 2

Cat. 404

Cat. 407

SCALE 1:2

Cat. 410

Cat. 413

Cat. 415

Cat. 419

Cat. 421

Cat. 424

Cat. 425

Cat. 430

Cat. 431

Cat. 432

DSP07-H10-01

Cat. 433

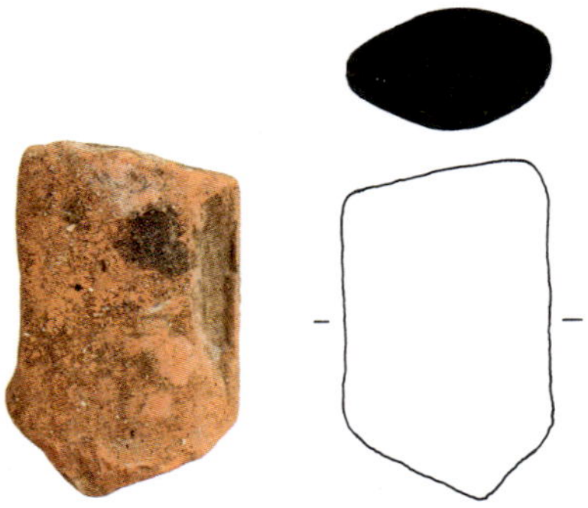

Cat. 434

SCALE 1:2

DSP07-H10-01

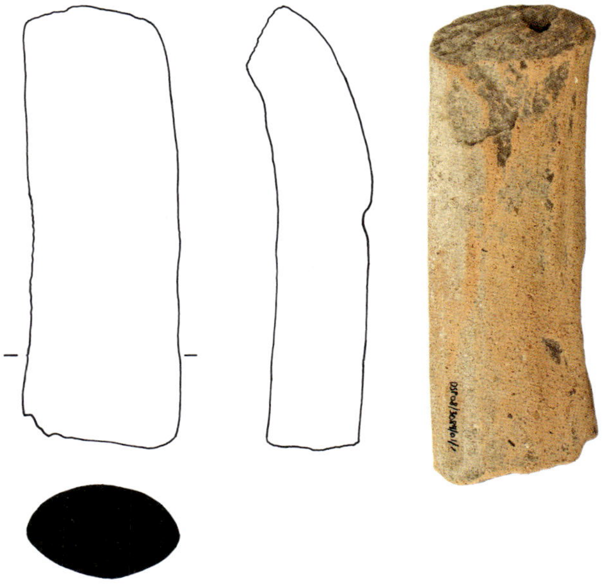

Cat. 435

Cat. 436

FEATURE 5 S. 404

Cat. 439

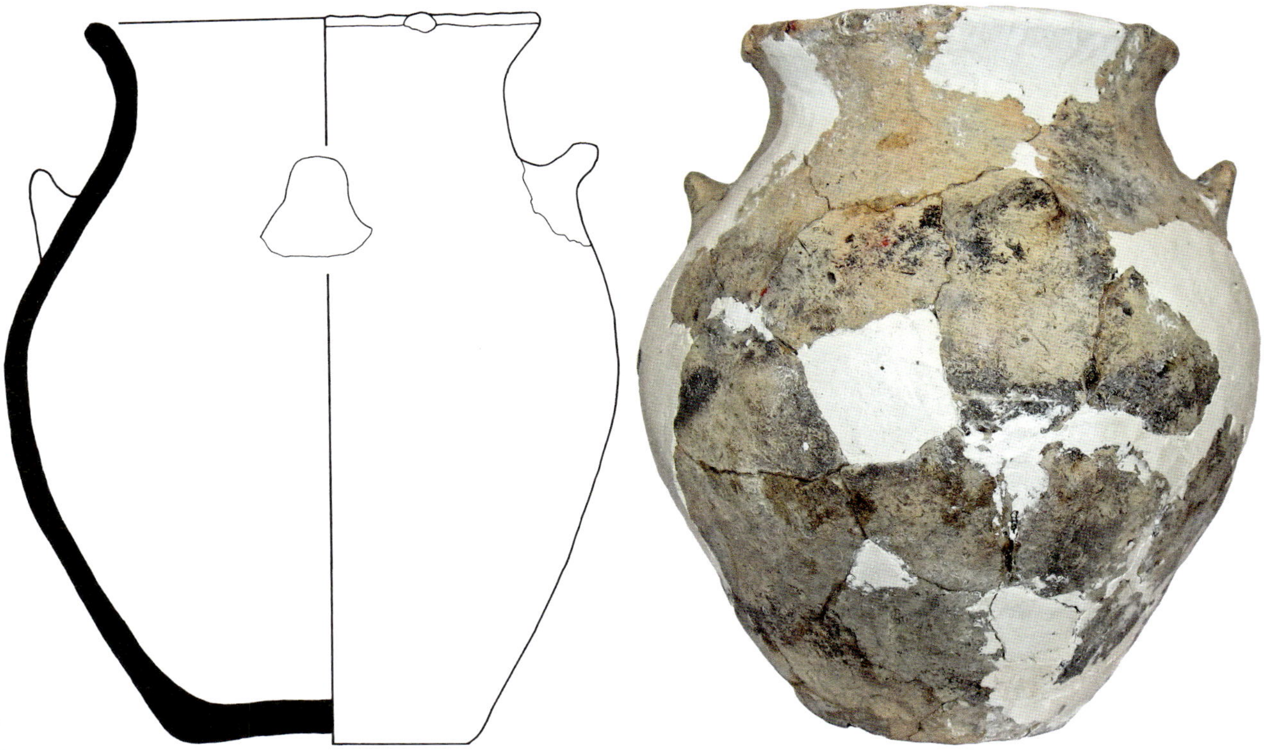

Cat. 442

SCALE 1:2

FEATURE 5

Cat. 445

Cat. 446

Cat. 449

Cat. 450

FEATURE 5

Cat. 448

DSP07-H11-01

Cat. 451

Cat. 452

DSP07-H11-01

Cat. 454

Cat. 455

Cat. 456

SCALE 1:2

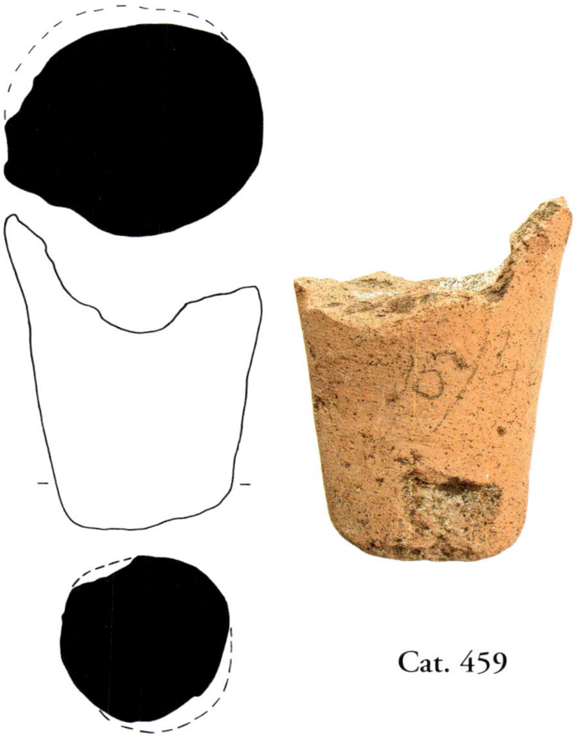

Cat. 459

Cat. 461

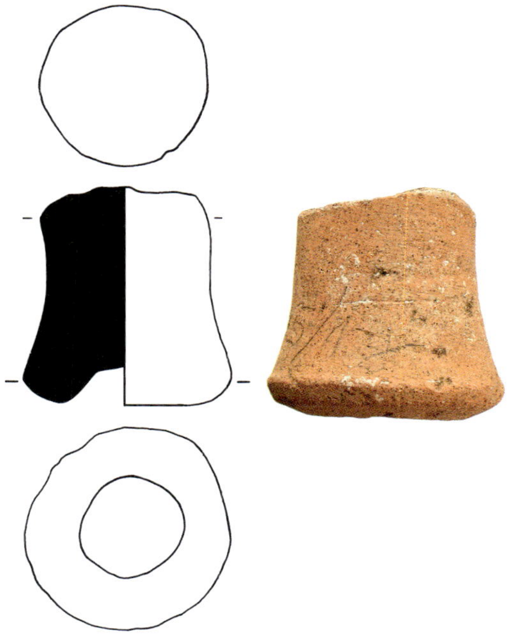

Cat. 462

Cat. 464

Cat. 466

SCALE 1:2

DSP07-H11-01

Cat. 467

Cat. 468

Cat. 469

Cat. 474

Cat. 477

Cat. 483

Cat. 485

Cat. 489

Cat. 491

SCALE 1:2

DSP07-H11-02

Cat. 492

Cat. 493

Cat. 495

Cat. 497

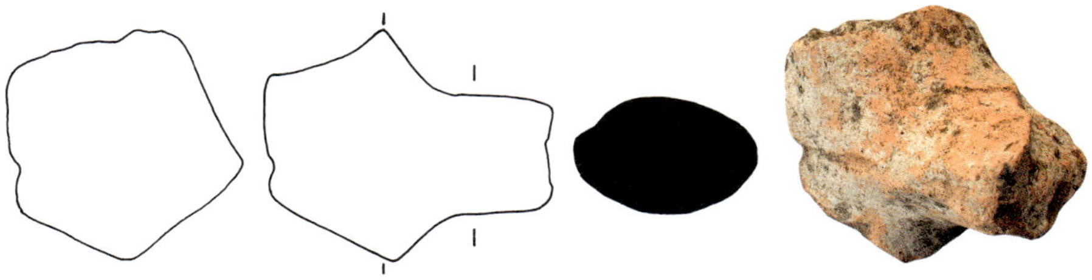

Cat. 500

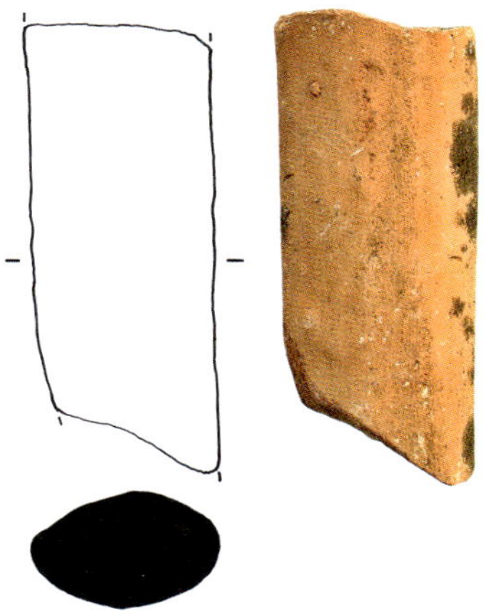

Cat. 502

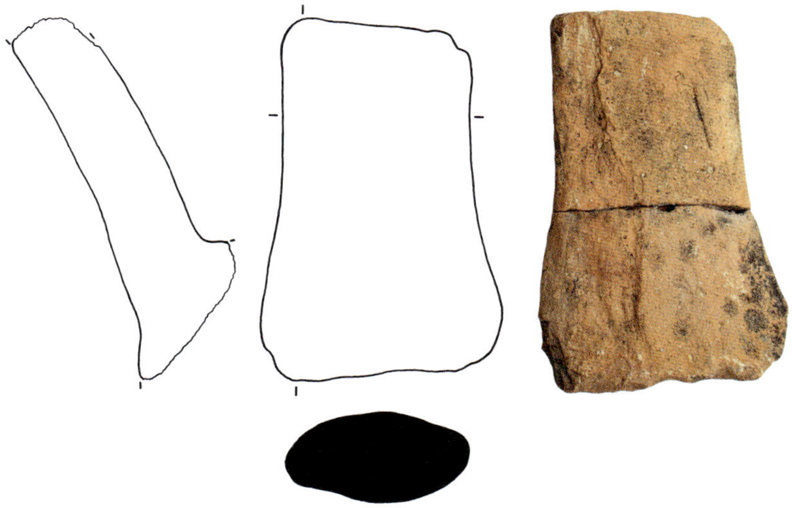

Cat. 505

SCALE 1:2

DSP07-H11-02

Cat. 506

Cat. 507

FEATURE 9

Cat. 508

Cat. 510

Cat. 511

Cat. 512

Cat. 513

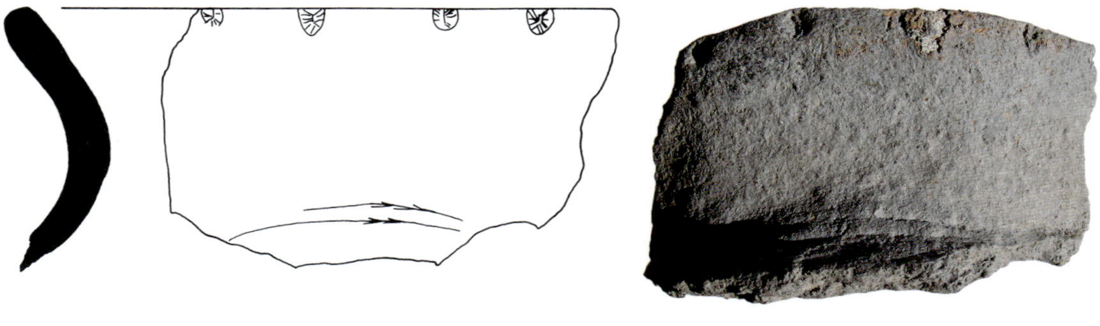

Cat. 514

SCALE 1:2

FEATURE 9

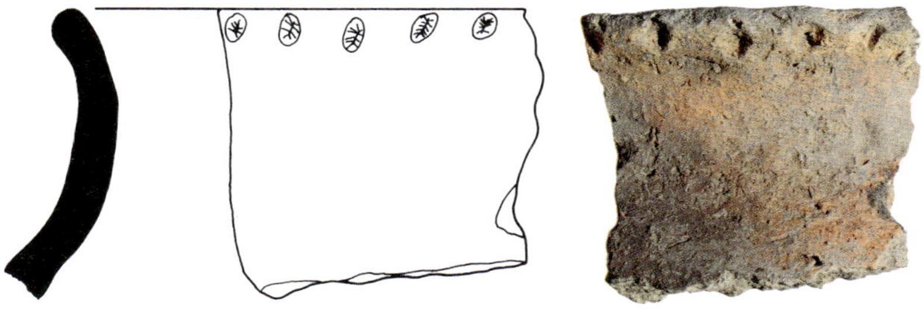

Cat. 519

Cat. 524

Cat. 527

Cat. 528

DSP08-H12-01

Cat. 530

Cat. 536

DSP08-H12-01

Cat. 537

DSP08-H13-01

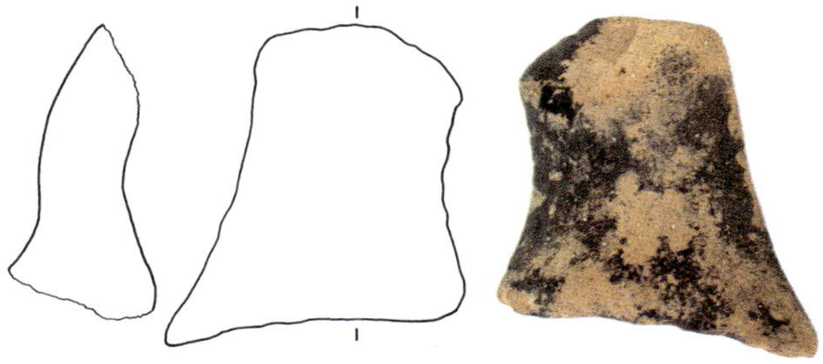

Cat. 538

SCALE 1:2

FEATURE 2 S. 414

Cat. 542

Cat. 543

Cat. 547

Cat. 550

Cat. 552

Cat. 556

Cat. 557

Cat. 561

DSP08-H17-02

Cat. 565

DSP07-H19-01

Cat. 566

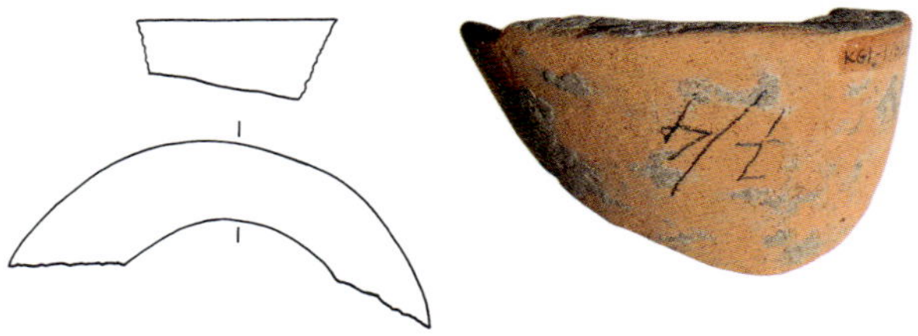

Cat. 567

DSP07-H19-03

Cat. 568

Cat. 569

DSP07-H19-07

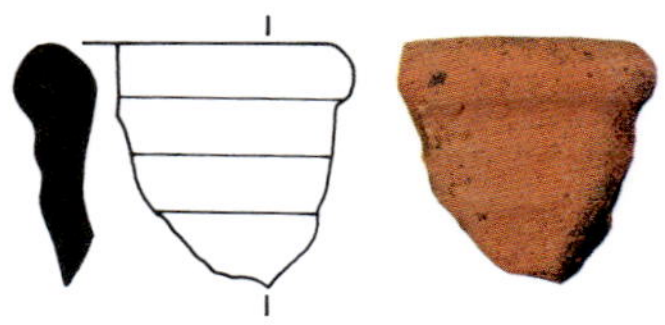

Cat. 570

SCALE 1:2

DSP08-H22-02

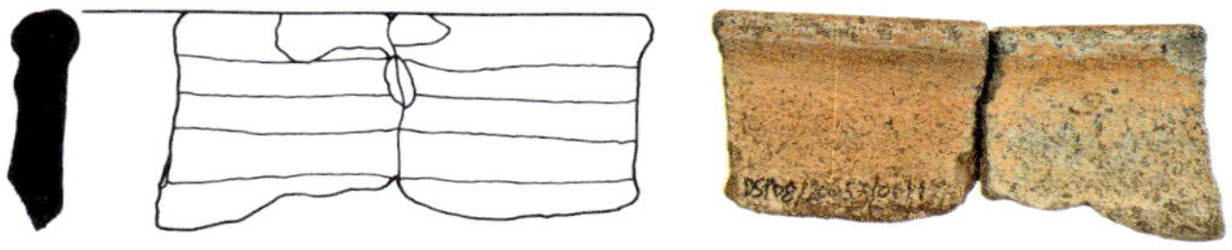

Cat. 572

DSP08-H24-01

Cat. 574

DSP08-H24-03

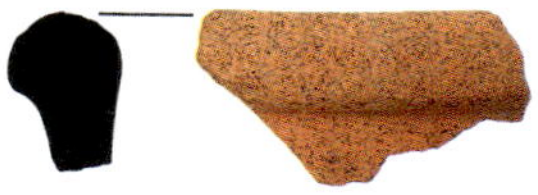

Cat. 575

DSP08-H26-01

Cat. 576

BAJ KIJAT 1 (AND 2)

Cat. 578

Cat. 579

Cat. 581

Cat. 582

Cat. 583

Cat. 584

BAJ KIJAT 3

Cat. 585

Cat. 587

Cat. 589

SCALE 1:2

BURUN ELI 2

Cat. 590

BURUN ELI 6

Cat. 592

Cat. 594

Cat. 596

GROTY I

Cat. 598

Cat. 600

Cat. 603

Cat. 606

Cat. 607

Cat. 608

Cat. 609

KURAMA KOSTEL

Cat. 610

Cat. 611

Cat. 613

SCALE 1:2

Cat. 615

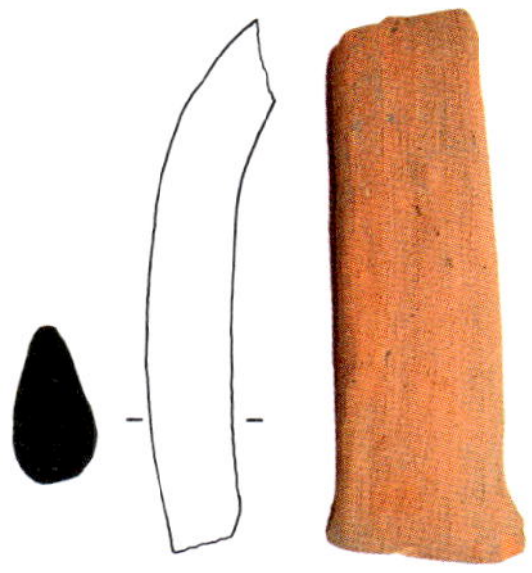

Cat. 616

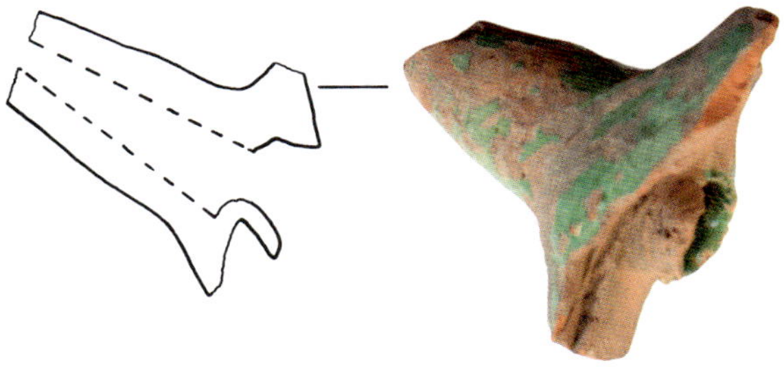

Cat. 618

Cat. 619

Cat. 620

Cat. 621

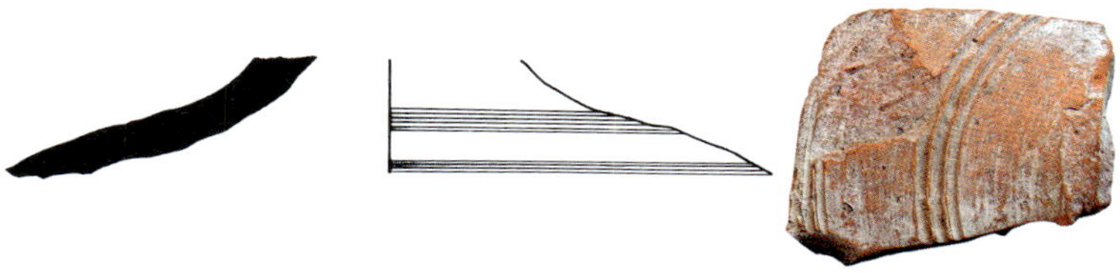

Cat. 622

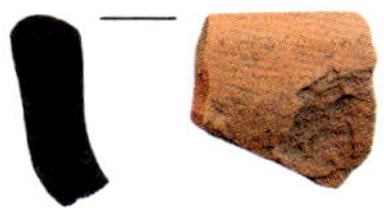

Cat. 624

Cat. 625

Cat. 626

Cat. 628

Cat. 629

SCALE 1:2

KURAMA KOSTEL

Cat. 633

Cat. 630

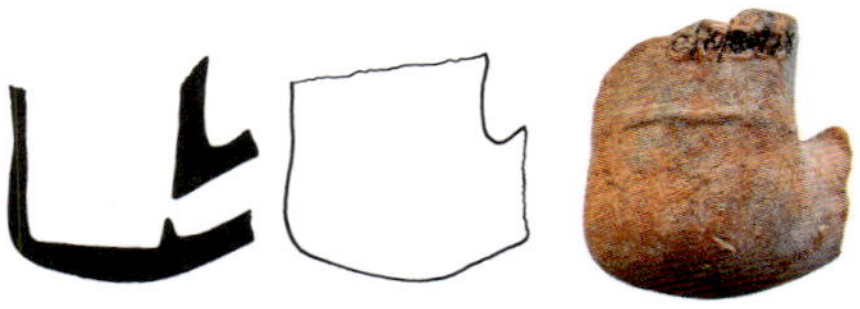

Cat. 634

Cat. 636

Cat. 637

MASLINY I

Cat. 640

Cat. 641

PANSKOE II/DSP06-F03-01

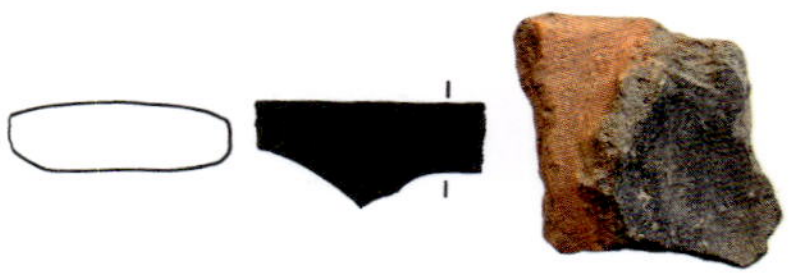

Cat. 643

Cat. 644

Cat. 645

Cat. 647

SCALE 1:2

PANSKOE II/DSP06-F03-01

Cat. 649

Cat. 650

Cat. 652

Cat. 653

Cat. 654

Cat. 655

Cat. 658

PANSKOE V/DSP07-F01-02

Cat. 659

Cat. 661

Cat. 662

SKALISTOE 2/DSP08-F03-02

Cat. 664

Cat. 665

Cat. 666

SCALE 1:2

Cat. 668

Cat. 670

Cat. 672

Cat. 673

Cat. 674

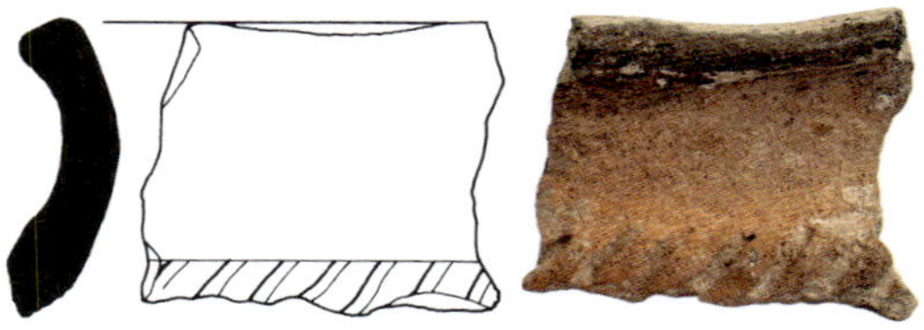

Cat. 675

Cat. 676

Cat. 679

Cat. 681

Cat. 682

Cat. 685

Cat. 688

Cat. 690

Cat. 692

Cat. 694

Cat. 695

Cat. 697

Cat. 703

Cat. 704

Cat. 705

SCALE 1:2

SKALISTOE 2/DSP08-F03-02

Cat. 706

Cat. 708

SKALISTOE 5

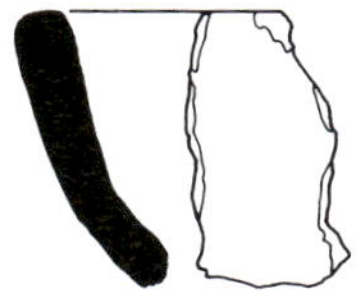

Cat. 713

Cat. 714

SKALISTOE 5

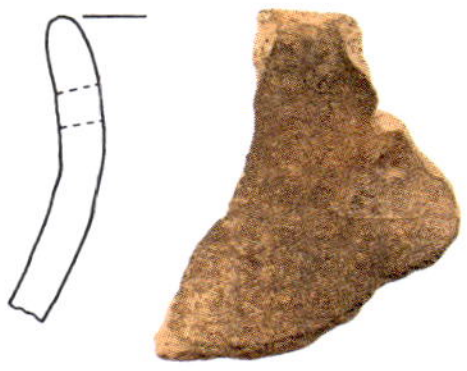

Cat. 720

Cat. 724

Cat. 727

Cat. 728

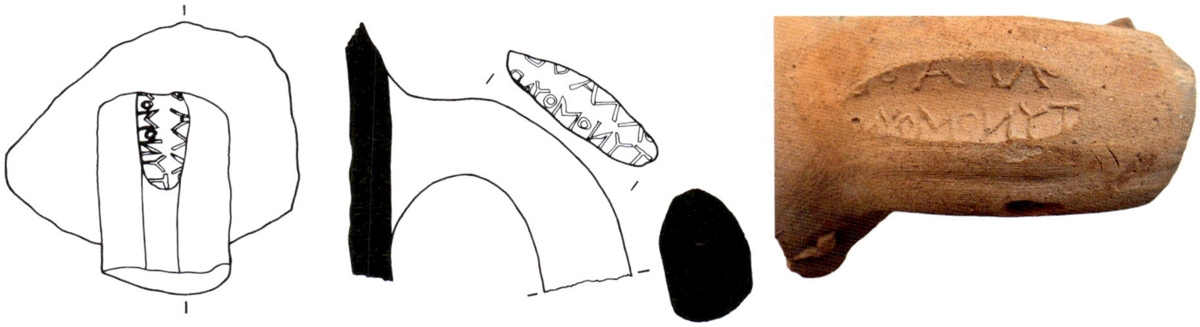

Cat. 735

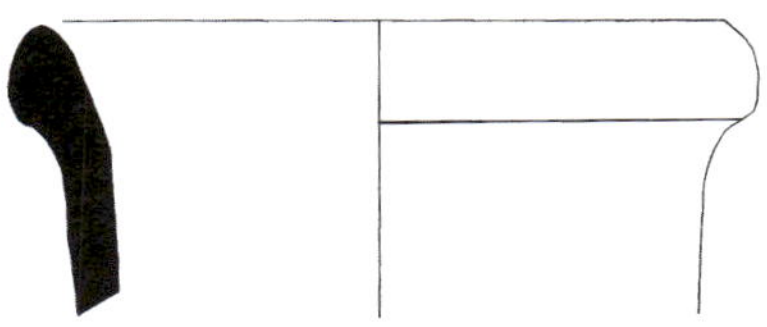

Cat. 736

Cat. 738

Cat. 740

SCALE 1:2

SKALISTOE 5

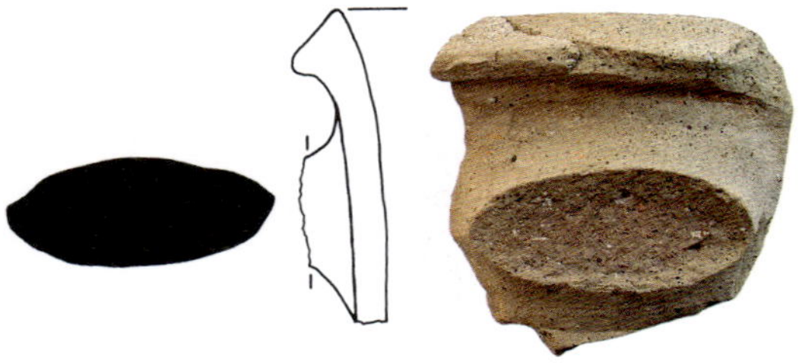

Cat. 741

Cat. 747

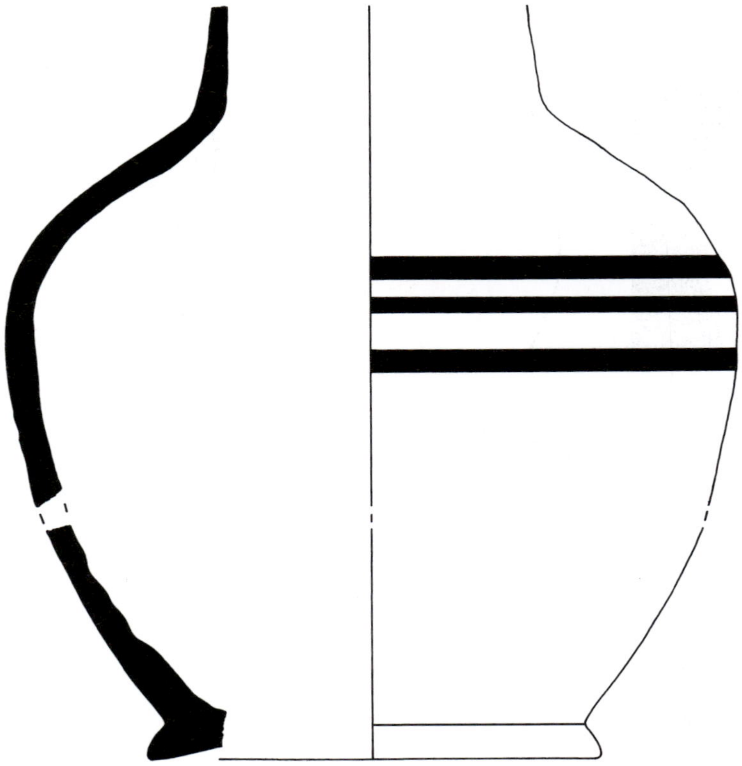

Cat. 749

Cat. 766

Cat. 768

Cat. 767

SKALISTOE 10

Cat. 775

VODOPOJNE 3A + B

Cat. 778

Cat. 779

Cat. 780

Cat. 781

SCALE 1:2

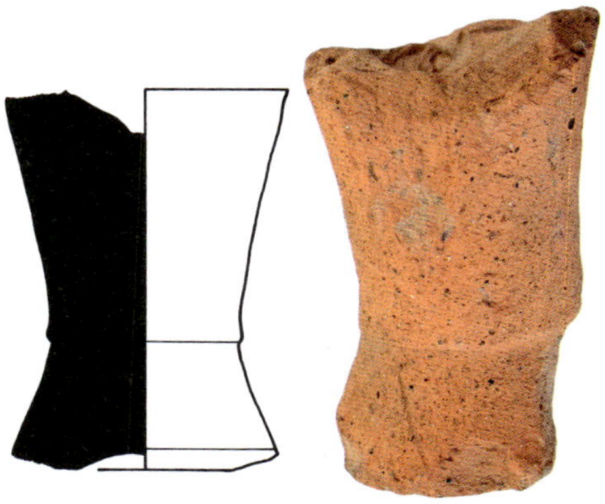

Cat. 782

Cat. 783

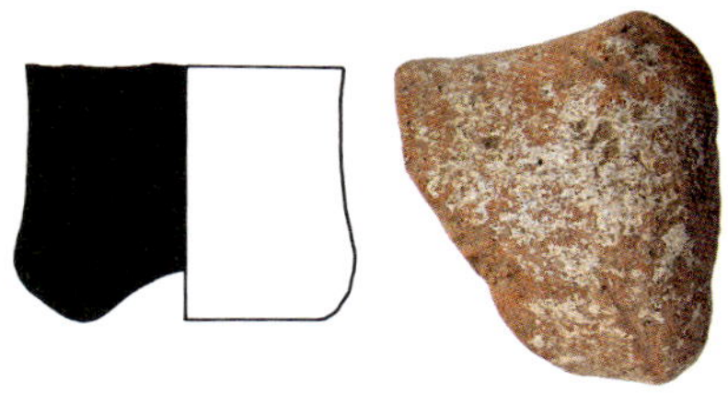

Cat. 784

Cat. 785

Cat. 786

Cat. 787

Cat. 788

Cat. 789

Cat. 792

Cat. 794

Cat. 795

Cat. 796

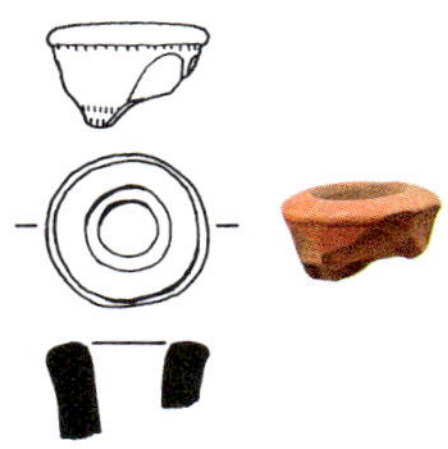

Cat. 799

SCALE 1:2

Cat. 801

Cat. 804

Cat. 806

Cat. 807

Cat. 809

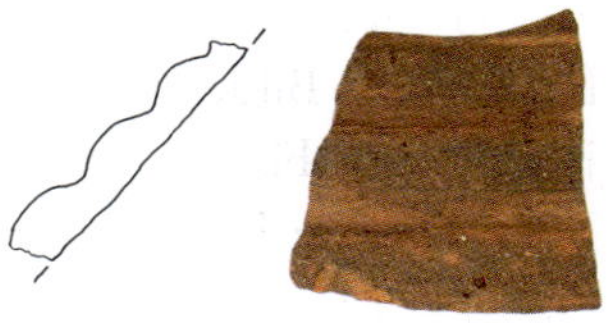

Cat. 810

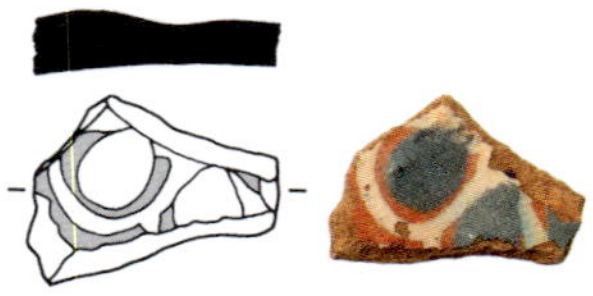

Cat. 811

Cat. 812

Cat. 813

Cat. 815

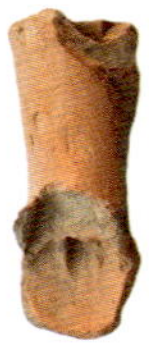

Cat. 816

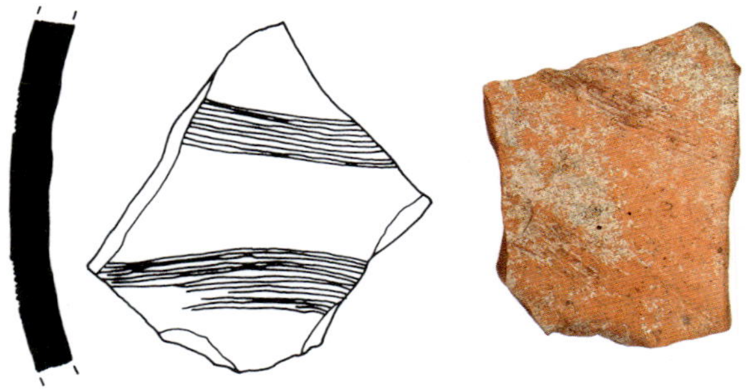

Cat. 818

Cat. 820

Cat. 822

Cat. 823

SCALE 1:2

Illustration credits

P. Attema: Figs. 3.5, 8-10, 12; 4.161, 183-184, 187, 198
P. Guldager Bilde: Figs. 2.10; 3.13; 4.1-2, 4-12, 15, 21, 25-26, 30, 38, 150, 162; 6.9, 12, 14-15
P. Guldager Bilde, T. Smekalova: Figs. 4.33-36
J. Delvigne: Figs. 3.2, 6-7, 11
Eastern Atlas: Figs. 4.153, 238, 301-303, 306-307
Google Earth: Fig. 4.185
Google Earth, P. Guldager Bilde: Figs. 4.148, 160, 167, 234; 6.11, 16, 21-23, 25, 27, 30, 35, 39
T. de Haas: Figs. 0.1-2; 2.1, 9, 13; 3.1, 3-4; 4.39-47, 49-51, 53, 55-58, 60-66, 68-72, 74, 76-78, 80-87, 89-92, 94-96, 98-103, 105-106, 109-113, 115, 117, 119-120, 122, 124-125, 127, 129-137, 139, 145-147, 159, 165-166, 168, 170, 172-174, 178, 182, 195, 200, 202, 208-212, 228-231, 233, 239-240, 243, 246-247, 257, 259, 261-262, 264, 266-267, 269-270, 273-275, 277-278, 285, 289, 295; 6.1-8, 10, 13, 17-20, 24, 26, 28-29, 31-34, 36-38, 40-41
T. de Haas, Eastern Atlas: Figs. 4.149, 232, 237, 241, 245, 268, 271, 296, 304
T. de Haas, T. Smekalova: Figs. 4.163, 177, 180-181, 196, 204-207, 248, 280, 290
T. de Haas, T. Smekalova, Eastern Atlas: Fig. 4.249
A. Karjaka: Figs. 4.155, 157-158, 189-191, 193-194, 253-254, 256, 283-284, 293-294.
S. Koltuchov: Figs. 0.3-4
S. Lancov: Figs. 4.188, 192, 242, 252, 255, 282, 292
S. Lancov, V. Stolba: Figs. 4.16-20
J. Munk Højte: Figs. 4.141, 144, 244, 258, 263, 265, 272, 276, 287, 297, 298, 300; 5.2-5
W. de Neef: Fig. 4.142
D. Pilz: Figs. 2.12
T. Smekalova: Figs. 2.2-4; 4.13-14, 108, 151-152, 164, 175-176, 179, 186, 197, 199, 201, 235-236, 251, 281, 291; 5.1
V. Stolba: Figs. 4.154, 156
S. Tiebackx: Figs. 4.213-227
M. Unger: Fig. 4.27
E. Weistra: Fig. 2.7
C. Williamson: Figs. 2.8; 4.75, 171
K. Winther-Jacobsen: Figs. 4.48, 52, 54, 59, 67, 73, 79, 88, 93, 97, 104, 107, 114, 116, 118, 121, 123, 126, 128, 138, 140, 143, 169, 250, 260, 279, 286, 288, 299, 305, 308
H. Zöllner: Fig. 2.11